Why Banks Fail

Sebastián Royo

Why Banks Fail

The Political Roots of Banking Crises in Spain

Sebastián Royo
Suffolk University
Boston, MA, USA

ISBN 978-1-349-95991-4   ISBN 978-1-137-53228-2   (eBook)
https://doi.org/10.1057/978-1-137-53228-2

Cover illustration: © Alex Linch shutterstock.com

This Palgrave Macmillan imprint is published by the registered company Springer Nature
America, Inc.
The registered company address is: 1 New York Plaza, New York, NY 10004, U.S.A.

All truths are easy to understand once they are discovered. The point is to discover them.
Galileo Galilei

To my daughters—Andrea, Monica, and Abigail. You are my light and my motivation, and you have taught me the most important lessons of life.

Previous Publications

Portugal: Forty Years After the Revolution, 2018.
O Dilema da Espanha, with Sonia Alonso, Cristobal R. Katwasser, Manuel Maldonado, Ignacio Molina, Sebastián Royo, Ilke Toygür, 2015.
Lessons from the Economic Crises in Spain, 2013.
Portugal in the 21st Century: Politics, Society and Economics, 2011.
Varieties of Capitalism in Spain: Remaking the Spanish Economy for the New Century, 2008.
Do Isolamento a Integaçào, 2004.
Spain and Portugal in the European Union: The First 15 Years, 2003, coedited with Paul C. Manuel.
"A New Century of Corporatism?" Corporatism in Southern Europe: Spain and Portugal in Comparative Perspective, 2002.
From Social Democracy to Neoliberalism: The Consequences of Party Hegemony in Spain 1982–1996, 2000.

Preface and Acknowledgments

This book has taken me too long to write. I started it back in 2014 as a follow up to my 2013 book *Lessons from the Economic Crisis in Spain*. By the time that I finished that book, the effects of the 2008 economic crisis in Spain were not yet fully clear. As I wrote that book felt as if I was in the middle of a hurricane, pulled in all directions, sinking from the weight of the storm, and never quite in control. When I finished it, there were still many uncertainties and open questions about what was going to happen in Spain. I still stand by the analysis and conclusions of that book but it was clearly incomplete. In particular, the causes and effect of the banking crisis were just beginning to show. This book gives me the opportunity to come full circle and to finish the story. Moreover, as I was half-way through this manuscript I was appointed Acting Provost at Suffolk University. What was supposed to be a few months assignments, turned into a three-year commitment, and as much as I tried, my administrative responsibilities kept me away from the manuscript. Regardless, while the book arrives at a time of relative financial stability in Spain and around the world (just before the COVID-19 pandemic) after the country was unwinding from its financial bailout, and banking supervision has been tightened and capital requirements strengthened, it is still very much relevant because rather than merely reacting to the latest crisis, this book looks at deep and long-term causes of financial instability, and it focuses on long-term patters and structural factors of the Spanish financial system that have made the country more vulnerable to financial crises, which are as relevant today as they were a few years ago.

The following pages are the product of these struggles. The crisis was devastating for Spain, but the country has recovered. The solid foundations of the country are still there: is human capital with of the most educated generations in Spain's history; its extensive connections abroad, particularly in the EU and Latin America; its cost advantages compared to other western European countries; its impressive infrastructure network; as well as the success of Spanish firms and banks abroad. At the same time, while the crisis has not had the full cathartic effect that I expected, and new challenges have emerged (particularly with regard to political stability caused by the erosion of the two-party system and the increasing political polarization and fragmentation, as well as the Catalan crisis), the country has been purging some the excesses and mistakes that led it to the 2008 crisis in the first place, and is now growing under a much healthier model based on exports. Still, as I mentioned on the *Lessons* book, it remains essential that we understand what has happened and why, and that learn from the mistakes in order to avoid them again. While researching for both books I have sometimes been surprised by the attempts to blame others (the United States, Germany, Brussels, and even the euro) for the crisis and by the tendency to avoid any self-criticism or assume responsibility for the domestic decisions that contributed to the crisis. If we persist in that pattern, we are bound to repeat the same mistakes. It is my hope that we do not. This book is, I hope, another modest contribution toward that effort.

As I discussed in the 2013 *Lessons* book, the economic crisis in Spain was severely intensified by the global financial crisis that spread throughout the world in 2008 after the collapse of Lehman Brothers. The crisis was also the by-product of the institutional and political deficiencies of the European Monetary Union project. There is no question about it. Yet, that crisis also has deep domestic roots that can be attributed only to Spain. Much can be written about the global financial crisis, or the Eurozone crisis. That is not the objective of this book either. While acknowledging the international dimension and causal components of the crisis, this book is particularly interested in examining the domestic political causes of the banking crisis. It focuses on particular on domestic banking crises and looks at the political bargains that have made the Spanish banking system fragile. When I first started this research, I was planning to focus exclusively on the most recent banking crisis that started in 2008. As I worked on it, however, it was clear that that crisis was not an isolated event, but rather a continuation of a pattern. Hence, I decided that it was essential to place that banking crisis in a historical context and I proceeded to examine the overall evolution and of the Spanish banking system to explain why it had been so crises prone.

The foundations of this book are rooted in my interest for my country of origin, Spain, and in particular for its political economy. I am fortunate that one of my responsibilities at Suffolk University is to oversee its campus in Madrid. This has allowed me to travel to Spain very regularly and to witness firsthand the effects of the crisis.

Over the years, my thinking about the political economy issues has been influenced by many people. I would like to acknowledge again the insight of the following people: Joaquín Almunia, Ángel Berges, Nancy Bermeo, Katrina Burgess, Cesar Camisón, Michele Chang, William Chislett, Carlos Closa, Rachael Cobb, Xavier Coller, Francisco Conde, Alvaro Cuervo, Roberto Domínguez, Omar Encarnación, Enrik Enderlein, Miguel Angel Fernández Ordoñez, Bonnie Field, Mauro Guillén, Peter Hall, Kerstin Hamann, Iain Hardie, Diego Hidalgo, David Howarth, Elena Llaudet, Andrew Martin, Cathy Jo Martin, Felix Martin, Fernando Moreno, Carlos Mulas, Rafael Myro, Emilio Ontiveros, Andrés Ortega, Sofía Pérez, Charles Powell, Lucia Quaglia, Marino Regini, Álvaro Renedo Zalba, Joaquín Roy, Vivien Schmidt, Philippe Schmitter, Ben Ross Schneider, Kathleen Thelen, Pablo Toral, Mariano Torcal, José Ignacio Torreblanca, and Amy Verdun. In particular I want to acknowledge and express my utmost gratitude to Robert Fishman. He has provided feedback on many chapters, which has been instrumental. Robert's seminal work on Spain and Portugal has always been a major source and an inspiration for my research. He has been an intellectual mentor and a great friend.

Drafts of chapters of this book have been presented at universities and academic conferences, including Harvard University as well as meetings of the American Political Science Association, the Conference of Europeanists, and the European Union Studies Association. I want to thank all the people who were a part of those panels for their valuable insight and comments. I also want to express my gratitude again to David Howarth, Iain Hardie, and Amy Verdun for the workshops that they organized at the University of Edinburgh and the University of Victoria on market-based banking. They were instrumental in my research regarding the impact of the crisis on the Spanish financial sector, and their feedback, and that of all the participants in that project, was invaluable.

Over the years, my research in Spain has been greatly facilitated by the active collaboration of people including former cabinet members, business leaders, entrepreneurs, union leaders, scholars, and national and regional administration officials. They were all extremely generous with their time and interest. I am indebted to them. In particular, I would like to acknowledge the help I received from Emilio Ontiveros and Ángel Berges, who have been mentors and inspiration throughout my academic career. A lot of what I know about the Spanish financial system I owe to them. The *Real Insitituto Elcano* in Madrid where I am a Senior Research Fellow has also been a wonderful supporter of my work, and Charles Powell, Carlota García Encina, Miguel Otero-Iglesias and Federico Steinberg have been wonderful colleagues and important intellectual influences.

This book would not have been possible without the help of and inspiration from a number of people. I owe a great debt of gratitude to the many institutions and people who have supported my research over the years. In particular, I would like to thank all my colleagues and students in the Department of Political Science & Legal Studies at Suffolk University for providing a cordial and supportive environment, as well as my colleagues and students at the Suffolk University, Madrid Campus. Working with them has been an extraordinary experience. As I was about to finish this book, Agnes Bain, a former chair of the Government Department at Suffolk University, and the person who brought me to Suffolk University passed away. Agnes was a mentor and a role model. I have had very few people that have been as influential in my career as she was. Her life stands as testament and an example to many of us.

The Minda de Gunzburg Center for European Studies (CES) at Harvard University, where I am currently a Visiting Scholar, as well as a local affiliate and co-chair of the *Europe in the World Seminar*, has also proved to be an exceptionally supportive institution for my research throughout the years. I want to thank Peter Hall for his constant guidance and inspiration. My greatest academic debt is to him. I also want to thank Grzegorz Ekiert and Elaine Papoulias for the continuing support and inspiration. They do an incredible job running the Center and have built an amazing intellectual and personal community. The entire staff, Michael Barrio, Filomena Cabral, Laura Falloon, Elizabeth Johnson, Gila Naderi, Anna Popiel, and Peter Stevens have been immensely helpful and have provided a very supportive environment to finish this book. Vassilis

Coutifaris has been an amazing host and a wonderful friend. He has always been ready to help and we have shared great moments together. Finally, I want to thank my fellow 2019–2020 visiting scholars at Harvard CES. Special thanks to Sean McGraw, not only we have shared an office at CES but many stories to fill another book! He has been an inspiration. It was an extraordinary group and I had the tremendous privilege to spend a sabbatical at Harvard with them. Fridays' Visiting Scholars Seminar, chaired by the inestimable Arthur Goldhammer, also provided a wonderful opportunity to share our research and learnt from each other.

I also want to thank Kenneth Greenberg former Dean of the College of Arts and Sciences at Suffolk University. I had the privilege to work as associate dean for him for seven years. He has been a mentor and a great role model, and I have learned a lot from him. Ken has been incredibly supportive throughout my career at Suffolk, and I am forever indebted to him for all the opportunities that he has given me. I feel enormously lucky to have worked for him and to count him as a dear friend. I would also like to thank President Marisa Kelly. I have had the privilege to work with her for over five years. Her passion for our students, her professionalism, her integrity, and her work ethic have inspired me all along. She has been not only my boss, but also my mentor. We have gone through so much together and I have learnt so much from her. Her son Jack is inspiring us from heaven. I cannot be more grateful for all the opportunities that she has given me. Under her leadership, Suffolk University is reaching new heights. Finally, I also want to recognize two of my closest friends and colleagues at Suffolk University: Gary Fireman, and Rich Miller. They are wonderful colleagues and even better friends. There is no way I would have finished this book without their unrelenting support. I am very fortunate to work with an amazing group of people at Suffolk University and, at risk of leaving some people out, I want to recognize the unwavering support of my dear colleagues Linda Brunjes, Alina Choo, Rachael Cobb, Brian Conley, Ann Coyne, Sara Dillon, Tom Dorer, Greg Gatlin, Donna Grand Pre, Mary Lally, Tom Lynch, Elena Llaudet, Brian McDermott, Raffi Muroy, Jessica Murray, Abraham Peña, Andrew Perlman, Marilyn Plotkins, Colm Renehan, Jennifer Ricciardi, Antonia Rizzo, Carlos Rufin, Joyya Smith, Susan Spurlock, Maureen Stewart, Bryan Trabold, and Valerie Ventura. They are wonderful colleagues and even better friends. I could not be more grateful to them.

As I publish this book I have just started a new adventure at Suffolk University as Vice President of International Programs. And I could not have started under more difficult circumstances. As this book goes to press, the world is in the midst of an unprecedented pandemic caused by the COVID-19. The United Spain and Spain, as many other countries, are suffering immensely with thousands of cases of coronavirus, hundreds of deaths and in the case of Spain the entire country in an estate of emergency. I follow the news from the distance in fear for my parents, friends, and loved ones. Boston is not far behind. Harvard University has in effect been shut down, and many of my fellow visiting scholars at CES have returned to their home countries. At Suffolk we have been working around the clock to minimize the impact of the crisis and to support the members of our community. We are taking actions and seeing things that I never thought we would have to see. I am proud to be a member of a community at Suffolk and Harvard where people put the greater good above their own self-interest. It is very hard to anticipate what we will will face in the weeks and months ahead, but everyone knows enough to understand that COVID-19 will test our capacities. I truly hope that this crisis will bring the best of who we are, and that we will proceed with the wisdom and generosity that this crisis demands. My heart goes out to all the people who have suffered losses from this devastating crisis.

On a more personal note, I would like to thank all the members of my family. My parents, José Antonio and María del Valle, have always been incredibly loving and supportive. They are my role models and the sources of the best in me. I would have never achieved anything without their love and support. My brother Borja has always been a champion of my work. I love them very much. I also want to give a shout to my NH crew: Steve, Karen, Phil, Maureen, Shawn, Shannon, Jeff, and Laura. Misty Harbor is my haven of sanity and I wrote many sections of this book there. They have always been incredible supportive, and they have kept me grounded and sane!

As I finish this book, we are close to the tenth anniversary of the death of my twin brother Pepe, after bravely fighting a battle against cancer. As I have said before, he was the smartest, kindest, and the most generous person I have had the privilege to know, and he was, and still is, the most influential person in my life. I would not be who I am without him, and I would never have accomplished anything without his love, leadership, inspiration, and constant support. I still miss him so much, every single day. The pain has not abated. I still feel like I am missing a limb, and I struggle every day to cope with the loss. There is not a single day that I do not think of him, and that he is not with me. He is a central part of my life, of my memories, and of who I am. He also lives through his beloved widow Sue, his daughter Zoe, and son McKenzie. They are a delight, and there is so much of my brother in them. We always cherish the times that we spend together. Poet Pablo Neruda once said: "To feel the love of people whom we love is a fire that feeds our life." I have been inspired by my twin brother Pepe, and I feel I should carry on those things that were so important to him. I still try every day.

Finally, I also want to thank my family who is everything that truly matters. My greatest debt is to my wife Cristina. We have shared 26 extraordinary years. She is an exceptional wife, mother, and professional. Cristina is the best thing that ever happened to me. Nothing that I have achieved, including this book, would have been possible without her love, patience, dedication to our family, incredible hard work, and unwavering support. Writing a book also has distributional consequences and the ones who have suffered the most from my absences have been my daughters Abigail, Andrea, and Monica. Regardless, they have been a joy and a constant source of happiness. Their achievements make me so proud. They have taught me the most important lessons of life. I love them so much and I am eternally grateful for the patience and support. I have dedicated this book to them.

Boston, MA, USA   Sebastián Royo
March 2020

CONTENTS

1 Introduction: Spanish Banking—How Do We Explain
a History of Fragility? 1

2 The Origins of the Spanish Banking System 53

3 Spanish Banking in the Twentieth Century 79

4 From Boom to Bust: The Economic Crisis in Spain
2008–2013 119

5 The Global Financial Crisis and the Spanish Banking
System: Explaining Its Initial Success (2007–2010) 141

6 A 'Ship in Trouble': The Spanish Banking System
in the Midst of the Global Financial System Crisis
(2010–2012) 173

7 Bank Bargains and Institutional Degeneration 213

8 Conclusions: The Implications of 'Bank Bargains'
for Democratic Politics 247

Bibliography 313

Index 331

ABOUT THE AUTHOR

Sebastián Royo is a Visiting Scholar at Harvard University's Minda de Gunzburg Center for European Studies (2019–2020). He is a Professor in the Department of Political Science & Legal Studies and Vice President of International Affairs at Suffolk University in Boston, USA, where he served as Acting Provost between August 2016 and August 2019. Royo's articles and reviews on comparative politics have appeared in *Comparative Political Studies, European Journal of Industrial Relations, PS: Political Science and Politics, West European Politics, South European Society and Politics, Democratization, Mediterranean Quarterly, SELA, FP, Perspectives on Politics*, and other publications. His books include *From Social Democracy to Neoliberalism: The Consequences of Party Hegemony in Spain, 1982–1996* (2000), *A New Century of Corporatism? Corporatism in Southern Europe: Spain and Portugal in Comparative Perspective* (2002), *Spain and Portugal in the European Union: The First Fifteen Years* (ed. with P. Manuel, 2003); *Portugal, Espanha e a Integração Europeia: Um Balanço* (ed. 2005); Varieties of Capitalism in Spain (2008); *Lessons from the Economic Crisis in Spain* (2013), and *Portugal, Forty-Four Years After the Revolution* (ed. 2018). Royo is a Senior Research Associate at the Elcano Royal Institute in Madrid, and an and Local Affiliate at the Minda de Gunzburg Center for European Studies at Harvard University, where he is the co-chair of the *Europe in the World Seminar*. He is the co-founder and co-chair of the *American Political Science Association's* Iberian Studies Group, and serves in the editorial boards of *South European Society & Politics* and esglobal.org.

LIST OF FIGURES

Fig. 1.1 Number of systemic banking crises, World Bank
classification, OECD, 1976–2011 (*Source* World Bank,
Global Financial Development Database. From Copelovitch
and Singer [2020, p. 26])   6

Fig. 1.2 Total banking crises (Reinhart and Rogoff classification).
OECD, 1976–2009 (*Source* Allen and Rogoff 2011. From
Copelovitch and Singer [2020, p. 27])   7

Fig. 1.3 Worldwide Governance Indicators (*Source The Worldwide
Governance Indicators.* http://info.worldbank.org/govern
ance/wgi/#home)   16

Fig. 1.4 Long-term interest rates. Total, % per annum, March
2000–October 2006 (*Source* OECD Main Economic
Indicators: Finance)   31

Fig. 1.5 Institutional structure of the Spanish financial system in
2005 (*Source* AFI 2005)   38

Fig. 4.1 Unemployment rate (2000–2020) (*Source* OECD Labor
market statistics)   131

Fig. 5.1 Spanish credit institutions: wholesale funding   152

Fig. 6.1 Spain: savings vs. commercial banks, 1980–2010 (*Source*
IMF 2012)   199

Fig. 7.1 Business environment. Selected ranking within OECD,
2013 (*Source* IMF, *2013 Article IV Consultation*, p. 22)   234

Fig. 7.2 Spanish democracy in decline (*Source* Freedom House.
Freedom in the World 2020, p. 10)   243

Fig. 8.1   Short-term interest rates. Total, % per annum, Jan 2011–Feb 2020 (*Source* OECD Main Economic Indicators: Finance)   271

Fig. 8.2   Labor compensation per hour worked in Spain and Germany. Total, Annual growth rate (%), 2000–2019 (*Source* OECD Productivity Statistics: GDP per capita and productivity growth)   278

Fig. 8.3   Income Inequality. Euro Area (2017) Gini coefficient, 0 = complete equality; 1 = complete inequality (*Source* OECD Social and Welfare Statistics: Income distribution)   297

Fig. 8.4   Gross Domestic Product (GDP) Total (US dollars/capita, 2019) (*Source* OECD. Aggregate National Accounts, SNA 2008 [or SNA 1993]: Gross domestic product)   298

Fig. 8.5   Unit labor costs by persons employed/by hours worked, Percentage change, previous period, 2016 (*Source* OECD. Labor: Unit labor cost—quarterly indicators—early estimates)   301

Fig. 8.6   Labor productivity and utilization. Labor productivity/Labor utilization, Annual growth rate (%), 2016 (*Source* OECD Productivity Statistics: GDP per capita and productivity growth)   302

LIST OF TABLES

Table 1.1	Financing forms' initiatives in spain in the 1990s	12
Table 3.1	The Spanish economy (1980–1985)	104
Table 4.1	The boom years (2000–2008)	122
Table 4.2	The economic crisis: 2008–2013	130
Table 5.1	The provisions of Spanish banks (In millions of Euros)	169
Table 6.1	A chronology of crony capitalism among *cajas*	179
Table 6.2	The financial sector and the exposure to real estate risk	197
Table 6.3	Restructuring of the financial sector with FROB assistance (as of May 8, 2012)	202
Table 7.1	Consideration of different institutions	222
Table 7.2	Spaniards opinion about the degree of representability of their political system	224
Table 7.3	WEF competitiveness ranking. Spain 2006–2010	233
Table 7.4	The justice system treats wealthy and poor people the same (%)	236
Table 7.5	What is the main problem that currently exists in Spain? (% of the population that selects each option as the main problem)	238
Table 8.1	State aid to banks ten years after the crisis (FROB, FGD and support to Sareb)	269
Table 8.2	Economic performance Spain. 2014–2020	299

CHAPTER 1

Introduction: Spanish Banking—How Do We Explain a History of Fragility?

INTRODUCTION

This book examines the "fragility" and historical crises proneness of the Spanish banking system, with a particular focus on the most recent banking crisis that followed the global financial crisis of 2008. It borrows and builds on the arguments developed by Calomiris and Haber (2014) who have argued that banking systems arise from a process of political bargaining: Every country's banking system is the result of what they call a *Game of Bank Bargains*—a game in which actors with differentiated interests come together to form coalitions that will determine how banking systems are created and how they will operate: "Banks' strengths and shortcomings are the predictable consequence of political bargains and those bargains are structured by a society's fundamental political institutions" (p. x).[1] Each country is different: It has different rules of the game, specific players with distinctive interests, governments who get different shares of the benefits, and coalitions that forge specific "*bank banking*

---

[1] In their magisterial comparative analysis of banking systems that focuses on the United States, England, Canada, Mexico, and Brazil, Calomiris and Haber invite other scholars to test their framework "against additional country cases" (2014, p. 25) hoping that other scholars would evaluate their interpretations by comparing them with detailed narratives of other countries. This book takes that invitation and applies their framework and arguments to the Spanish case.

© The Author(s) 2020
S. Royo, *Why Banks Fail*,
https://doi.org/10.1057/978-1-137-53228-2_1

bargains." According to them, banking systems are the result of political struggles ("bank bargains") between government, voters, and interest groups. The *Game of Bank Bargains* determines the rules that define how banks are regulated, how they are chartered, and how they interact with the state. These outcomes, in turn, determine banking system's performance along two dimensions: propensity for banking crisis and the degree of private access to credit (p. 477). These authors focus on the structure of the banking system and the rules for allocating credit as the key explanation for banking instability, and in order to explain why banking systems vary in structure, they identify two key variables, regime type and the presence or absence of a populist coalition.

Calomiris and Haber (2014, p. 454) define abundant credit as a ratio of private bank credit to GDP of about 83% over the 1990–2010 period, and a stable banking system as one that has been crises-free since 1970. According to that definition, only six out of 117 countries meet that threshold: Australia, Canada, Hong Kong, Malta, New Zealand, and Singapore. This book aims to explain outcomes in this *"Game of Bank Bargains"* in a particular country, Spain, which has suffered two major banking crises since the 1970s.[2] It seeks to answer one fundamental question: *Why has Spain not been able to construct a banking system that avoids banking crises?* In answering this question, the main focus will be on political factors, and in particular in the interplay between politics and banking, which has been crucial to account for the performance of the Spanish banking system.

The book examines the historical circumstances that have shaped the formation of political institutions and coalitions that control banking outcomes in Spain. Those circumstances are crucial to determine the extent to which Spain has suffered banking crises and suffered scarce credit, and they depend on the historical context, which changes as a result of external influences (for instance, banking crises have arisen from political struggles). Yet, these circumstances are mediated by political institutions (for instance, they may be able to insulate banking systems from populist policies). Following Calomiris and Haber (2014, pp. 488–89), it looks at political preconditions on banking-system outcomes, but accepts that banking systems also shape politics as well: Banking systems are not just an outcome of politics, they also shape the coalitions that bargain and affect

---

[2] I also adopt Calomiris and Haber's definition of banking crises. They regard them as "either systemic insolvency crises or systemic illiquidity crises" (p. 5, footnote 1).

the bargaining power of the parties that participate in the bargain. Finally, political circumstances influence innovations, which shape the development of new financial services and instruments. The main argument of the book is that political circumstances influence *bank banking bargains*, and that they, in turn, define the types of banks that emerged in the country. It seeks to explain banking goals and how they are shaped by political bargains, not the extend of regulation, but rather on the goals that give rise to regulation (p. 14).

In regard to the most recent Spanish banking crises of 2008, the main hypothesis is that this financial crisis was first and foremost the outcome of a political bargain. A central argument of this book is that politics matter, and that political factors are central to understanding why Spain has suffered repeated banking crises. It starts from the premise that banks are not just institutions run and controlled by technocrats who make mistakes, and also that banking crises are not merely the results of such mistakes, or incompetence, bad luck, or moral shortcomings. On the contrary, this book shows how politics (understood as the political institutions that structure the incentives of economic and social actors) influence bankers' decisions, their operations, and the regulatory framework in which they operate. Indeed, political institutions and politics structure the incentives of actors involved in banks, from bankers to shareholders, depositors, debtors, to regulators. In other words, banking crises are the consequence of banks' characteristics and the political circumstances in which they operate (see Calomiris and Haber 2014).

As we will see below, financial crises are devastating. Despite this, and this is the puzzle that this book seeks to address, in countries like Spain these crises have been persistent throughout history and governments have been unable, or unwilling, to make sure that banks either limit their risk exposure and/or have sufficient capitalization. When governments deal with banking decisions, they often face conflicts of interest, and different groups will have distinctive preferences and leverage (based on their wealth and power) to influence those decisions. The way(s) in which governments resolve or mediate those conflicts, and the resulting coalitions that emerge from the bargaining process, will largely determine the strength (or fragility) of the resulting banking structure.

Governments are indeed essential because they regulate banks, they enforce contracts, they use banks as source of finance, and in the case of bailouts, they allocate losses among creditors. But other groups (e.g., bankers, shareholders, depositors, debtors, taxpayers) also have 'skins in the game' and have a stake in the performance of banks. The interplay of these actors (what Calomiris and Haber call the "*Game of Bank Bargaining*") will lead to coalitions that will determine bank entry rules, the degree of competition in the sector, credit flows, credit conditions, banks' activities, and the allocation of loses. Different bank structure configurations have different beneficiaries and losers among those stakeholders, and the outcome of the political struggle among them will determine the role that they will play in the financial system. This book seeks to explain how this struggle/bargaining process (what they call the "*banking games*") has played out in Spain in order to explain how banking rules have emerged in the country, and how the players have operated under those rules (Calomiris and Haber 2014, pp. 12–15).

To do so, the book traces the coevolution of banking and politics in Spain over time, with a particular focus on the two decades that preceded the 2008 crisis; and it seeks to test Calomiris and Haber's hypotheses (2014, p. 453): Democracies are more conductive to a broad distribution of bank credit than autocracies; democracies with liberal institutions are more conductive to broad distribution of credit and the absence of banking crisis than those in which bankers form coalitions with populists; and government safety nets tend to destabilize banking systems, and they arise as a result of political bargains, not for economic efficiency reasons. In other words, it will test the *hypothesis that stable democracies, like Spain's, with weak institutions that cannot prevent opportunities for the development of rent-seeking coalitions between populist and bankers tend to have less stable banking system (i.e., more crisis prone) and tend to provide less abundant credit.* On the contrary, democracies with strong institutions that prevent rent-seeking tend to have stable banking system that provides higher levels of credit.

BANKING CRISES

Financial crisis has very serious and lasting consequences: Banking collapses impact borrowers and can make credit more scarce, thus forcing businesses to cut back on investment and possibly reduce their workforce; individuals may also have to hold back on purchases and reduce their consumption; and they often lead to bailouts in which taxpayers have to pay to rescue the banks. As a result, unemployment increases, economic growth collapses, deficits and debt spiral, and currencies crash. Often, they also lead to political crises, as governments collapse. Reinhart and

Rogoff (2009) have looked carefully at the consequences of financial crises. According to their seminal study, stock prices fall by more than 50% on average over more than three years; real estate markets fall by 35% over six years; economic output declines on average 9% over two years; and unemployment rises by 7% over more than four years (Reinhart and Rogoff 2009, p. 466). Moreover, financial crises also have serious political consequences on domestic politics, political stability, and governance survival: In Spain, for instance, the Socialist government that was in power at the time of the crisis was voted out in 2011, and subsequent elections have led to the collapse of the two-party system and increasing fragmentation of the party system, which has made the formation of governments far more difficult, and consequently, the country has had 4 general elections in 4 years (between 2015 and 2019). Finally, they can also lead to domestic or international conflict, as it happened with the European Union (EU) during the 2008 Euro crisis. Therefore, it is crucially important to explain why crises occur.

Yet, financial crises have become an unfortunate recurrent fact of life, and no country seems immune. Following the collapse of the Bretton Woods system in 1971, one of the most consequential economic developments of the world economy has been the globalization of finance. According to the Organization for Economic Cooperation and Development (OECD), gross cross-border capital flows rose from about 5% of world GDP in the mid-1990s to about 20% in 2007, or about three times faster than world trade flows. While there have been many befits from this growth (e.g., in economic growth, investment, efficiency, or living standards), it has also led to increasing number of financial crises. According to the World Bank's Global Financial Development Database between 1976 and 2011, there have been 131 crisis years for OECD countries. During that period, only 3 countries, Australia, Canada, and New Zealand, have not experienced a systemic banking crisis. And Spain is one of the OECD countries with the highest number of crisis (Fig. 1.1).

However, the World Bank's classification is considered restrictive because it only measures systemic crises that result in multiple bank failures, or a systemic pattern of distress across a large number of financial institutions, and thus does not account for all episodes of financial in stability (see Copelovitch and Singer 2020, pp. 25–28). For this reason, scholars also use the classification of banking crises developed by Reinhart and Rogoff (2011), which is broader as they consider two types of events:

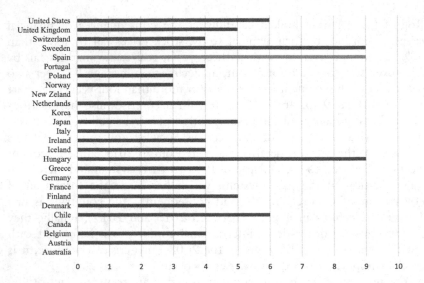

Fig. 1.1 Number of systemic banking crises, World Bank classification, OECD, 1976–2011 (*Source* World Bank, Global Financial Development Database. From Copelovitch and Singer [2020, p. 26])

"bank runs that lead to the closure, merging, or takeover by the public sector of one or more financial institutions," or "if there are no runs, the closure, merging, or takeover, or large scale governance assistance of an important financial institution (or group of institutions), that marks the start of a string of similar outcomes for other financial institutions" (Reinhart and Rogoff 2011, p. 1680). Under this classification, Spain also stands out for the number of banking crises (see Fig. 1.2).

While the focus of this book will be on banks[3] and banking crises defined as either "systemic insolvency crises or systemic illiquidity crises" (Calomiris and Haber 2014, p. 5), when looking at the more recent performance of the Spanish banking system, it is very important to

---

[3] The focus is on chartered banks rather than financial markets because in Spain (as in many other countries) banks have been, by far, the main sources of credit and access to capital. Financial markets are sustained and created by banks. For this reason, I use the terms 'financial system' and 'banking system' indistinctly in the book.

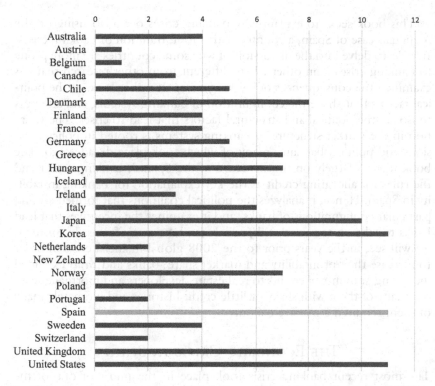

Fig. 1.2 Total banking crises (Reinhart and Rogoff classification). OECD, 1976–2009 (*Source* Allen and Rogoff 2011. From Copelovitch and Singer [2020, p. 27])

distinguish between the large banks and the *cajas* (savings and loans): The large banks performed relatively well while the *cajas* suffered from a somewhat traditional financial crisis. According to their argument, these crises occur when banking systems are vulnerable by virtue of the way in which they have been constructed (they are "*fragile by design*"). An analysis of banking crises shows that they typically take place either when banks have exposed themselves through high risks in their lending and investment, or when they have inadequate capital balance to absorb losses. Examining the case of Spain, this book confirms that such crises happen as a result of political choices, and that banks' crises reflect the structure of Spain's political institutions.

This book seeks to explain why banking crises occur, focusing on the particular case of Spain, a country with a long tradition of banking crises. It seeks to delve into the discussion of why some countries are more prone to banking crises than others. It is different from other research that has examined the consequences of banking crises. It rather looks at the political factors that shape the configuration of banking systems, and attempts to isolate the political and structural factors that lead to crises. The variation in the market structure of banking systems is rooted in policy decisions and political bargains, many of which took place decades ago. The book focuses largely on the structure of the Spanish banking sector and the rules for allocating credit as the key explanation for banking instability in Spain. Hence, it analyzes the political coalitions that formed around particular configurations of banks, and it examines the incentives that lead banks to take on too much risk, which can take many forms. In Spain, as we will see, in the years prior to the 2008 global financial crisis in order to increase their profitability and market share, banks and *cajas* extended increasing amounts of credits to real estate developers and mortgage seekers, many of them with shaky or little credit histories and ultimately many of them required a massive bailout.

THE ECONOMIC CRISIS IN SPAIN

The most recent banking crisis took place in the midst of one of the worst economic recessions in Spanish modern history (prior to COVID-19). Indeed, the financial and the economic crises were so intertwined that is impossible to explain one without the other. Between 1996 and 2007, the Spanish economy was one of the fastest growing and the most successful economies in Europe. High levels of immigration, low interest rates, and the liberalization and modernization of the Spanish economy all contributed to this spectacular performance. This success, however, came to a halt in 2008, and in the winter of 2013, Spain was still suffering the effects of a very painful economic crisis.

As we will examine in Chapter 4, the great recession that began in 2008 has been blamed on global capital imbalances, inadequate central bank policies, greed, the failure of institutions, faulty risk assessment models, or the pervasive belief that 'this time was different,' that we had found the cure against the business cycle, and that housing prices would go on rising forever. However, while the global economic crisis has been a significant contributing factor in this downturn, this book shows that domestic factors largely help account for Spain's current economic problems.

The crisis in Spain took place in the context of two larger transformations. First, a socioeconomic transformation was driven by globalization (with the downward pressure that it has been exerting on wages), as well as the introduction of new technologies (with its impact on the labor market). Changing family structures, the decline of unions, the erosion of civil society, and educational gaps also impacted the socioeconomic fabric of the country. In addition, the economic crisis took place in the context of the unraveling of the national fabric. After decades of authoritarianism, the democratic transition of the late 1970s that culminated with the 1978 Constitution seemed to settle many of the divisions that had torn the country for centuries. The crisis, however, exposed the fragility of that settlement and battered institutions, political parties, the establishment, and national elites.

Indeed, the crisis had an earth-shattering effect in Spain at all levels: economic, political, institutional, and social. There have been different interpretations of the causes and culprits for the crisis. Most of the analyses on the economic crisis in Spain have concerned themselves with phenomena like mismanaged banks, excessive debts, the bubble in the real estate sector, or the loss of competitiveness (Royo 2013; Ortega and Pascual-Ramsay 2013; Quaglia and Royo 2015; Molinas 2013; Barrón 2012). Others have sought to explain the crisis as a by-product of globalization and/or European Monetary Union (EMU) integration because it eliminated exchange rate risks in a way in which investors in the southern countries accepted lower yields (Bermeo and Pontusson 2012; Kahler and Lake 2013; Armigeon and Baccaro 2012; Cameron 2012). This led to massive capital inflows toward the periphery that fuelled booms that turned into bubbles (in the case of Spain, particularly in the real estate sector), causing massive current account deficits. When the global financial crisis hit these countries, the bubble burst and investors refused to continue financing these deficits, which exposed these uncompetitive economies.

By focusing on a particular dimension of the crisis, the financial crisis, this book seeks to contribute to the debate about the causes of the crisis in Spain and seeks to complement those analyses (Royo 2013). It moves beyond the above explanations and seeks to focus in particular on the crisis of the Spanish financial sector. In particular, it focuses on the interplay between politics and banking.

OVERVIEW OF THE BOOK

This book seeks to explain why the Spanish banking system has been historically so unstable, 'fragile,' and crises prone. It focuses on the intertwined connection between politics and banking in Spain, and it looks at the historical relationship between the economic power of Spain's largest financial institutions and the political power of successive Spanish governments. Building on the approach developed by Calomiris and Haber (2014), the book analyzes the coalitions, institutions, and regulations that have shaped banking in Spain. It shows how, through strategic bargains with successive governments and central bankers, large Spanish banks came to dominate the Spanish economy and how they retained their power even as the political system changed dramatically. Indeed, banks in Spain obtained regulatory concessions and privileges from governments in exchange for providing funding and channeling credit to politically popular causes. This led to a rent-seeking system in which politicians and bankers shared the spoils extracted from other sectors of society.

Furthermore, this historical analysis will help contextualize the changes in Spain's national financial system (NFS) and the extent to which it was affected by the international financial crisis of the mid-2000s. It also addresses the following question: *What factors explain the way in which the Spanish financial system has been affected by the crisis and has reacted to it?* In Spain, banking exposure to wholesale markets has increased significantly over the last decade and, therefore, has become "market-based banking," with significant implications for the character of credit provision and the nature of the Spanish NFS . The lower reliance on bank lending proved to be beneficial at the beginning of the crisis. While most market-based systems faced further problems in the wake of the Lehman Brothers collapse, Spanish banks performed relatively well. However, the crisis eventually had a major impact, once more "traditional" problems with government debt and lending emerged. This book analyzes this reversal and explains its causes.

Prior to the crisis, Spanish regulators placed emphasis upon preexisting regulatory and supervisory frameworks, which initially shielded the Spanish financial system from the direct effects of the global financial crisis. This contributed to their initial positive performance compared with their European counterparts. However, as in Ireland, the collapse of the real estate markets eventually led to a traditional banking crisis fueled by turbocharged lending on the liability side of the Spanish bank's balance sheets. The global financial crisis and the subsequent credit crunch had a simultaneous effect: On the one hand, it reduced banks' ability to borrow and therefore to continue lending, thus leading to the collapse of the real estate sector; on the other, the property collapse contributed to reduce banks' ability to borrow. In order to explain the recent Spanish financial crisis, this book also focuses on the following factors: first, the regulatory framework; second, the institutional features of the banking system, including the role of the Bank of Spain (BoS); and third, the impact of macroeconomic developments, notably the real estate bubble. This book argues that the outcome is historical and contingent: There have been other causal independent variables of Spanish political and economic life that can be wielded to explain the crises.

Moreover, the Spanish banking system has not only been crisis prone, but it also has a troubling record of providing relatively small amounts of credit to business enterprises. While the main focus of the book will be on banking crises, it also examines another weakness of the Spanish banking system, namely its traditional inability system to provide enough credit relative to the size of the country's economy, and argues that this shortcoming is also the result of political choices (for comparative data, see Calomiris and Haber 2014, pp. 7–8). Not surprisingly, Spanish Small and Medium Enterprises (SMEs) often complained that banks were indifferent to their credit needs and not disposed to lend them money.[4] Fishman (2010) highlights that the average annual rate of increase in credits to the economy during the period 1990–1998 was a mere 8.77% in Spain (as a point of contrast, it reached more than twice in Portugal—18.41%). And as a result small Spanish manufacturing firms were being forced to finance their growth and development largely through their own internal resources: The internal funds ratio for small manufacturing firms in the

---

[4] According to Fishman, such concerns from small businesses about the availability of credit had been voiced prior to the recent crisis. For a sample of representative press reports based on a survey of small businesses Fishman references, *El Pais*, February 5, 2009, p. 17; *Público*, February 5, 2009, p. 3; and *La Vanguardia*, February 5, 2009, pp. 50–51.

Table 1.1   Financing forms' initiatives in spain in the 1990s

	Average annual credit increase 1990–1998[a]	Internal funds ratio[b]	New equity ratio[c]	Long-term debt 1997[d]
Spain	8.77%	67.2%	7.3%	10.0%
EU median	n/a	[c]40.9%	[c]27.8%	11.7%

Source Fishman (2010, Table 5)
[a]Data reflect annual average for 1990–1998; *Source* OECD Historical Statistics 1970–1999 (2000, p. 92)
[b]Data refer to sources and application of funds for small manufacturing firms 1996–1997; Data from: Enterprises in Europe, Sixth Report, Eurostat/European Commissions (2001, p. 161)
[c]Ibid
[d]Data refer to long-term debt ratio as percent of total assets in small manufacturing firms; *Source* Ibid., p. 158
[e]EU data are based on figures from Austria, France, Germany, Italy, Portugal, and Spain, for these two measures

late 1990s stood at 67.2% in Spain (vs. 21.2% in Portugal). Table 1.1 shows aggregate-level data on credits and financing for firms.

Fishman also notes that comparatively, lower access to funding also had an impact on innovation expenditures in Spain during the 1990s, a factor that hindered productivity growth, as noted in Chapter 4. For instance, innovation expenditures as a percentage of total "Turnover" between 1996 and 1997 were only 1% by Spanish manufacturing firms vs. 2.5% average in the EU-15, and the average annual increase in gross fixed-capital formation between 1989 and 1999 was 2.8% vs 2.1% in the EU-15. This was aggravated by the fact that Spanish policy-makers preferred to use the EU funds for other purposes such as infrastructural improvements, and the Spanish state's contribution to gross domestic capital formation was also concentrated in public infrastructure (Fishman 2010).

The modern roots of this shortcoming are in the legacies of the authoritarian regime (Fishman 2010). According to Calomiris and Haber (2014, p. 47), under such regimes returns to equity holders are high, loans to insider firms are subsidized, governments and banks insiders extract significant rents, and periodic fiscal firms result in expropriations. Banks in autocratic regimes tend to allocate credit to insiders (which leads to what they call a "rent-distribution system"), and depositors are reluctant to put much of their liquid wealth into the banking system because they earn less than they would through other investments. This leads to a system of underdeveloped banks, a crony banking system that allocates credit narrowly and leads to scarce credit, which is self-reinforcing because it largely lacks the institutions that enforce contract rights: For instance, these countries and banks tend to lack adequate credit analysis, efficient property and commercial registers, and reporting services (they develop them when they have to make loans to unrelated parties and when they need to be able to sanction debtors for non-payments). In such autocratic systems, banks are less willing to incur the costs to develop these institutions/resources that can facilitate arm's-length lending, and may even oppose them if they would erode their rents and/or facilitate such lending by others. Finally, these systems tend also to be unstable and lead to frequent crises because of the lack of constraints on the autocrat's authority (which may give him/her incentives to expropriate the banks when he/she has financial needs) and the tendency on the part of bank insiders to lend to their nonfinancial enterprises and rescue them, particularly during economic crises.

Indeed, Spain was characterized by comparatively high costs of credit and high return on financial assets throughout the 1980s and 1990s (see Pérez 1997, pp. 12–18). According to the European Commission's BACH project, the average cost of investment finance for Spanish firms was high compared to other countries. Prior to the full establishment of the Single Market and the EMU, and in the absence of perfect capital mobility across borders, these higher costs were attributed to the comparatively scarcity of capital in Spain. But the problem was even deeper. Investment costs not only were higher but they were also unresponsive to economic conditions (even when macroeconomic variables like public debt, inflation, or public deficit improved, which they did throughout the 1980s and 1990s); the costs were still high; and the cost of investment finance was also high in relation to the rate of return (ROR) on investment: Spanish firms experienced a very negative leverage effect,[5] which meant that Spanish firms were decreasing the profitability of their equity when they borrowed to invest in productive capacity, job creation, or technological innovation, and therefore, it was a strong disincentive for investment. In Spain, the costs of finance were largely determined by the rates charged by banks on credit, because banking loans were the principal source of external funding for Spanish firms.

---

[5] It measures difference between the average rate of return (ROR) of firms and the costs of external financing: when positive firms can increase their ROR by borrowing, but when negative it would decrease the profitability of their equity.

The comparatively higher costs of investment finance behind the strong negative leverage effect in Spain have also been attributed to macroeconomic policies, and particularly the high levels of public spending and public deficits, because monetary policy had to compensate for the lack of financial restraint, which had a "crowding-out" effect (see González-Páramo et al. in Pérez p. 17, footnote 14). Pérez (1997) challenges this interpretation and shows that fiscal deficits had a structural component, that they were not comparatively high, and that there was also a limited degree of variation in credit rates: When deficits rose, the level of credit rates barely changed.

Other scholars (Torrero 1989) have attributed banks' capacity to charge higher interest rates in the 1980s to the lack of leverage of Spanish firms who were desperate for credit at a time of crisis and high indebtedness. When economic conditions improved in the second half of the 1980s, they decreased their borrowing and increased their self-financing. Yet, banks were still able to continue raising interest rates (and profit margins) by focusing on creditors who did not face external competition and could pass the higher costs to their consumers (i.e., utilities and service ventures). However, Pérez (1997, p. 20) shows convincingly that the reason for this behavior was "the existence of an oligopolistic financial structure" that allowed Spanish banks to increase their costs and earnings in the middle of a severe recession, which had a significant impact on the high financial costs borne by Spanish firms in the 1980s. This oligopolistic structure persisted for over a decade, despite a process of financial liberalization undertook by the financial system. Pérez (1997, pp. 21–22) shows how the credit deregulation process that started in the late 1970s was not accompanied by the introduction of any significant competition into the Spanish financial market, even when foreign banks and domestic savings and loans were allowed to expand their activities, because they still faced operational constraints during the 1980s that prevented them from exerting downward pressure on credit rates.

Not surprisingly firms responded in the late 1980s by reducing their recourse to financing. An *under-bank* situation has significant social and economic costs for any country, as several studies have shown that lower levels of financial development impact physical capital accumulation, economic growth, technological progress, job creation, and social mobility (i.e., King and Levine 1993; Taylor 1998). They also impact the investment behavior of firms. As a result, countries suffer lost investment opportunities and competitive disadvantages: Potential entrepreneurs and investors are starved for credit, while bank cronies have plenty of access to funds. In the case of Spain, the high costs of capital shifted the burden of adjustment in the 1980s toward Spanish firms and intensified a bias against investment in productive capacity. The book shows that the scarcity of credit (or under-banking) has also been a result of the structure of Spain's political institutions.

In addition, this book shows that banks' crises happen as a result of political choices, and that they reflect the structure of Spain's political institutions. It argues that a fundamental reason for both the 2008 financial crisis and the economic crisis is rooted in the process of institutional degeneration that preceded the crisis (see Ferguson 2013). It seeks to explore the consequences that the institutional degradation has brought to the Spanish economy. The World Bank's *World Governance Indicators* provide relatively detailed, survey data-based assessments of the operation of political systems. The database contains six variables: (1) control of corruption, (2) government effectiveness, (3) political stability and absence of violence, (4) regulatory quality, (5) rule of law, and (6) voice and accountability. Out of the six, three—(1), (5), and (6)—can be considered direct measures of democratic quality, whereas the other three—(2), (3), and (4)—can be rather seen as measures of government efficiency. In the Spanish case, all three democratic quality variables showed little improvement between 1996 and 2008, and the corruption one even deteriorated. Meanwhile, the three government efficiency variables—political stability and absence of violence, government effectiveness, and regulatory quality—also exhibited deterioration between 1996 and 2013 (see Fig. 1.3). Moreover, although Freedom House still scores Spain 4 out of 4 for an independent judiciary, it also highlights deficiencies in the country's safeguards against official corruption (scoring it 3 out of 4) and stresses that:

Although the courts have a solid record of investigating and prosecuting corruption cases, the system is often overburdened, and cases move slowly. For instance, among other high-profile proceedings during the year, in May 2018, after 10 years of investigation, the courts handed down convictions for 29 of the 37 people indicted over their alleged involvement in the illegal financing of the PP from 1999 to 2005. The party itself was found to have benefited from the schemes and was ordered to pay a €240,000 ($280,000) fine.[6]

[6] https://freedomhouse.org/country/spain/freedom-world/2019.

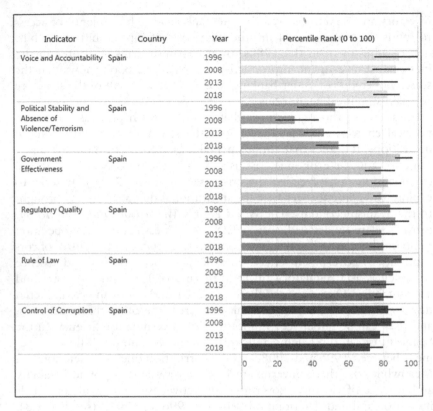

Fig. 1.3 Worldwide Governance Indicators (Source *The Worldwide Governance Indicators.* http://info.worldbank.org/governance/wgi/#home)

This book argues that the process of institutional degeneration led to a Spanish version of crony capitalism characterized by the misgovernment of the public; an outdated and inadequate policy-making process; an inefficient state; and an often corrupt and inefficient political class. Mismanaged banks, excessive debts, bubbles in the real estate sector, or competitiveness losses are all symptoms of an institutional malaise that intensified in the years prior to the crisis. Indeed, we cannot understand the bubble in the real estate sector, the loss of competitiveness, or the financial crisis without referencing the institutional divergence in the rule of law between Spain and the EU core. As analyzed in great detail in

Chapter 7, the real estate bubble, the competitiveness divergence, and the financial crisis are all rooted on the rule of law divergence. Moreover, it was this institutional divergence that made it difficult to implement the reforms that EMU demanded (and that the financial bailout ended up imposing in the country). Chapter 7 seeks to examine the causes of Spain's 'stationary state' and argues that they are in large part the result of 'laws and institutions.' They are the problem, and the economic crisis was a symptom of the institutional degeneration (Ferguson 2013, pp. 8–10).

Therefore, this research also builds on and develops further the institutional approach by Ferguson (2013), and Acemoglu and Robinson (2012), identifying the distinctive institutional features that underline countries' economic success (or lack thereof). It shows that the economic success spurred by the country's modernization and European Union membership was not sustained because the governments (at all levels, local, regional, and national) became less accountable and responsive to citizens. In terms of causal mechanisms between institutional degradation and economic crisis, it shows (following Acemoglu and Robinson's terminology) that institutions across the country became more 'extractive' and concentrated power and opportunity in the hands of only a few. Indeed, political and economic institutions came short in empowering and protecting the full potential of Spanish citizens to innovate, develop, and invest. They did not foster the degree of 'creative destruction' that is vital for innovation and sustainable growth, and instead, they promoted an unsustainable model based on a real estate bubble. Finally, institutional deterioration made it difficult to implement the economic reforms required as members of a monetary union. Political and economic institutions resisted reforms because they would jeopardize the existence of the extractive rent-seeking mechanisms that became the main source of rent for the economic and political elites that controlled them.

In sum, the book seeks to examine the structure of the Spanish financial system with particular emphasis on the last two decades (1994–2014) to answer the question: *Why has Spain not been able to construct a banking system that avoids banking crises?* In doing so, it analyzes in particular the basis of the financial system's apparent success during the 1997–2007 period, and then, it examines the imbalances that made that success unsustainable, and ultimately led to the financial bailout. The book also looks at its consequences of the financial crisis. Finally, based on the analysis of the Spanish experience (both within the EU/EMU and during the crisis), it draws some lessons that would be of interest from a normative standpoint.

This book contends that the performance of the Spanish financial system was largely driven by institutional and political factors. It examines the rules and institutions that have shaped Spanish banking, focusing, in particular, on the role that politicians and governments (Calomiris and Hager's "banking games") have played. It offers an alternative to state- and market-driven conceptions of financial regulation and reform, and examines the accommodation between private bankers, politicians, and central bankers in Spain. It examines the political coalitions that have shaped the structure of the Spanish financial system, and the consequences that these trade-offs had during the recent economic and financial crisis.

In addition, the book also examines the development of the sovereign debt crisis in Spain. Initially, it had a 'good' banking crisis and it was only hit by the sovereign debt crisis as late as mid-2011. Spanish banks suffered a traditional banking crisis from late 2010 onward. As the crisis intensified, Spain's banking sector could not escape its dramatic effects. Deteriorating economic conditions, the implosion of the real estate market, the dependence on wholesale funding, weaknesses in the regulatory framework, and the role of the BoS all help to explain this reversal. In the end, the Spanish government resorted to European Union (EU) financial aid in order to recapitalize its banks. The book argues that the distinctive features of bank business models and of national banking systems in Spain have considerable analytical leverage in explaining the different scenarios of the crises. This 'bank-based' analysis contributes to the flourishing literature that examines changes in banking with a view to account for the differentiated impact of the global banking crisis first and the sovereign debt crisis in the Eurozone later.

The book also addresses the impact of European integration on a Southern European/peripheral economy. The recent experience of the country shows that EU and EMU membership have not led to the implementation of the structural reforms necessary to address the country's economic weaknesses (i.e., dependency on the constructions sector and the erosion in competitiveness). On the contrary, the book will make the case that EMU contributed to the economic boom, which was fueled by consumption and record-low interest rates, thus facilitating the postponement of necessary economic reforms (Royo 2013). Indeed, the

Spanish experience shows that the process of economic reforms has to be a domestic process led by domestic actors willing to carry them out.

Finally, the book explores one of the core questions facing the new Europe, namely the sustainability of the EMU in the context of sharp differences in economic performance and levels of competitiveness. Spain shows the pitfalls of monetary integration for less competitive economies, used previously to high inflation and high interest rates. These countries are likely to experience an explosion in consumer spending and borrowing, because lower interest rates and the loosening of credit will likely lead to a credit boom. This development resulted in further losses in external competitiveness, together with a shift from the tradable to the non-tradable sector of the economy, which had a negative impact on productivity. How can these issues be addressed to make EMU sustainable?

What Is Unique About This Book?

First, the focus is on Spain. There are very few books published in English that examine the Spanish financial sector, and almost none that examine the crises proneness of the sector from a historical perspective looking at institutional and political factors. Second, the book examines the distinctive performance of Spain's financial sector during the recent 2008 global financial crisis. Indeed, initially during the first years of the global financial crisis, one of the few unexpected surprises was the positive performance of the Spanish financial sector. Contrary to their counterparts all over the world, at the beginning of the crisis, Spanish banks appeared to have escaped the direct effects of the global financial crisis, and it is important to understand why. There is consensus that the stern regulations of the BoS and its countercyclical provisions played a key role in this outcome. It also made it so expensive for financial institutions to establish off-balance sheet vehicles, which had sunk banks elsewhere, that Spanish banks stayed away from such toxic assets. However, the deepening economic crisis would eventually catch up with Spanish banks and *cajas* and Spain was ultimately forced to request a financial bailout. This reversal needs to be explained.

Furthermore, the book offers an alternative to state- and market-driven conceptions of financial regulation and reform, and examines the accommodation between private bankers, governments, and central bankers in Spain. Indeed, by looking at the latest crisis from a historical perspective and examining the historical fragility of the Spanish banking system, the book offers an alternative explanation to the recent financial crisis in Spain. A lot of the focus in explaining the crisis has been on the corrupt business practices and the greed of bankers (particularly in the *Cajas de Ahorro*). Others have focused on the inherent characteristics of banks that make them vulnerable to panics. Following Calomiris and Haber (2014), this book focuses on the politics of banking and it shows that the crisis was a result of human agency. The structure of Spain's financial system was the product of the country's culture, history, and the political system, yet the crisis has been the outcome of domestic political conditions and domestic political bargains between competing (and sometimes cooperating) interests mediated in the political arena. It shows how EMU and cheap wholesale financing encouraged bankers (particularly in the *cajas*) to engage in aggressive lending to the real state sector. Despite the BoS's countercyclical regulatory framework, the capital requirements proved insufficient and banks were able to rely heavily on borrowing. Ultimately, that proved to be fatal for many of them.

The book also integrates the Spanish financial European integration experience and analyzes the effects of financial integration in (relatively) new democracies in Europe. The Spanish experience within the European Union (EU) and the EMU offers one of the few instances in which integration took place in an economic, political, and institutional context markedly different from that of the other European states. At the same time, the book evaluates the impact of EU/EMU accession on the Spanish economy and financial sector. While initially successful, EMU membership brought its own problems for the Spanish economy. The Spanish experience with EU/EMU integration illustrates the economic, social, institutional, and cultural challenges of this undertaking and will provide useful lessons for other countries. While integration has had very positive effects, the process of integration has also brought significant costs in terms of economic adjustment, loss of sovereignty, and cultural homogenization. Furthermore, EU/EMU integration does not guarantee success. Indeed, Spain suffered a severe economic downturn between 2008 and 2013 and experienced serious budgetary and fiscal problems that hampered economic growth and are likely to impact the Spanish economy and financial sector for the foreseeable future. The financial sector is a crucial actor to account for these challenges.

Finally, at a time in which there are increasing doubts about the future of EMU, this book sheds light on the impact of monetary integration on a 'peripheral' economy, with a particular focus on the financial system. It shows that in the case of Spain, EMU membership contributed to prevent capital flight and helped to avoid a repetition of the attacks on the *peseta*. Without the Euro, the huge trade deficit that countries like Spain experienced prior to the 2008 crisis would have led to massive capital flight, a devaluation of the *peseta*, an inflationary spiral, increases in interest rates and the risk premium (with the associated impact on the cost of the debt), and the implementation of more restrictive monetary and fiscal policies. However, the Spanish experience also shows the pitfalls of monetary integration for less competitive economies with an inflationary history. These countries are likely to experience an explosion in consumer spending and borrowing, because lower interest rates and the loosening of credit will likely lead to a credit boom, driven by potentially over-optimistic expectations of future permanent income, which in turn may increase housing demand and household indebtedness; and lead to overestimations of potential output and to expansionary fiscal policies. The boom also led to higher wage increases, caused by the tightening of the labor market, higher inflation, and losses in external competitiveness, together with a shift from the tradable to the non-tradable sector of the economy, which had a negative impact on productivity. In the end, the insufficient responsiveness of prices and wages, which did not adjust smoothly across sectors, led to accumulated competitiveness losses and large external imbalances. In this regard, the book stresses the need to implement supply-side reforms to bring labor costs down, through wage restraint, payroll tax cuts, and productivity increases, in order to make the country's economy more competitive.

LITERATURE REVIEW

Banking crises theories have focused on three main factors: the structure of banks (they focus on liquidity risks arising from the mismatch between illiquid and long-term loans a liquid and short-term liabilities); interbank connections (they focus on the problem of externalities, when banks fail to take into account the spillover effects created by their integration in a banking system); and human nature (human errors motivated by excessive optimism or excessive fear). Calomiris and Haber (2014), however, show that the problem with these theories is that they fail to "explain why banking crises are equally likely across all countries and all of recent history" (pp. 480–81). According to them, the problems that these theories identify are not *sufficient conditions* for causing banking crises, and the crucial factor that determines whether the problems identified by these theories will result in a banking crisis is political: the choices made by politicians, often motivated by their short-term interests. These political choices will determine the degree of safety nets and prudential regulation, which in turn will influence banks' decisions. In addition, the degree of generosity of the safety net, and/or the level of appropriate prudential regulation, will influence whether banks are more or less cautious in how they manage their risk. For instance, if the safety net is too generous, and/or prudential regulation is insufficient, it is more likely that banks will be less cautious in their management of risk and vice versa. Finally, politics can also lead to shocks (from war to political transitions) that will influence approaches to bank risk. In the end, the structure of the banking system, and its susceptibility to adverse shocks, is an outcome of politics and political outcomes, which shape the rules under which banks operate and determine the context to which they are subject (2014, pp. 483–84).

Furthermore, the economic literature has paid considerable attention to the economic factors associated with sovereign debt crises. One of the most extensive studies of financial crises highlights a strong causal link between banking crises and sovereign defaults in developed and developing countries (Reinhart and Rogoff 2009). Kaminsky and Reinhart (1999) also highlight the link between banking crises and balance of payment crises (the so-called twin crises). They found that financial liberalization often precedes banking crises, and also that problems in the banking sector typically precede a currency crisis—in other words, the currency crisis often deepens the banking crisis, activating a vicious spiral. According to these analyses, economic crises occur as the economy enters a recession, following a prolonged boom in economic activity fueled by credit, capital inflows, and accompanied by an overvalued currency. These accounts, written by economists, draw our attention to banking crises and balance of payments crises as the precursors of sovereign debt crises. The accounts are, however, devoid of political analysis and do not explain the initial causes of banking crises and balance of payment crises.

In the political economy literature, several complementary explanations account for the occurrence of sovereign debt crises, mainly in developing countries. These explanations can be articulated either at the international level, whereby the focus is on external factors that caused the crisis, or alternatively at the national level, whereby the focus is on the domestic factors that caused the crisis (for a similar macro-meso distinction, see

Germain 2012). At the international level, different authors have pointed out a variety of factors that can fuel sovereign debt crises, namely the impact of capital liberalization (Stiglitz 2000); the activity of bond markets (Mosley 2003) and financial innovation, especially financialization (Hardie 2011); the spread of neoliberal ideas (Major 2012); and the lack of a proper hegemon in the international system (Kindleberger 1973)—more recently in the European Union (Mabbet and Schelke 2015). Finally, Skocpol (1979) seminal comparative work on social revolutions argued for structural analysis and emphasized the effects of transnational and world-historical contexts upon domestic political conflicts.

While some elements of an economic and/or financial crisis can certainly be explained by international factors—such as financial liberalization—these factors are basically the same for all countries, especially within the relatively homogenous regional block of the Euro area. Hence, they are not well suited to explain differences in the outcomes of the sovereign debt crises across countries. These background factors fueled the sovereign debt crisis, but domestic factors better explain different dynamics of the crisis across countries (see Jabko and Massoc 2012; Haggard and Mo 2000).

This book is also situated within the broader literature about the comparative political economy of Spain's financial system. Some scholars have sought to place the politics of financial reform in Spain in a historical context. Sofía Pérez (1997) and Arvid Lukauskas (1997) have already challenged the widespread assumption that international market forces alone explain domestic financial reforms in Spain. Both claim that in that country domestic politics played an even larger role than international pressures. Pérez contends that domestic elites, particularly a group of reformers within the central bank, seized on liberal economic arguments and developed new patterns of accommodation with private bankers to promote reforms. She emphasizes the oligopolistic nature of the sector, arguing that this system generated significant costs for Spanish firms outside the finance sector.

Lukauskas also examines the role of public officials in the evolution of financial regulation in Spain, but he is more favorable to the Spanish political class. He claims that they undertook financial liberalization, despite opposition from powerful groups, to achieve political goals; democratization gave them a strong incentive to improve economic performance through financial reform in order to compete for votes. He attributes their banking policies to an electorally based desire to secure economic outcomes, such as growth, pleasing the median voter.

More recently, in another seminal piece, Robert Fishman (2010) has also analyzed the economic consequences of Spain's banking system and has shown how that country's poor employment performance has also been rooted in the financial system. Fishman argues that the Spanish policy-makers' approach was molded by the country's path to democracy, which shaped the political handling of banking and of financing for SMEs.

In addition, this book is positioned within the Varieties of Capitalism (VoC) debate about liberal market economies and coordinated market economies (Hall and Soskice 2001). According to the VoC literature, Spain is characterized by strong strategic coordination in financial markets, but not so in the field of labor relations (Royo 2008). In this regard, some have claimed that Spain is moving toward the Anglo-Saxon model because the lack of articulation of the Spanish institutional model precludes the evolution toward a Coordinated Market Economy–CME (see Molina and Rhodes 2007). The analysis of the Spanish experience during the crisis confirms the thesis that coordination is a political process. It shows that institutional change is a political matter and how successful coordination depends not only on the organization of the social actors but on their interests and strategies. Indeed, in a context of structural changes, we have to examine the political settlements that motivate the economic actors, and we need to look at the evolving interests of capital and the structural and political constraints within which economic actors define and defend their interests (Royo 2000, 2008).

Furthermore, theoretically, the book builds on and develops further the literature on VoC, highlighting the bank bargains and the role of certain domestic political economy institutions in the development of the crisis. While financial capitalisms have converged toward deregulation as a result of the combined processes of globalization and European integration, this book shows that differences persist. Indeed, in the case of Spain, the crisis has led to extensive regulatory intervention that has served to reinforce the pre-existing model. Methodologically, it 'dissects' the sovereign debt crisis into three interconnected crises: the banking crisis, the fiscal crisis, and the balance of payment crisis, examining the 'anatomy' of each crisis paying particular attention to the interplay of domestic political economy institutions therein.

According to the standard typology of the varieties of financial capitalism, NFSs can be divided into 'bank based' and 'market based' depending on the dominant sources of nonfinancial-company (NFC) finance (Zysman 1983). This literature focuses on national systems as governments and/or NFCs as agents of change. Indeed, the transformation of NFSs is largely seen in terms of the rise of shareholding capitalism and a move away from bank-financed capitalism—an analysis which points to the implications of the end of 'patient capital' for other elements of the national variety of capitalism. According to this view, banks are static, so any changes in their activities, and the implications of those changes, are largely neglected. In other words, it tends to overlook the important role of banking and banks and their contribution to change (Hardie and Howarth 2013). This book seeks to address this shortcoming by focusing on agency (the operation of Spanish banks) in order to explain changes in national financial (and specifically banking) systems during the global financial crisis (Quaglia and Royo 2015). Yet, while the focus is upon the activities of banks, it is also necessary to understand the institutional framework that has shaped banking activities in Spain.

Moreover, an obvious additional starting point to investigate the importance of national-level institutional factors in how the sovereign debt crisis played out is also the literature on VoC (for comprehensive analyses, see Amable 2003; Hancké et al. 2007; Hall and Soskice 2001; Schmidt 2002; for a review, see Jackson and Deeg 2008). These works have examined the main components of VoC, namely industrial relations institutions; education and training systems; corporate governance and systems of corporate financing; product markets; social protection systems and welfare states; and public intervention in the economy.

This literature has generally focused on the Anglo-Saxon and continental VoC, characterizing them, respectively, as 'liberal market economies' and 'coordinated market economies.' The Southern European countries have been overlooked or placed in a residual category of 'Southern European' or 'State-led' model of capitalism (Schmidt 2002; Della Sala 2004; Royo 2008). The literature on VoC is of limited utility in explaining the dynamics of the economic crisis in Spain, because it inaccurately predicts similar outcomes among countries with similar institutional frameworks.

For instance, in the academic literature, Italy and Spain are often grouped together in the same variety of 'Southern European' capitalism. Consequently, one could have expected similar outcomes from the crises in the two countries. However, Spain experienced a full-fledged sovereign debt crisis resorting to *Euroarea* financial assistance for its banks, whereas Italy did not. Indeed, Spain experienced a severe banking crisis including the bailout of a number of the country's banks which substantially increased the public deficit and debt. By contrast, Italian banks, with one exception, did not experience such significant losses and they needed no recapitalization by the government (Quaglia and Royo 2015). To be fair, this literature was not developed to explain sovereign debt crises, but rather to tease out institutional complementarities that can enhance or hinder the competitiveness of national economic systems in the world economy.

From a political economy standpoint, this book also challenges the interpretation according to which the responses of European countries to the pressures associated with globalization and the process of European integration are uniform. Contrary to this prediction, this book shows that in Spain globalization and European integration have promoted rather than undermining alternative domestic responses. While technological changes, capital market integration, and post-industrialization have affected the balance of power between governments and private actors and have triggered new political realignments, they have also influenced the interests and strategies of the actors and have led to new strategies and patterns of change. These developments have led to particular economic policies and preferences, but they have also precluded the implementation of alternative policy options, and they have often hindered the necessary reforms.

The literature on domestic political economy institutions and economic policies (e.g., Thelen 2004; Mahoney and Thelen 2009) is also of relevance for this research, in particular those works that have looked at the interaction between domestic and external factors in economic policy. Some authors have examined the effects of 'internationalization' and 'liberalization' on domestic political economy institutions, such as financial markets (Pérez 1997; Deeg and Luetz 2001; Thatcher 2007) and systems of corporate finance and governance (Deeg and Pérez 2000). Other authors have highlighted the role of the state and domestic political institutions in framing globalization (Weiss 2003), and more generally in political economy (Schmidt 2009). This book is grounded in the comparative historical analysis literature. It seeks to offer a historical grounded explanation of the 2008 economic crisis in Spain, and it is concerned with causal analysis (the causes of the economic crisis) while emphasizing processes over time.

This book builds on these contributions and stresses the agency of domestic actors in shaping NFS change. It uses a comparative historical analysis approach to explain the financial crisis. This holistic analysis will contribute to a more complete understanding of the national varieties of financial capitalism.

Lastly, this book is also placed within the financial stability literature. A large body of literature has examined the impact of capital flows on financial crises. Many have concluded that global capital is the most likely culprit of banking crisis: Reinhart and Rogoff (2009) look at the historical pattern that connects capital inflows, large current account deficits, asset bubbles, excessive indebtedness, sovereign borrowing, and financial crises. Chinn and Frieden (2011) examine the causes of financial crisis, like the 1994 Mexican crisis, the 1997 Asian crisis, or the 2008 global financial crisis, and attribute to capital inflows the economic boom that fueled the financial exuberance that ultimately ends in a financial crisis. Portes (2009) emphasizes distortions in the allocation of capital as it is channeled through financial intermediaries and banking institutions. Schularick and Taylor (2012) analyze long-terms patterns of financial instability and find that the single most important determinant of banking crisis is domestic credit growth, and that capital inflows go hand in hand with credit booms. Other scholars, however, have found a weak link between capital inflows and domestic credit growth (Amri et al. 2016), while others have found a weak causal link between capital inflows and banking crises and no significant correlation between capital inflows and bank credit, showing that credit booms appear to be a separate channel of financial instability (Copelovitch and Singer 2020). They show that, while sustained capital inflows can be dangerous by linking current account deficits and large capital inflows to distortions in the allocation of capital, there are many cases (like Australia and New Zealand) in which large capital inflows, asset bubbles, and macroeconomic imbalances do not result in banking crises. Furthermore, capital does not always flow from surplus to deficit countries (e.g., the United States where two-thirds of capital inflows arrive from countries with which the United States does not run a large current account deficit) (pp. 185–86). These authors show, as we will see below, that larger capital inflows are associated with reduced bank capital levels only at high levels of market/ban ratio (p. 68). Finally, Caballero (2016) has also examined the association between capital inflows and banking crisis and found out that the impact is driven by debt flows and portfolio equity.

Another stream of this literature led by Kindleberger and Minsky's classic financial stability model looks at the stages of financial crises. Their model can be summarized in five different stages: displacement, boom, overtrading, revulsion, and tranquility (see Kindleberger 1996; Minsky 1982a, b). According to this model, the crisis starts with an exogenous shock to the economic system that impacts profits and opportunities in at least one sector (*displacement*). Economic actors take advantage of this opportunity(ies), which lead to the second stage, a *boom*, that in turn increases the money supply. The economy accelerates, which leads to the third stage, *overtrading*, caused by over-borrowing, over-consumption, and/or over-investment. When some insiders try to capitalize from this boom and sell out to take their profits, it leads to the fourth stage, *revulsion*. At that stage, when economic actors realize that the market cannot continue growing, any shock can turn into a *stampede*. Finally, the last stage, *tranquility*, takes place when investors move to less liquid assets, trade is cut off, or a lender of last resort provides sufficient liquidity (Kindleberger 1996, p. 15). This model provides a systematic interpretation of the anatomy of the Spanish financial crisis. As we will see throughout the book, the stages of the model can be traced in five episodes of the turmoil: financial liberalization policies; credit growth; over-borrowing, over-consumption, and over-investment; financial distress and crisis; and financial rescue.

Furthermore, the financial stability literature also examines how we can achieve financial stability in a world of cross-border banking, and whether governments can still produce this public good at the national level in today's globalized financial markets. In this regard, the examination of the Spanish case confirms Schoenmaker's financial stability trilemma (2008, 2011). Schoenmaker examines the trade-off between financial integration and national financial autonomy and argues that financial stability, financial integration, and national financial policies are incompatible. Any two of the three objectives can be combined but not all three; one has to give. Spain tried to achieve all three and failed. In order to address this trilemma, Schoenmaker supports transferring powers for financial policies (regulation, supervision, and stability) further to the European level. This would imply a European-based system of financial supervision.

Avgouleas and Goodhart (2015) have analyzed bail-in processes that involve the participation of bank creditors in bearing the costs of restoring a failing bank to health as an alternative to the unpopular bailout approach. The bail-in tool involves replacing the implicit public guarantee with a system of private penalties. They show the benefits and the disadvantages of this approach. The analysis of the Spanish case confirms the downsides of the bail-in approach and shows how bail-in regimes do not eradicate the need for injection of public funds where there is a threat of systemic collapse because a number of banks (in the Spanish case *cajas*) simultaneously entered into difficulties.

The book is also situated in the literature on financial instability caused by defaults on mortgages. Goodhart et al. (2010) explore financial instability due to a housing crisis and defaults on mortgages. They build a model that shows how tight money reduces prices and quantities traded. While government support to banks in crises stabilizes the economy, when banks become risk-loving, a subsequent crisis becomes even more extreme. Their model helps account for the risk-seeking behavior of Spanish financial intermediaries in 2004–06 that ultimately led to the combined collapse of the housing and (nearly) financial markets. Jiménez et al. (2014) and Maddaloni and Peydró (2011) have also shown how low interest rates may lead to a search for higher yields, encouraging banks to soften their credit standards, thereby increasing both the volume and average riskiness of supplied loans; and Maddaloni and Peydró (2011) have found that lower overnight interest rates induce lowly capitalized banks to grant more loan applications to risky firms and "to commit larger loan volumes with fewer collateral requirements to these firms, yet with a higher ex post likelihood of default."

Moreover, Goodhart et al. (2012) look at regulatory design. They compare different potential financial externalities and examine how to regulate them. They recognize that "trying to lean against the wind to reduce the credit expansion and house prices in the boom via regulation is not easy" (p. 42) because the boom (as shown in the Spanish case) brings large increases in asset prices, and higher prices deliver higher capital gains to owners of the assets. At the same time, while these home price gains improve the equity of mortgage holders and lower the loan to value ratio on their mortgages, they also improve bank capital ratios because the mortgages are less risky and because the home price gains raise bank equity. All these effects show the difficulty during boom periods, as it happened in Spain before the crisis, of trying to impose higher loan to value requirements, to raise capital standards, or to lift margin requirements on repo loans enough to slow down credit expansion and try to reduce house price appreciation. These authors show that the most effective ways to address these issues are through dynamic provisioning rules and liquidity requirements that may help slow mortgage credit growth and thus lower the relative price of house prices. However, their model shows, as confirmed by the Spanish experience, that changes to capital rules, progress on revising other regulations such as liquidity, margin requirements, or time-varying provisioning rules, have a limited impact, as capital alone is unlikely to be sufficient to contain the problems arising during a crisis, as it happened in Spain.

Many studies have also examined how monetary policy drives bank risk-taking and have focused specifically on the overall failures of the ECB's monetary policy. Jiménez et al. (2014) have shown how monetary policy affects the composition of the supply of credit, in particular with respect to credit risk. They have analyzed the impact of lower interest rates on banks' risk-taking and have found that a lower overnight interest rate induces banks to engage in higher risk-taking in their lending and encourages lowly capitalized banks to grant more loan applications to risky firms than highly capitalized banks. Moreover, they have found that when the overnight rate is lower, applications granted by lowly capitalized banks also have a higher ex post likelihood of default.[7] Their analysis suggests that when the monetary policy rate is lower, the intensity of risk-taking is not simply the result of more lending by capital-constrained banks, but is also consistent with risk-shifting (see also Rajan 2006; Allen and Rogoff 2011). Their findings are consistent with other studies that have shown that monetary and macroprudential policies may not be independent (Goodhart 1988; Stein 2012), while others have found, as noted before, that credit booms have the highest ex ante correlation with banking crises (see Schularick and Taylor 2012; Gourinchas and Obstfeld 2012). These findings support the attribution of new responsibilities to central banks in the realm of macroprudential supervision (Diamond and Rajan 2012).

While it is unquestionable, as examined in Chapters 3 and 7, that the loose monetary policies of the ECB (see Fig. 1.4) had an impact on the housing bubble, it is important to emphasize that monetary policy alone did not cause the crisis. Indeed, the initial growth of mortgage risk and the decline in prudential regulation preceded the ECB's loose policies,

---

[7] On the contrary, they found that a lower long-term interest rate and other key macrovariables such as securitization and current account deficits (which entail capital inflows) have no such effects.

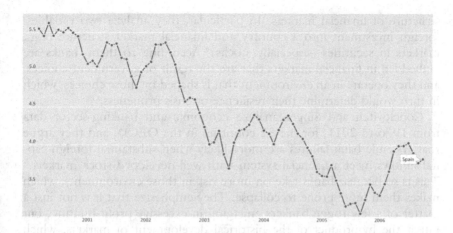

Fig. 1.4 Long-term interest rates. Total, % per annum, March 2000–October 2006 (*Source* OECD Main Economic Indicators: Finance)

and the BoS could have countered the effect of the ECB loose monetary policies by increasing capital requirements from banks and *cajas*. Furthermore, although monetary policy can contribute to the overpricing of real estate assets and the development of bubbles, banking crises require first that banks invest in those overpriced and risky assets, and also that they back those investments with insufficient capital (see Calomiris and Haber 2014, p. 271). Therefore, it is not enough to just account for weaker lending standards and prudential regulatory failures; these banks' decisions need to be explained. Moreover, we also have to explain differences among banks/*cajas* because not all banking institutions behaved the same way: Most *cajas* were far more aggressive and reckless in their lending practices than banks. These differences also need to be accounted for. This book builds from those analyses, to show that "banks bargains" created the institutional framework that allowed *cajas* to lend recklessly and provided the incentives to do so.

Finally, in their recently published book, Copelovitch and Singer (2020) examine why some countries are more prone to systemic bank crises than others, and what makes banks vulnerable under contemporary conditions. They also study banking focusing on politics, but rather than studying the political and public policy preferences of bankers or their political contributions, they focus on the political decisions that shape the structure of financial markets. In particular, they analyze two variables: foreign investment into a country and financial market structure (e.g., markets in securities—especially stocks). According to them, banks are embedded in financial markets that are the result of government choices, and they operate in an environment that is shaped by those choices, which in turn would determine their resilience or crisis proneness.

Copelovitch and Singer analyze economic and banking-sector data from 1976 to 2011, for the 32 countries in the OECD, and they argue that systemic bank failures are more likely when substantial foreign capital inflows meet a financial system with well-developed stock markets.[8] This is so because banks take on more risk in those environments, which makes them more prone to collapse. They emphasize that it is not just a matter of a few rogue bankers engaging in excessive profit-hunting, but rather the by-product of the historical development of markets, which creates conditions ripe for crisis. They show that while some countries like Canada (which has always had small, regional stock markets and is the only OECD country without a national stock-market regulator) have been able to accommodate and channel capital inflows productively throughout the economy, in others like Germany these capital inflows have led to their banking system to go awry. According to them, this difference is based on the fact that a large securities market is a form of competition for the banking sector, to which banks respond by taking greater risks because they both compete for the business of firms that need to raise money. In countries where the stock markets are small and unsophisticated, there's not much competition and firms go to their banks to financing. However, when stock markets are well developed, banks do not want to lose customers and they assume larger risks. They show that Canadian banks do not face a competitive threat from stock markets the way banks in the United States do, and hence, they can still be Conservative and remain competitive and profitable; by contrast, German national-scale banks have been feeling pressure from a thriving set of regional banks, and they looked at stock markets to improve their competitive advantage and bolster their profits through securities investment, leading to many banking blowups in the last two decades. In order to address these sources

---

[8] The collapse of the Bretton Woods system of international monetary-policy cooperation led to a significant increase in foreign capital movement. From 1990 to 2005 alone, international capital flow increased from $1 trillion to $12 trillion annually. (It has since slid back to $5 trillion, after the 2008 global financial crisis.)

of instability, they advocate for macroprudential regulations for banks to ensure that they are holding enough capital to absorb any losses they might incur.

As it relates to Spain, Copelovitch and Singer claim that the Spanish experience validates their hypothesis. First, they confirm that large capital inflows measured by either net or gross portfolio flows are not always associated with banking instability: Spain was a country with low levels of gross inflows, and yet it experienced frequent banking crises (p. 34). Second, they use the case of Spain in 2003 (with a market/bank ratio of 0.39 in 2003) and compare it with Sweden in 1981 (with a market/bank ratio of −3.75) in order to illustrate the substantive significance of their results and to calculate the predicted effect of a 1.5 standard deviation increase in gross portfolio flows on the probability of a banking crisis. They find that "for Spain 2003 the baseline predicted probability of a crisis is 3.2 percent and the estimated first difference effect of a 1.5 standard deviation in gross portfolio flows is extremely large (48.0 percent) and significant at the 95 percent confidence level" (p. 56).[9] These results confirm their theory that capital inflows are destabilizing only in more securitized financial systems. In other words, capital inflows influence the propensity of banks to take on greater risk, through the assumption of greater insolvency risk and/or through a reduction of capital reserves, depending on the country's domestic financial market structure: at low levels of market/bank ratio. And capital inflows have no correlation with bank capital levels and vice versa. In sum, according to them, capital inflows trigger banking crises not because they cause credit booms, but because they lead banks to reduce the capital holdings and lend to riskier lenders (p. 73).

METHODOLOGY

This book has chosen an historic and institutional approach that considers the objectives of policy-makers and social actors, as well as the way that they interpret existing economic and political conditions. This approach allows the researcher to examine the ways institutions structure the relations among actors and shape their interests and goals, thus constraining political struggles and influencing outcomes (Steinmo et al. 1992, p. 2).

---

[9] For Sweden, they find that the estimated first difference is 3.0%, which is significantly large given that the baseline-predicted probability of a crisis is only 1.6% (p. 56).

From a methodological standpoint, and following Calomiris and Haber (2014, p. 452), the book is a case study that studies a sequence of events in a particular country, Spain, over long periods of time. It emphasizes the role of narratives in causal inference. This historical analysis allows us to explore the roots of the country's banking problems as well as the antecedents and timing of key banking decisions. This is an appropriate approach to identify causal patterns and develop "'structural narratives' which combine the logic of economics and political bargaining with a careful examination of the specific historical events in individual countries." The aim is to show evidence of the key factors that contributed to economic, political, and financial history. While this approach has limitations (i.e., it is based on causal inferences, and being a case study it may not be necessarily representative), it builds on Calomiris and Haber's comparative analysis by emphasizing the role of narratives in causal inference, insisting on the value of narrative evidence, and reinforcing the applicability of the identified patterns to other cases.

In this book, the focus is on the institutions that shape the incentives of individuals and groups. *Institutions* are crucial because they set the "rules of the game." They determine the capacity of coordination among political and economic actors, and they encourage them to form coalitions to advance common interests. Institutions are also responsible for establishing standards and setting rules, and also for monitoring, rewarding, and/or sanctioning behavior (depending on the case). While banks operate within existing institutional constraints, these institutions are constructed as political solutions to political problems, they are the result of political deals, and they evolve as a result of shifting political and economic circumstances. The problem arises when these political institutions create incentives to develop coalitions that may be detrimental to the well-being of a country.

The book also examines *coalitions*: It looks at domestic coalition (cleavage) formation (Lipset and Rokkan 1967; Moore 1967). The benefited group will gain more political power by way of economic leverage and strategic incentives (Rogowski 1990). But *individual actors* are also important because they shape outcomes and they have the ability to identify opportunities for forming coalitions and implementing innovations. Indeed, individuals in central positions of power (for instance, in government and/or regulatory agencies) can make a difference for banking outcomes.

Government leaders should remember that banks and capital markets are also mediated by public institutions and political decisions. Politics are not only desirable but also inevitable in dealing with the political and economic effects of banking crises. This book wants to contribute to this discussion by taking recent changes in the Spanish political economy as a point of departure. It seeks to explore the relationship between institutions and the interests of economic actors and their experiences.

Coalition bargaining is a political process, and strategic actors with their own interests design institutions (Thelen 2004). Institutional change is a political matter because institutions are generated by conflict, they are the result of politics of distribution, and hence they are politically and ideologically construed and depend on power relations (Becker 2009). Institutions are important for banks because they influence interests and impact coalitions, and in a context of structural changes, we have to examine the political settlements that motivate the economic actors.

Finally, most of the research material for this project has been gathered in Spain. In order to pursue this analysis, the author conducted an extensive review of the secondary literature and has interviewed scholars, economic actors, and policy-makers. The author has conducted interviews with leaders of the Spanish banking sector, as well as leaders and representatives of the main political parties. In addition, the author has interviewed former and current high-ranking officials from state agencies: ministry of economics, the BoS, the European Commission, and the European Central Bank (ECB). Finally, the author has conferred with leading scholars and specialists in politics, finances, and economics in Spain and the United States. Finally, primary and secondary sources of data come from the libraries and records of selected Spanish government departments and international institutions, like the ECB, the World Bank, the OECD, and the International Monetary Fund (IMF).

STRUCTURE OF THE SPANISH
FINANCIAL SYSTEM 1986–2005[10]

Since Spain joined the European Union (EU) on January 1, 1986, the Spanish Financial System (SFS) has been adapting to the European context of which it is part.[11] The regulatory organs of the SFS are all the institutions with competences to dictate legal norms (i.e., the government, Congress, and the Ministry of Economy), while the supervisory organs are the BoS and the National Stock Market Commission (*Comisión Nacional del Mercado de Valores*, CNMV). The BoS is the central supervisory institution. Regulated by the *13/1994 Law of Autonomy of the Bank of Spain* (which has been subsequently partially modified), the Bank is in charge of the supervision of all credit institutions. It shared some competences with the autonomous communities, which also had supervisory power over the savings banks (*Cajas de Ahorro*) and credit cooperatives. Finally, the *Comisión Nacional del Mercado de Valores* (CNMV), created by the 24/1988 Law, has been the institution in charge of supervision and inspection of stock markets and the activities of legal and physical people involved in these markets.

Most of the *cajas* were established in the nineteenth century as pawnshops with the support of the Catholic Church and/or local municipalities, with the aim to redistribute their profits through social work (they dedicated a significant portion of their provisions, typically over 20% to social causes) (see Güell 2001). Overtime most of them came under the control of the regional and local governments who often used them to advance their political agendas. They were regulated by both the national government (in charge of the basic norms) and the autonomous communities' governments (in charge of the application and development of the rules established by the central government). Political institutions (parties and unions) participated in the governing bodies of the *cajas*.

As noted throughout the book, the distinctive regulatory framework that separates *cajas* from commercial banks was a crucial factor in explaining the differences in performance between the *cajas* and commercial during the 2008 crisis. Indeed, in the last decades *cajas* made a push to

---

[10] From: "Why the Spanish Financial System Survived the First Stage of the Global Crisis?" *Governance* (Article first published online: November 6, 2012. Volume 26, Issue 4. October 2013, pp. 631–56).

[11] From Royo (2013, pp. 179–81).

increase their market share, and they expanded aggressively beyond their traditional markets competing with traditional commercial banks to offer real estate loans. When the real estate bubble collapsed after 2007, many of them accumulated billions of euros of loans at risk of default and were forced to require state support.

Prior to the liberalization of the late 1980s and 1990s, the SFSs (like those of Greece and Italy) was typical credit-based Mediterranean system, characterized by extensive interventionism, state control over the banking system, and underdeveloped capital markets (Pérez and Westrup 2010). The role of financial institutions was to provide funding to contribute the process of economic development and industrialization (Pérez 1997; Lukauskas 1997), and bank deposits were turned into low-interest credit for industrial enterprises and the government (Deeg and Pérez 2000). The oligopolistic nature of the sector generated significant costs for Spanish firms outside the finance sector (Pérez 1997).

The liberalization of the sector started in the second half of the 1980s, driven by the country's integration in the European Community and the subsequent European Single Market program. In Spain's case, the process of financial liberalization was also part of the BoS's effort to achieve effective disinflation (Pérez and Westrup 2010). The European integration process, and particularly the creation of the Single Market and the European Monetary Union, has been a driving factor in subsequent developments. While financial regulations were still the responsibility of the national governments, in reality over the last two decades there has been a harmonization process throughout all the member states. This process led to the liberalization, modernization, consolidation, and opening up of the system.

In Spain, the regulation and supervision of the SFS seek two main objectives: to guarantee the correct functioning of its markets and to protect the consumers of financial services. Regulation, defined as the development of the norms that rule the activities of financial markets and institutions, has to be distinguished from supervision: the guardianship of the fulfillment of the norms. The regulatory organs of the SFS are all the institutions with competences to dictate legal norms, i.e., the government, Congress, and the Ministry of Economy, while the supervisory organs are the BoS and the National Stock Market Commission (*Comisión Nacional del Mercado de Valores*, CNMV) (see Fig. 1.5).

Financial institutions were very important for the Spanish economy because they accumulated approximately 36% of the total financial assets

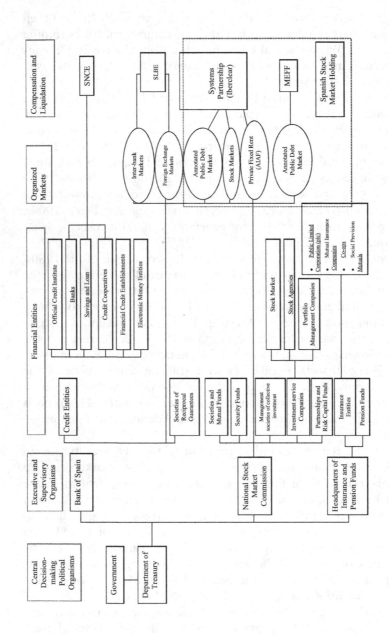

Fig. 1.5 Institutional structure of the Spanish financial system in 2005 (*Source* AFI 2005)

and liabilities prior to the crisis. Credit institutions concentrated approximately 94% of the credit process, which illustrates the high relative weight of financial intermediation in the Spanish economy. Indeed, the SFS was considered as highly bankarized (*bancarizado*) given the strong direct or indirect weight of credit institutions, particularly banks and *cajas*. Yet, the degree of financialization[12] of the sector was relatively low prior to the crisis and the majority of banks' *assets* were loans to customers, and a significant part of bank assets involved Spanish government securities, which were considered among the safest possible asset investments. This relatively low degree of financialization of Spanish banks (like Italian and Greek ones) can be explained by the slow evolution of the NFS and Spanish banks' reluctance to change a business model that has been consistently successful for decades.

Based on Hardie and Howarth's banking typology (2013), corporate finance in Spain cannot be considered 'market based.[13] Indeed, European Central Bank (ECB) data show that assets are really very much not market based. According to the ECB, data customer loans as a percentage of assets were 67.31% at the end of 2007, a much higher percentage compared to other countries, such as the UK, Germany, or France. This would place Spain very low on the market-based banking category.

Furthermore, securitization provided collateral for about 20% of bank lending in Spain prior to the crisis. This was actually moderately high in comparative terms (for instance, in France it was only 1%). Thus, a lot of the Spanish credit boom was fed by the securitization of liabilities, which allowed Spanish banks to lend more. This was not off-balance sheet, but the result (increased lending) was similar to what we saw in other countries like the UK (although about 50% less in GDP terms). Thus market-based banking was still pretty important to Spain. According to the IMF, at the end of 2007, cross-border liabilities as a percentage of GDP represented only 60.6%, a much lower figure than that of countries such as

---

[12] Financialization is defined as the trading of risk, and it is operationalized by looking at Spanish banks' assets (i.e., the size of the trading book and the presence of toxic assets) and liabilities (i.e., the funding base of banks), their reliance on wholesale market rather than retail deposits for funding, the securitization of lending, and the use of structured investment vehicles (SIVs).

[13] The concept of "market-based" banking (MBB) considers the extent to which the banks' ability to supply credit to the economy is driven by market pressures on both the asset and liability sides of their balance sheets.

France (157.6%) or Germany (143.3%). International liabilities as a percentage of total liabilities represented 24.1% (versus 66.1% in the UK or 77.7% in Ireland).

Finally, prior to the crisis, according to the BoS data, on the liabilities side Spanish banks were traditionally very active in capital markets, relaying particularly on the interbanking markets for their funding (on average about 20% of the total), while *cajas* and credit cooperatives, which were more successful capturing resources from their customers were not so dependent on those markets (only around 7%). Spanish banks lent more than what they got in deposits, which forced them to rely on wholesale markets to fill the gap. This forced them during the crisis, as we discuss throughout the book, to rely heavily on the ECB.

THE FINANCIAL BAILOUT

As mentioned above, the crisis of the financial system in Spain led to a European bailout in 2012 (see Royo 2013). As part of its historical analysis of the Spanish banking sector, this book seeks to explain how that banking crisis happened. The main detonator that led to the bailout was Bankia's rescue, which highlighted concerns that the level of provisions that Spanish banks had taken against distressed property portfolios was too low. At the same time, concerns about the fiscal situation of the country (public debt had doubled since the crisis started from 35.5% of GDP at the end of the first quarter of 2008 to 72.1% at the end of the first quarter of 2012) and the interconnectedness between sovereign issuer and banks (the infamous "doom loop") pushed Spanish yields up and led to growing concerns about the need for a bailout. In the end, the intensification of the financial crisis led to a deepening sovereign crisis, and since Spain's banking system was perceived as systemically important for the Eurozone because the country had about 450 billion euros of deposits from foreign companies and individuals, the European (and IMF) leaders decided to intervene.

Prior to the bailout, investors perceived those concerns and they were punishing Spanish banks and Spain's sovereign. The performance of Spanish banks' stock reflected those worries as well: between mid-March and the end of June of 2012, Santander and BBBA's stock declined about 30%. At the same time, Spain's cost of borrowing kept increasing and the spread on Spanish ten-year bonds over German Bunds hit new European highs in late May 2012, climbing to 511 points, while yields of ten-year bonds moved above 6.5% (reaching 6.9% on June 14), and moving closer to the 7% level that was widely considered unsustainable, and which led to bailouts for Greece, Portugal, and Ireland.

Yet, despite these alarming signs, as late as Spring 2012, the Spanish government and PM Rajoy insisted that the country would not need international bailout for its banks. On April 12, Mr. Rajoy stated that "talking about a rescue makes no sense... Spain is not being rescued; Spain can't be rescued. There is no intention and no need and so Spain will not be rescued." Rajoy was also on record stating that "we are not going to let any regional government fall, or any bank fall, because they can't ... if that happens the country will fall."[14] This followed the repeated statements of the members of the government that Spain would not take any form of international rescue.

By early June 2012, however, it was becoming clear that despite all the denials, a rescue was inevitable and that Spain would in fact need a financial bailout. The government's attempts to force the EU and BCE's hands (the budget minister, Cristobal Montoro, responded to a question about whether Spain needed a bailout with a veiled threat: "those with the most interest in whether Spain does all right are the holders of debt, who have to be repaid in full and have that right," and Minister Guindos was on the record stating that the battle for the euro was going to be waged in Spain[15]), and its insistent demand for help from its European partners only exasperated them and intensified concerns in European capitals that it was only considerations over the political stigma associated with a bailout, rather than the policy constraints that Brussels would include on it, that was keeping the Spanish government from accepting aid that was on the table.

Spain's risk premium—the difference between its bond yields and those of Germany—continued soaring after the Bankia nationalization, adding pressure to the government, and Bankia's shares plunged. By June 5, the severity of the situation was finally creeping in and the government was already admitting (in the words of Mr. Montoro) that given the high perceived risk of its sovereign debt, Spain did "not have the doors of the markets open," and this despite the fact that it was planning to auction

---

[14] See "Doubts Emerge over Spain's Leaders," *Financial Times*, April 13, 2012.

[15] See "Defiant Spain to Test the Bond Market," *Financial Times*, June 1, 2012.

up 2 billion euros of bonds that same week. Mr. Rajoy, for his part, continued insisting on the need for a banking union and Eurobonds, stating during a senate session that Europe "needs to support those that are in difficulty ... The most important thing is we have a problem of financing, of liquidity and debt sustainability."[16] Germany still refused to provide aid unless there was a formal request from the Spanish government.

Finally, on Saturday, June 9, following a few days of fierce official negations, the government asked the EU for funds to recapitalize its struggling banking sector at a conference call between the Eurozone's 17 finance ministers. It agreed to accept a bailout of up to $125 billion, nearly three times the $46 billion in extra capital that the IMF said it was the minimum the banking sector needed to guard against the deepening of the country's economic crisis. The decision also aimed at quelling rising financial turmoil ahead of the Greek parliamentary election scheduled for June 17. This decision made Spain the fourth and largest European country to agree to accept emergency assistance (albeit in this case only for the country's financial sector).

The European statement on the aid gave few details, which initially allowed the government to claim that it was not a "rescue" package, and that it was not subject to the conditionality and supervision by the troika (EU, IMF, and ECB) that characterized the other three rescue packages of Greece, Ireland, and Portugal. In Spain, the announcement was made by Minister Guindos, which caused a political storm and forced PM Rajoy to give an impromptu press conference the day after, on Sunday, June 10, before he departed to watch the Spanish football national team play a EuroCup game in Poland. Guindos announced that "what we are asking is financial support, and this has absolutely nothing to do with a full bailout," and added that the terms of the emergency loan would be "very favorable," and that "the problem that we face affects about 30% of the Spanish banking system."[17] The amount of the financing was expected to be completed after the two consulting firms had been hired to look at the bank accounts published their audit report on June 21. The funds would be channeled through the Spanish bank bailout fund, the FROB, and the

---

[16] See "Spain Makes Explicit Plea for Bank Aid," *Financial Times*, June 5, 2012.

[17] See "Spain to Accept Rescue from Europe for Its Ailing Banks," *New York Times*, June 9, 2012; and "Spain Seeks Eurozone Bailout," *Financial Times*, June 10, 2012.

Spanish government would ultimately be responsible and had to sign the memorandum of understanding and the conditions that came with it.

The market's response to the bailout package was initially sanguine. While observers praised the decisive pre-emptive action of European leaders (something relatively unusual during the crisis), and the fact that the aid was directed to the banks and that the amount was much larger than estimated to give some margin in the case of further need, there was disappointment regarding the failure to inject the money directly into the banks as equity.[18] The model proposed for the aid failed to recognize the crucial link between sovereign debt and the banks. Spanish banks accounted for a third of Spanish sovereign bonds, nearly double the tally before the crisis started (they had purchased 83 billion euros between December 2011 and June 2012). As in Greece where the sovereign debt dragged down the banks, this made them very vulnerable to a potential sovereign debt crisis, which was becoming increasingly more likely in Spain, thus intensifying the "doom loop." Consequently, after an initial market rally, the Spanish bond continued jumping higher and reached a new record, very close to 7%, just four days after the announcement of the bailout, demonstrating that investors were growing increasingly anxious about Spain's ability to pay back its debts.

Main Argument of the Book

The 2010 banking crisis in Spain confirms a long-standing tenant: Banks or banking systems collapse when they meet two conditions: They take on too much risk in their loans and investments, and they do not have sufficient capital on reserve to absorb the losses associated with their risky investments and loans (Calomiris and Haber 2014, p. 207). Indeed, the cause of the 2012 crisis in Spain was rooted in policies that eroded underwriting standards and weak prudential regulation.

As we will see later in the book, the banking crisis affected in particular a set of financial institutions, the *cajas* (savings and loans, S&L). Indeed, with a few relatively small exceptions, the Spanish financial crisis has been

---

[18] See Patrick Jenkins, "'Doom Loop' Takes the Fizz Out of Madrid's Brief Euphoria," *Financial Times*, June 12, 2012, 2; and Andrew Ross Sorkin, "Why the Bailout in Spain Won't Work,' *New York Times*, June 12, 2012, B1.

a crisis of the *cajas*. These institutions borrowed short term from depositors and then lend long term on fixed-rate mortgages.[19] The success of this model, however, was based on two conditions: low inflation volatility and discipline from depositors who would withdraw their deposits if the managers acted imprudently. However, by the 2000s, the context in which these institutions operated had changed markedly and the 2007 global financial crisis hit them hard because they were very dependent on wholesale funding.

Yet, government protection of *cajas* (part of the *bank bargains* that we will examine) had insulated them from the consequences of their own risk-taking and facilitated the reckless decisions that led to their downfall. When the real estate market collapsed after 2007, wholesale funding dried up and their funding costs skyrocketed, which caused significant problems because they had to pay more for capital, and they held mortgages that (many of which went into default as a result of the crisis) still earned only low fixed interest rates of return. This brought several of them to the point of insolvency. If their vulnerability, driven by their high reliance on the real estate market and wholesale funding, had been recognized and addressed in the years prior to the crises through *cajas'* closures, shrinkage, or consolidation, the crisis for the *cajas* would have been significant but not as devastating as it ended up being. As losses started to pile up the Spanish government, supervisory agencies should have shut down insolvent ones or forced them to raise additional capital. Yet, they ignored or minimized the signs and looked the other way, postponing the day of reckoning. But in doing so, they ensured that the final outcome would be much worse.

In many Spanish *cajas*, there was a failure of risk management, which led to an increase in risky lending and to inadequate levels of capital cushions. The question remains *how it was that so many cajas ended up making so many risky loans while maintaining insufficient capital to protect themselves against insolvency*. What were the processes by which *cajas'* portfolios became increasingly risky, and by which increased risk in bank assets was not adequately matched by increasing amounts of capital in reserve.

This book will argue, following Calomiris and Haber (2014), that institutional and regulatory frameworks favored both the government and other privileged actors' access to finance at the expense of an environment

---

[19] See Calomiris and Haber (2014, pp. 199–201).

conducive to a stable banking system. Indeed, political institutions have structured the incentives of bankers and political and economic actors to form coalitions that shaped regulations and policies in their favor This institutional framework was the result of political choices that made it vulnerable, because prudent lending practices continued being influenced by the desires of the groups that were in control of the government, who often channeled credit to groups that were considered politically crucial. Therefore, it is not surprising that banks have been fragile and crises prone.

ORGANIZATION OF THE BOOK

The book continues in Chapter 2 with an overview of the origins of the Spanish banking system prior to the twentieth century. It shows the role of the state in the development of the Spanish banking system, and how financial mismanagement contributed to the decline of the country prior to the twentieth century. It also examines the 'bank bargains' that took place during that era, which ultimately led to the development of a fragile banking system and recurrent banking crises.

Chapter 3 examines the evolution of the Spanish banking system in the twentieth century and analyzes the causes of the mid-1970s and early 1980s banking crisis. It shows how the banking sector was transformed and modernized, growing from an underdeveloped structure into a comparatively modern sector. However, successive governments continued establishing institutional and regulatory frameworks that favored both the government and other privileged actors' access to finance at the expense of an environment conductive to a stable banking system which ultimately led to a systemic baking crisis in the 1970s.

Chapter 4 analyzes the overall economic crisis that started in 2008 in Spain. It is impossible to disentangle the 2008 banking crisis from the overall economic crisis that affected the country at the same time. As the economic crisis intensified, Spain's banking sector could not escape its dramatic effects. This chapter looks at the performance of the Spanish economic throughout the 1990s and the first decade of the twentieth century. It examines the reasons for the success of the Spanish economy in the 1990s and provides an overview of the main causes of the 2008–2013 crisis and the governments' responses.

Chapter 5 analyzes the impact of the global crisis on the SFS between 2008 and 2010. It shows that, overall, the performance of the largest

Spanish financial institutions was positive. The chapter examines why and outlines some lessons from the Spanish experience. It contends that this response was largely driven by institutional, political, and cultural factors. Finally, the chapter considers the Spanish experience within the framework of the VoC literature. While financial capitalisms have converged toward deregulation as a result of the combined processes of globalization and European integration, this chapter shows that differences persist. Indeed, in the case of Spain, the crisis led to extensive regulatory intervention that served to reinforce the pre-existing model.

Chapter 6 examines the subsequent impact of the economic crisis on the SFS, as well as the banking bargains that ultimately led to the financial bailout. In addition, it focuses on the following variables: deteriorating economic conditions; the implosion of the real estate market; the dependence on wholesale funding; weaknesses in the regulatory framework; and the role of the BoS.

Most of the analyses on the crisis in Spain have concerned themselves with phenomena like mismanaged banks, excessive debts, the bubble in the real estate sector, or the loss of competitiveness. Others have sought to explain the crisis as a by-product of EMU integration. Chapter 7 moves beyond those explanations and argues that a fundamental reason for the economic and financial crisis is rooted in the process of institutional degeneration that preceded the crisis. It analyzes the deteriorating performance of Spanish institutions to explain that reversal.

The concluding chapter provides an overview of the main arguments to explain the fragility of the Spanish banking system. It shows how domestic social, political, and economic factors have been crucial to understand coalition formations and policy choices in Spain. These coalitions have not been neutral, and they have influenced the stability and resilience of the Spanish banking system and its ability to provide credit. It also outlines some lessons from the Spanish experience and analyzes the main implications of the financial crisis.

REFERENCES

Acemoglu, Daron, and James Robinson. *Why Nations Fail: The Origins of Power, Prosperity and Poverty.* New York: Random House, 2012.

AFI (Analístas Financieros Internacionales). *Guía del Sistema Financiero Español.* Madrid: AFI, 2005.

Allen, Franklin, and Kenneth Rogoff. "Asset Prices, Financial Stability and Monetary Policy." In *The Riksbank's Inquiry into the Risks in the Swedish Housing*

Market, edited by Per Jansson, and Mattias Persson, 189–218. Stockholm: Sveriges Riksbank, 2011.

Amable, Bruno. *The Diversity of Modern Capitalism.* Oxford: Oxford University Press, 2003.

Amri, Puspa D., Greg M. Richey, and Thomas D. Willet. "Capital Surges and Credit Booms: How Tight Is the Relationship." *Open Economics Review* 27, no. 4 (2016): 637–70.

Armigeon, Klaus, and Lucio Baccaro. "The Sorrows of Young Euro." In *Coping with the Crises,* by Nancy Bermeo and Jonas Pontusson, 162–98. New York: Russell Sage Foundation, 2012.

Avgouleas, Emilios, and Charles Goodhart. "Critical Reflections on Bank Bail-Ins." *Journal of Financial Regulation* 1, no. 1 (March 2015): 3–29.

Barrón, Iñigo. *El Hundimiento de la Banca.* Madrid: Catarata, 2012.

Becker, U. *Open Varieties of Capitalism: Continuity, Change and Performance.* NY: Palgrave, 2009.

Bermeo, N., and Jonas Pontusson. *Coping with the Crises.* New York: Russell Sage Foundation, 2012.

Caballero, Julián A. "Do Surges in International Capital Inflows Influence the Likelihood of Banking Crises?" *The Economic Journal* 126, no. 591 (2016): 281–316.

Calomiris, Charles W., and Stephen H. Haber. *Fragile by Design: The Political Origins of Banking Crises & Scarce Credit.* Princeton: Princeton University Press, 2014.

Cameron, David. "European Fiscal Responses to the Great Recession." In *Coping with the Crises,* by Nancy Bermeo and Jonas Pontusson, 91–129. New York: Russell Sage Foundation, 2012.

Chinn, Menzie D., and Jeffry A. Frieden. *Lost Decades: The Making of America's Debt Crisis and the Long Recovery.* New York: W.W. Norton & Company, 2011.

Copelovitch, Mark, and David A. Singer. *Banks on the Brink: Global Capital, Securities Markets, and the Political Roots of Financial Crises.* New York: Cambridge University Press, 2020.

Deeg, Richard, and Sofía Pérez. "International Capital Mobility and Domestic Institutions: Corporate Finance and Governance in Four European Cases." *Governance* 13, no. 2 (2000): 119–53.

Deeg, Richard, and Susanne Luetz. "Internationalisation and Financial Federalism: The United States and Germany at the Cross Roads?" *Comparative Political Studies* 33, no. 3 (2001): 374–405.

Della Sala, Vincent. "The Italian Model of Capitalism: On the Road between Globalization and Europeanization?" *Journal of European Public Policy* 11, no. 6 (December 2004): 1041–57.

Diamond, Douglas W., and Raghuram G. Rajan "Illiquid Banks, Financial Stability, and Interest Rate Policy." *Journal of Political Economy* 120 (2012): 552–91.

Eurostat/European Commission. *Enterprises in Europe: Sixth Report.* Luxembourg: Office of Official Publications of the European Communities, 2001.

Ferguson, Neil. *The Great Degeneration.* New York: Penguin Press, 2013.

Fishman, Robert. "Rethinking the Iberian Transformations: How Democratization Scenarios Shaped Labor Market Outcomes." *Studies in Comparative International Development* 45, no. 3 (2010): 281–310.

Germain, Randall. "Governing Global Finance and Banking." *Review of International Political Economy* 19, no. 4 (2012): 530–35.

Goodhart, Charles A. E., Anil K. Kashyap, Dimitrios P. Tsomocos, and Alexandros P. Vardoulakis. "Financial Regulation in General Equilibrium." NBER Working Papers Series. Working Paper 17909 (March 2012). http://www.nber.org/papers/w17909.

Goodhart, Charles. *The Evolution of Central Banks.* Cambridge, MA: MIT Press, 1988.

Goodhart, Charles A. E., Tsomocos, P. Dimitrios, and Alexandros P. Vardoulakis. "Modeling a Housing and Mortgage Crisis." In *Financial Stability, Monetary Policy, and Central Banking,* edited by Rodrigo A. Alfaro. Santiago, Chile: Central Bank of Chile, 2010.

Gourinchas, Pierre-Olivier, and Maurice Obstfeld. "Stories of the Twentieth Century for the Twenty-First." *American Economic Journal: Macroeconomics* 4 (2012): 226–65.

Güell, Joan. "Las Cajas de Ahorro en el Sistema Financiero Español. Trayectoria Histórica y Realidad Actual." Paper presented at the Universidad de Zaragoza, during the *Jornadas sobre La singularidad de las cajas de ahorros españolas.* May 28, 2001.

Haggard, Stephan, and Jongryn Mo. "The Political Economy of the Korean Financial Crisis." *Review of International Political Economy* 7, no. 2 (2000): 197–218.

Hall, Peter, and David Soskice. *Varieties of Capitalism.* New York: Oxford University Press, 2001.

Hancké, Bob, Martin Rhodes, and Mark Thatcher, eds. *Beyond Varieties of Capitalism: Contradictions, Complementarities, and Change.* Oxford: Oxford University Press, 2007.

Hardie, Ian. "'How Much Can Governments Borrow?' Financialization and Emerging Markets Government Borrowing Capacity." *Review of International Political Economy* 18, no. 2 (2013): 141–67.

Hardie, Iain, and David Howarth. "Market-Based Banking and the Financial Crisis." Mimeo: Paper present at the University of Victoria, 2013.

Jabko Nicolas, and Elsa Massoc. "French Capitalism under Stress: How Nicolas Sarkozy Rescued the Banks." *Review of International Political Economy* 19, no. 4 (2012): 562–95.

Jackson, Gregory, and Richard Deeg. "How Many Varieties of Capitalism? From Institutional Diversity to the Politics of Change." *Review of International Political Economy* 15, no. 4 (2008): 679–708.

Jiménez, Gabriel, Ongena, Steven, Peydró, José Luís, and Saurina, Jesús, "Hazardous Times for Monetary Policy: What do Twenty-Three Million Bank Loans Say about the Effects of Monetary Policy on Credit Risk-Taking?" *Econometrica* 82, no. 2 (2014): 463–505.

Kahler, Miles, and David Lake. *The Great Recession in Comparative Perspective.* Ithaca: Cornell University Press, 2013.

Kaminsky, Graciela, and Carmen Reinhart. "The Twin Crises: The Causes of Banking and Balance-of-Payments Problems." *The American Economic Review* 89, no. 1 (1999): 473–500.

Kindleberger, Charles. *The World in Depression 1929–1939.* Berkeley: University of California Press, 1973.

Kindleberger, Charles. *Manias, Panics, and Crashes: A History of Financial Crises.* New York: Wiley, 1996.

King, Robert G., and Ross Levine. "Finance and Growth: Schumpeter Might Be Right." *Quarterly Journal of Economics* 108 (1993): 717–37.

Lipset, Seymur Martin, and Stein Rokkan. *Party Systems and Voter Alignment.* New York: Free Press, 1967.

Lukauskas, Arvid. *Regulating Finance.* Ann Arbor: Michigan University Press, 1997.

Mabbet, Deborah, and Waltraud Schelke. "What Difference Does Euro Membership Make to Stabilization? The Political Economy of International Monetary Systems Revisited." *Review of International Political Economy* 22, no. 3 (June 2015): 508–34.

Maddaloni, Angela, and Peydró, José Luís. "Bank Risk-Taking, Securitisation, Supervision and Low Interest Rates: Evidence from the Euro-Area and the U.S. Lending Standards." *The Review of Financial Studies* 24, no. 6 (2011): 2121–65.

Mahoney, James, and Kathleen Thelen. *Explaining Institutional Change: Ambiguity, Agency, and Power.* New York: CUP, 2009.

Major, Aaron. "Neoliberalism and the New International Financial Architecture." *Review of International Political Economy* 19, no. 4 (2012): 536–61.

Minsky, Hyman. *Can 'It' Happen Again: Essays on Instability and Finance.* New York: M.E. Sharpe, 1982a.

Minsky, Hyman. "The Financial-Instability Hypothesis: Capitalist Processes and the Behavior of the Economy." In *Financial Crises: Theory, History, and*

Policy, edited by Charles Kindleberger and Jean Pierre Laffargue. New York: Cambridge University Press, 1982b.

Molina, Oscar, and Martin Rhodes. "Conflict, Complementarities and Institutional Change in Mixed Market Economies." In *Beyond Varieties of Capitalism*, edited by B. Hancké, M. Rhodes, and M. Thatcher, 223–53. Oxford: Oxford University Press, 2007.

Molinas, César. *Qué Hacer con España*. Madrid: Imago Mundi, 2013.

Moore, Barrington. *Social Origins of Dictatorship and Democracy: Lord and Peasant in the Making of the Modern World*. New York: Beacon Press, 1967.

Mosley, Layna. "Attempting Global Standards: National Governments, International Finance, and the IMF's Data Regime." *Review of International Political Economy* 10, no. 2 (2003): 331–62.

OECD. *OECD Historical Statistics 1970–1999*. Paris: OECD, 2000.

Ortega, Andrés, and Angel Pascual-Ramsay. *¿Qué nos ha pasado?* Madrid: Galaxia Gutemberg, 2013.

Pérez, Sofía A. *Banking on Privilege*. New York: Cornell University Press, 1997.

Pérez, Sofía A., and Jonathan Westrup. "Finance and the Macroeconomy: The Politics of Regulatory Reform in Europe." *Journal of European Public Policy* 17, no. 8 (2010): 1171–92.

Portes, Richard. "Global Imbalances." In *Macroeconomic Stability and Financial Regulations: Key Issues for the G20*, edited by Mathias Dewatripont, Xavier Freitas, and Riichard Portes. London: CEPR, 19, Vol. 19, 2009.

Quaglia, Lucia, and Sebastián Royo. "Banks and the Political Economy of the Sovereign Debt Crisis in Italy and Spain." *Review of International Political Economy* 22, no. 3 (2015): 485–507.

Rajan, Raghuram G. "Has Finance Made the World Riskier?" *European Financial Management* 12 (2006): 499–533.

Reinhart, Carmen M., and Kenneth Rogoff. *This Time Is Different: Eight Centuries of Financial Folly*. New York: Princeton University Press, 2009.

Reinhart, Carmen M., and Kenneth Rogoff. "From Financial Crash to Debt Crisis." *American Economic Review* 101 (August 2011): 1676–706

Rogowski, Ronald. *Commerce and Coalitions: How Trade Affects Domestic Political Alignments*. Princeton, NJ: Princeton University Press, 1990.

Royo, Sebastián. *From Social Democracy to Neoliberalism*. New York: St. Martin's Press, 2000.

Royo, Sebastián. *Varieties of Capitalism in Spain*. New York: Palgrave, 2008.

Royo, Sebastián. "A 'Ship in Trouble' The Spanish Banking System in the Midst of The Global Financial System Crisis: The Limits of Regulation." In *Market-Based Banking, Varieties of Financial Capitalism and the Financial Crisis*, edited by Iain Hardie and David Howarth. New York: Oxford University Press, 2013.

Schmidt, Vivien. *The Futures of European Capitalism*. New York: Oxford University Press, 2002.

Schmidt, Vivien. "Putting the Political Back into Political Economy by Bringing the State Back Yet Again." *World Politics* 61, no. 3 (2009): 516–48.

Schoenmaker, Dirk. "The Trilemma of Financial Stability." Paper prepared for CFS-IMF Conference: *A Financial Stability Framework for Europe: Managing Financial Soundness in a an Integrating Market*. Frankfurt, 26 September 2008.

Schoenmaker, Dirk. "The Financial Trilemma." *Economics Letters* 111 (2011): 57–59.

Schularick, Moritz, and Alan M. Taylor. "Credit Booms Gone Bust: Monetary Policy, Leverage Cycles, and Financial Crises, 1870–2008." *American Economic Review* 102 (2012): 1029–61.

Skocpol, Theda. *States and Social Revolutions*. New York: Cambridge University Press, 1979.

Stein, Jeremy C. "Monetary Policy and Financial Stability Regulation." *The Quarterly Journal of Economics* 127 (2012): 57–95.

Steinmo, Sven, Kathleen Thelen, and Frank Longstreth, eds. *Structuring Politics: Historical Institutionalism in Comparative Analysis*. New York: Cambridge University Press, 1992.

Stiglitz, Joseph. "Capital Market Liberalization, Economic Growth, and Instability." *World Development* 28, no. 6 (2000): 1075–86.

Taylor, Alan. "On the Costs of Inward-Looking Development: Price Distortions, Growth, and Divergence in Latin America." *Journal of Economic History* 58 (March 1998): 1–28.

Thatcher, Mark. *Internationalisation and Economic Institutions: Comparing the European Experience*. Oxford: Oxford University Press, 2007.

Thelen, Kathleen. *How Institutions Evolve: The Political Economy Skills in Germany, Britain, the United States and Japan*. New York: Cambridge University Press, 2004.

Torrero, Antonio. " La formación de los tipos de interés y los problemas actuales de la economía española." *Economistas*. Madrid no. 39, 1989.

Weiss, Linda (ed.). *States in the Global Economy: Bringing Domestic Institutions Back In*. Cambridge: Cambridge University Press, 2003.

Zysman, John. *Governments, Markets, and Growth: Financial Systems and Politics Industrial Change*. Ithaca, NY: Cornell University Press, 1983.

CHAPTER 2

The Origins of the Spanish Banking System

Introduction

Spain's banking history provides fertile ground to show how banking systems are the outcome of bargains struck by coalitions of market participants and the actors that control the government, and how they advance their interests and welfare. It offers a great case study of how politics have intruded into bank regulation, and how coalitions of politicians, bankers, and other interest groups generate policies that determine "who gets to be a banker, who has access to credit, and who pays for banks bailouts and rescues." The allocation of political power determines the composition of these coalitions and structures their bargains. These bargains take place and are structured by an existing set of political institutions, and they ultimately determine the laws and regulatory frameworks, the winners and losers, as well as the actors that will be in charge of enforcing them (Calomiris and Haber 2014, pp. 38–39).

The Spanish banking system has a long history characterized by steady progress and marred by an alternation of booms and busts, typically closely associated with political crises and/or the performance of the Spanish economy at large. Spanish finances staggered from crisis to crisis approximately every twenty years: 1557, 1575, 1596, 1607, 1627, and 1647 (Ehremberg 1928, p. 334). Indeed, it is not a coincidence that systemic banking crises have coincided with economic crisis: The three largest ones in modern history were the railroad crises of the 1860s; the oil crisis of the 1970s and 1980s; and the real estate/financial one

© The Author(s) 2020
S. Royo, *Why Banks Fail*,
https://doi.org/10.1057/978-1-137-53228-2_2

that started in 2008. As we will see throughout the book, this banking history has been marked by the need to support the state's funding needs. Indeed, governments had incentives to establish regulatory environments that favored both the government's (and other actors') access to finance at the expense of an environment conductive to a stable banking system. Often prudent lending practices conflicted with the desires of the group(s) that were in control of the government to channel credit to groups that were considered politically crucial, which in the end made the Spanish banking sector fragile.

From early times, the Spanish banking system was the result of political choices that made it vulnerable by construction; and it has been consistent with institutions that set the distribution of power. Indeed, political institutions in the country structured the incentives of politicians, bankers, shareholders, depositors, debtors, and taxpayers to form coalitions that shaped laws, regulations, and policies in their favor (and often at the expense of others) (Calomiris and Haber 2014, p. 4). Therefore, it is not surprising that bank insolvency and illiquidity crises became the pattern.

This chapter examines the origins of the banking system in Spain from the twilight of empire in the fifteenth century, through the end of the nineteenth century. During the earlier times (fifteenth-seventeenth centuries), Spain was under an absolute monarchy, an autocrat with absolute power that conformed to Calomiris and Haber's autocratic regime taxonomy (2014, pp. 41–50). That period was characterized, despite the wealth from the colonies, by constant funding crises that led Spanish kings to declare bankruptcies, which often led to financial crises. While taxpayer-funded bailouts of banks only started in the mid-twentieth century (and early on the costs of failure were borne mainly by bankers, shareholders, and depositors), these bankruptcies destroyed the credibility of the Crown, and they led to regular funding crises and to the absence of a meaningful banking system.

The banking system was the product of political deals that determined which laws, policies, and regulations were passed, which groups of people could become bankers, as well as their power to contract with whom, for what and under which terms. Once the power of the king was eroded, particularly in the nineteenth century, the Crown was forced to build a network of alliances that included bankers, in order to be able to finance expenditures that, given the precariousness of the Spanish Treasure's finances, most often exceeded tax revenues. These coalitions, as we will see, were unstable because bankers were fully aware of the potential risks of failure, which forced the Crown to raise the rate of return on their capital to compensate them. In addition, the Crown often gave bankers other political privileges to cement their alliances, including government positions, nobility tittles, loans to their business enterprises, or the right to collect taxes and serve as a government's agent. During this period, the number of bank charters was also restricted to minimize competition among banks, which allowed them to charge higher interest rates, and led to a banking system that restricted entry and granted privileges to bank insiders as part of the political bargain.

Under this system, the gains accrued as rents to the insiders and the Crown. It produced a coalition of interest that included three main actors: the Crown, the minority shareholders, and the bank insiders. The Crown received a steady source of public finance and whenever needed emergency funding; the bank insiders earnt rents as bank managers and privileged access to cheap credit (often for their own nonfinancial enterprises and/or those of their families); finally, minority shareholders earnt compensation in stock returns, which were often above normal to compensate them for the risk of expropriation. Under this compact, the government's role was to establish a framework of banking laws and regu-lations that preserved the coalition, which led to an unstable system of underdeveloped banks and also scarce credit. The instability of the system was caused by the weak rule of law and the lack of constraints on the Crown's authority and discretion, because periodically it needed finance in excess of tax revenues, and it retained the power to expropriate the banking system through some combination of mandated loans to the state, high taxation, high reserve requirements, and nationalization of bank ownership. In addition, instability was fostered by the propensity of the bank insiders to resort to inside lending through extending credit to their (or their families') nonfinancial institutions, in effect allowing them to loot the banks to save those enterprises in times of crises. In effect, the political deal was a 'rent-distribution system' whereby the Crown retained the power to 'expropriate' the banking system whenever its fiscal needs demanded it; bank insiders were potential sources or rent extrac-tion and hence were compensated for that risk; debtors were excluded from the credit system and suffered lost investment opportunities; and taxpayers were also a source of rent extraction when the government bailed out banks. This repressed banking system was far away from any model of efficient allocation of resources. On the contrary, competition in the credit market was limited, and banks did not allocate credit broadly because the bankers did not benefit directly from doing so. This created a situation in which the insiders of the banking political coalition had ample access to credit, whereas the majority of the population, including potential entrepreneurs, were starved for credit. As a result, the banking system was small and crises prone because occasionally the Crown expropriated the banks, and periodically, the insiders expropriated minority shareholders and depositors (for this framework, see Calomiris and Haber 2014, pp. 47–50).

ORIGINS OF THE BANKING SYSTEM IN SPAIN

From early times, Spain lacked a large stock of precious metals, like silver and gold, and was forced to trade with Africa, Eastern Europe, and the Middle East to get these metals. A new frontier was open with the discovery of America in the fifteenth century, which opened access to the wealth of Peruvian and Mexican gold and silver mines, and contributed to the country's enormous prosperity and the building of an unprecedented global empire. By the mid-sixteenth century, however, constant wars and empire overreach led to financial difficulties for the Spanish Crown that provoked recurrent suspensions of payments and debasing the currency (the so-called *vellón* episodes) that ruined Spanish bankers and ultimately brought about the decline of Spain in the seventeenth century.

In the medieval period, there were well-known bankers in the main commercial centers of the two largest kingdoms, Castile and Aragón, such as Burgos, Medina del Campo, Toledo, Valladolid, Barcelona, Valencia, or Saragossa (this chapter summarizes Tortella and García Ruiz 2013, as what follows is largely drawn from this work). They were mostly money changers very active in trade fairs and in financing trade and commerce, who charged interests and paid commissions to intermediaries who brought business to them.[1] Loans on collateral were also common. Barcelona was the host of one of the earliest municipal banks, the Table of Change, *Taula del Canvi* (1401), devoted to servicing the city's financial needs and to attracting deposits. It was later emulated in other Spanish cities.

The expulsion of the Jews in 1492 by the Catholic Kings was a major blow to the banking system, as their absence was not easily filled. Genoese

---

[1] There are references to a banking transaction in the famous *Poema del Cid* written around 1150.

bankers capitalized and became mayor players in cities like Seville, which played a crucial role in the trade with the Americas, financing the large flows of trade that went through the city and lending to the Crown. Indeed, medieval times set the pattern for a central element of bankers' role in Spain: to provide financing to the Spanish Crown, which at that time required massive funding for its military campaigns. They used *asientos*, contracts, to supply a certain amount of merchandise like specie at some predetermined location, payable against the silver shipments that arrived from the Americas. The country had to export a large amount of the bullion it imported from the Americas to finance its wars and commercial deficit. These *asientos* became very profitable (and also very risky) and attracted bankers from all over Europe. Payments, however, were often contingent on the arrival of the silver fleets, and any delay or disruption provoked market crisis and often threatened the solvency of bankers. Not surprisingly, from the time of Charles V (1500–1558) bankers' close links with the kings drove them to bankruptcy, and even more importantly, it prevented the development of a banking system comparable to the Northern European one (Tortella and García Ruiz 2013, pp. 4–6).

Indeed, Spain suffered regular bankruptcies. International bankers (like the Fuggers of Augsburg) lent regularly to the House of Hapsburg, starting with funding to ensure the election of Charles V as Holy Roman Emperor (who was also king of Spain), and continued funding him during the wars between Spain and Austria and Spain and France. The Fuggers borrowed in Antwerp to relend to Spain and were paid when the silver cargos arrived in Seville (Kindleberger 1984, p. 45). His son, Phillip II, was involved in constant military campaigns that increased the Crown's indebtedness fourfold during his entire reign had to declare three bankruptcies in 1557, 1575, and 1596. In addition, he introduced a new innovation, the debasement of the coinage to finance the deficits, that became a pattern under his successors. The financial strains of the Crown were compounded by other factors such as a regressive tax system that over-taxed and penalized the cities versus the countryside, and the commoners versus the church and nobility, and discouraged industry and trade (the once flourishing Castile textile industry lost its competitiveness as a result of bad regulation and taxes); the sale of patents to nobility and gentry, which reduced the number of taxpayers (it has been estimated that between 1591 and 1631, one million taxpayers "disappeared"); the insecurity of markets and property rights caused by state bankruptcies and confiscations, and the uncertainty about the value of money caused by the debasement of coinage; finally, corruption was also rampant (according to some studies as much as 40% of taxes were lost in collection at that time). For instance, Phillip IV (king between 1621 and 1665) tried to confront the financial needs of the Crown by lowering interest rates on loans to the state, confiscating silver remittances, issuing new copper money, and turning to the bankers for help with new *asientos*. But despite all these measures he had to declare bankruptcy in 1627! These decisions led to the impoverishment of the country, particularly Castile which was far more heavily taxed that Aragón. Following yet another suspension of payments in 1647, many bankers, who had been lured by the American silver and the trade/commercial opportunities, stopped extending credit to the Crown, as those disastrous decisions generated widespread mistrust and resentment among bankers (many of the them foreign ones from Genoa, Germany, or Portugal). In the end, the structural imbalances of Spain's finances stretched bankers' resources, and often ruined, those who dealt with the Crown. Consequently, in the eighteenth century, baking was largely carried out by traders who undertook banking operations as part of their business, and *gremios* (guilds) who accepted deposits and became specialized in banking operations (Tortella and García Ruiz 2013, pp. 4–6).

Similar to other European countries, Spain had adopted in the early modern period a bimetallic standard, but it was one of the few countries that in the nineteenth century (because silver was abundant in the country) decided to adopt the silver standard rather than the gold standard, which ended up isolating the Spanish economy, but helped the country during the Great Depression. The diffusion of banknotes only became widespread in the last decades of the nineteenth century.

From a comparative standpoint, Spain was late developing a modern banking sector. Indeed, it can be said that Spanish modern financial history started with the American War of Independence when Spain sided with the French against the British. This momentous decision had important consequences for Spain because in retaliation British ships cut off the country from the flow of silver in Mexico, which led King Charles III, who was looking for a way to finance the deficit without increasing taxes, to print for the first time in Spanish history paper money and issue a new type of public debt in 1870, the *vales reales*, or royal notes, which played the role of paper money and yielded a 4% interest rate (Kindleberger 1984, pp. 146–47). Unfortunately for the country, these *vales reales* quickly went to a discount, and in order to address the problem,

Francisco Cabarrus, a French banker established in Spain, came up with the project (modeled on John Law's *Banque Général* in France) to create the first join-stock bank in Spanish history, the *Banco Nacional de San Carlos* (the Bank of Saint Charles) in 1782, which was popularly known as the Bank of Spain (although that formal name was only adopted by a successor seventy-five years later). According to the charter, the bank had three main goals: to discount letters, *vales*, and bills; the supplying of the army; and to make foreign payments and transactions on account of the government. The bank had features typically associated with central banks such as the discount and negotiation of bills of exchange. In addition, for a 10% commission it was required to provide the army and the navy with provisions, to attract investors it was encouraged to trade on its own stock, and the bank was allowed to form the *Compañia de Filipinas* (the Philippines Company) for trade with that colony. However, all earnings were required to be paid as dividends, which did not allow the bank to build a reserve. Despite these generous provisions, the bank had a hard time. Wars with France (1793–1795) and England (1786–1802) led to budgetary difficulties, which were compounded by the fact that the country was cut off from the Americas as a result of another blockade that led to new discounts to the *vales reales*. These monetary difficulties were aggravated by the Napoleon's occupation of Spain (and later by the Duke of Wellington). The government stopped printing *vales reales* at the end of the Napoleonic Wars, despite the fact that they had been profitable and reliable for the treasury, which led to a sharp decline of their price. This was compounded by the scarcity of silver that followed the independence of Peru (1821) and Mexico (1822). The provisioning of the army and the navy, which included the purchase and transportation of diverse commodities and victuals to distant places, proved extremely complex and ended up overwhelming the bank, which was also accused of overcharging and profiteering. When the government delayed and rejected payments, it also led to mounting losses.[2] The chronic deficit of the Spanish state during and after the War of Independence made repayments of its debt to the Bank of Saint Charles not possible, and its shareholders had to

---

[2] The bank's note issue averaged close to 3% and had never exceeded 14% of its capital. The Bank had also made important mistakes: It invested heavily in a project to build a canal from Madrid to Sevilla that failed, it purchased French public debt bonds that turned out to be an almost total loss; and the Company of the Philippines ended up heavily indebted with the bank as well (Tortella and García Ruiz 2013, pp. 17–21).

accept a steep reduction of their claims (from 310 million *reales* to 40 million) that was to be used as the capital of a renewed Bank of Saint Charles. In the end, at the last minute and to please the sitting king, Ferdinand VII (1784–1833), by invoking his patron saint, the bank was reorganized in 1829 as the *Banco Español de San Fernando*—the Spanish Bank of Saint Ferdinand (Kindleberger 1984, pp. 146–47; Tortella and García Ruiz 2013, pp. 17–27).

The history of the Bank of Saint Charles confirmed a pattern in which Spanish banks had to support the state's funding needs in exchange for privileges, and it is another instance in which the structural imbalances of Spain's finances ruined those who dealt with it (in this case aggravated by the egregious errors of the bank's directors). It is another example of a coalition of interest composed by the king, the minority share-holders, and the bank insiders. The king received a steady source of public finance and emergency funding, while the bank insiders earnt rents and the shareholders earnt compensation in the form of above-normal returns in exchange for the real risk of expropriation (in this case they ended up accepting a steep haircut of 270 million *reales*!).

The Spanish Bank of Saint Ferdinand started with a capital of 40 million *reales* taken over from the Bank of Saint Charles, and it was designed as a bank of issue and discount and a lender to the government: 80% of its lending went to the government. Deposits became free of charge. For the first few years, the main goal of the bank's managers was to rebuild the trust that has been shattered by the disaster of the Bank of Saint Charles and it acted very conservatively keeping its assets as liquid as possible. Yet again, however, the bank operations were impacted by the fiscal effects of the collapse of the Spanish empire with the inde-pendence of most of its colonies, which deprived the country from the inflow of silver and had a very severe impact on trade and government revenue because for the previous centuries remittances from the Americas have helped fill the gap between expenditures and revenues (Vicens Vives 1969, pp. 713, 724, 727).

The government tried to shore up public finances by increasing taxes, which led to internal unrest that aggravated the situation, and left Spanish finances in a situation of permanent deficit. Some help came from repa-triated savings from the immigrants returning from the colonies, and the government resorted to selling civil, aristocratic, and Church land to raise revenue and accommodate the growing numbers of landless peasants

(Kindleberger 1984, p. 147).[3] Yet, it still remained very dependent on the banks, both domestic and foreign ones. Despite relief from some bankers (notably, Alejandro Aguado who became the court's banker as was made Marquis of the Marshes for his services), the financial challenges proved unsurmountable, and at the end of Ferdinand's reign, the country was again insolvent.

The dead of Ferdinand led to a war of succession (the Carlist civil war) between his brother Charles, who received support from the most traditional and Catholic sectors of the country, and Ferdinand's widow, Maria Cristina, who became queen regent on behalf of their daughter Isabella II and was supported by the more liberal and progressive sectors of the country. The war lasted a decade and it created further tensions over a treasury that was already exhausted. The Spanish government was able to capitalize on the resources from its limited assets (including land and remittances from the last colonies, Cuba, the Philippines, and Puerto Rico) and funding from foreign bankers, notably Rothschild of London that provided 15 million francs in 1834, in exchange for repayment, the appointment as official bankers of the court, and the recognition of the Triennium debt (an outstanding debt largely in foreign hands that Ferdinand had refused to recognize).[4] These were years of enormous turmoil in the country and constant cabinet changes. The appointment of Juan A. Mendizábal as prime minister in 1835 proved providential because he was able to carry out the *desamortización* (confiscation) of the properties of the monastic orders. It is important to note that despite the fact that the clergy only represented around 4.5% of the male population in Spain in 1700 (it went down to 3.5% by 1770 and less than 2% by 1840), the Spanish Church was the largest landowner in Spain: According to the famous *Catastro de Ensenada* (Cadastre of Ensenada) that took place between 1750 and 1760, the Spanish Church owned 15% of the country's agricultural land, which generated 24% of the agricultural income. It is estimated that between 1750 and 1780 the Church owned between 25

---

[3] Most observers agree that the disentailment process that run throughout the eighteenth (1830s–1850s) and nineteenth centuries had limited success because the aristocrats were able to acquire the bulk of the land (they were also largely unprogressive and absentee), and also because it slowed down economic growth by diverting savings away from industry and infrastructure (Kindleberger 1984, p. 147).

[4] The Rothschilds were also able to obtain, through bribes to the Queen and the finance minister, the Almadén contract to exploit mercury resources which they had pursued for years and turned out to be very profitable (Tortella and García Ruiz 2013, p. 31).

and 30% of all the country's properties including land, real estate, financial assets, etc. However, since the Reconquest, the Church paid the Crown a share of the income from the land in contribution to the funding of the saint war that took place between 718 and 1492 to expel Muslims from the peninsula. These payments represented between 10 and 25% of its income, but sometimes they reached up to 50%. They were extended afterward and were renegotiated over time (Piketty 2019, pp. 119, 122, 197–99). Hence, this explains the importance of the *desamortización*. The lands were sold in auction, and the Church was compensated in public bonds, and it accepted cash and public debt securities as payment, thus providing much needed income to the Crown to finance the ongoing war.

The government also pressured the Spanish Bank of Saint Ferdinand for funds offering all kind of securities as guarantees that the bank tried to resist. Mounting pressures often forced the bank to lend to the government, but typically it would only agree to a fraction of what was requested. This became a pattern throughout the initial war years, but subsequently the bank abandoned its original Conservative stand to lending to the government (it had learned from the mistakes of its predecessor, the Bank of Saint Charles), and credit to the government grew significantly after 1834 while it neglected credit to the private sector, which almost disappeared by 1939. This was just another instance of the government's financial needs crowing out the private sector, which has been so characteristic throughout Spain's banking history. In addition to providing credit to the government, the bank also acted as the government's agent, mediated with other bankers and suppliers, and discounted promissory notes (Tortella and García Ruiz 2013, pp. 33–35).

The end of the war in 1839 with the victory of the liberal government provided a boost to the country's economy, but its structural problems (large public debt and chronic budget deficit) and the political turmoil persisted. In the years that followed, new companies were established, and the industrial and textile sectors recovered. These companies demanded financial services and this led to the establishment of new 'merchant' and 'savings' banks (such as the *Sociedad Valenciana de Fomento, Caja de Descuentos Zaragozana* and the *Caja de Ahorros de Madrid*), as well as commercial banks like the Bank of Barcelona (*Banco de Barcelona*) and the *Bank of Isabella II* (*Banco de Isabel II*). These latter two banks could not be more different. While the former was very Conservative, the latter was very innovative (and aggressive).

Indeed, these two new banks emerged in Catalonia in the first half of the nineteenth century: The private banking firm of the Marquis of the Remisa was founded in 1827 and focused largely on dealing in bills of exchange with the owner's capital (Kindleberger 1984, p. 147). The Bank of Barcelona was the first private bank established in 1844 in response to the need for capital from merchants and also from the industrial firms that emerged in the 1830s that followed the economic revival after the Carlist War. Founded by Manuel Girona, the scion of a merchant family, it was a commercial bank of issue that also lent on merchandise (mostly cotton, as the cotton industry was very important in Catalonia) as collateral and was authorized to open branches in the Balearic Islands and Catalonia. The Bank of Isabella II was also founded in 1844. It was a commercial bank that could issue banknotes under a different name and was largely led by the José Salamanca, a savvy man of very questionable morals and practices who would become cabinet minister, and was modeled after the *Caisse Générale* of Jacques Laffitte of Paris. The new bank was aggressive trying to compete with the Bank of Saint Ferdinand by discounting, lending, and supporting industrial ventures. It was also very innovative introducing small-denomination banknotes, using its issuing license liberally, while opening a new branch in southern Spain (Cadiz) and placing notes in circulation in that city. The new bank also lent aggressively to the new industrial, textile, mining, and railroad companies (including to the Madrid-Aranjuez railroad company that belonged to Salamanca) accepting their shares as collateral, a practice that henceforth would become more widespread. This liberality with loans would ultimately prove its undoing (Kindleberger 1984, pp. 147–48; Tortella and García Ruiz 2013, pp. 36–40).[5]

The Madrid Savings Bank (*Caja de Ahorros de Madrid*) was founded in 1838 under the auspices of a 1838 royal order that sought to promote the establishment of savings banks in the country to promote a thrift spirit and the formation of investible funds that the state could use. It led to the establishment of *Cajas* in other cities between 1839 and 1845, like Granada, Seville and Santander, Coruña, Valencia, Barcelona, Burgos, and

---

[5] The new bank was opposed by the Spanish Bank of Saint Ferdinand, which even refused to accept its money, and they had a legendary confrontation. But the new bank forced its more traditional competitor to amp its game intensifying its operations with the private sector, not charging fees for deposits, facilitated the concessions of loans, expanded its note issue (Tortella and García Ruiz 2013, p. 38).

Cadiz. These *Cajas* built upon the model established by the *Montes de Piedad*, charitable organizations that originated in the fourteenth century to provide small loans at reasonable interest rates. Since they had no capital or shareholders, they had to cooperate with savings banks, which collected charitable funds but lacked investment outlets for those funds. In Madrid, the *Monte* and the *Caja* signed an agreement through which the *Caja* lent to the *Monte* all the funds deposited (Tortella and García Ruiz 2013, pp. 40–41). As we will see later in the book, these *cajas* ended up deviating from their original mission and were one of the main culprits of the 2008 banking crisis in Spain.

The boom of the 1840s also attracted money of former colonists in Spanish America, as well as European funds that were used to fund infrastructure (mostly railroad building) but also some industry (Kindleberger 1984, p. 148). The European crisis of 1847–1848 caused a sharp decline in Spanish exports and it led to an economic crisis, bankruptcies, and closedowns, severely impacting banks, and forcing the closing of many of them, with fatal consequences for the Bank of Isabella II, which was dragged down by companies that closed or suspended payments as a result of the crisis (including the Madrid-Aranjuez railroad, one of José Salamanca-projects financed by his own bank) as they had received credits from the bank and now were unable to repay. That risky and reckless investment in the private sector and Salamanca's own companies finally caught up with the bank. The Bank of Saint Ferdinand was also impacted by the crisis. In this case, the problem was not so much loans to the private sector, which the bank had largely neglected, but rather loans to the government and the public sector in fulfillment of a contract that it signed with the state in 1844 which made the bank the state's cashier, whereby it would collect some taxes and provide the government advances on those taxes. The crisis had also impacted government's revenues and forced it to borrow from the bank, making the bank vulnerable to the government's solvency and willingness to repay. To address the crisis, the government ended up forcing the merge of the two banks in 1847, a process led by Salamanca (!) who "indebted for enormous amounts to the Bank of Isabella II, could find no other means of getting out of this morass than by becoming Finan1966ce Minister" and used his political clout to convince the queen to appoint a new government with the help of a general she was infatuated with (Santillán 1966, pp. 311–12). He appointed a commission to evaluate the assets and liabilities of both banks that ended up admitting them at face value, to the detriment of the

Bank of Saint Ferdinand (which also had a majority in the commission). The commission made a decision that was patently against the interests of the Bank of Saint Ferdinand because "the government menaced them with stopping the payments it owed the bank" (Tortella and García Ruiz 2013, pp. 42–43), which would have caused the banks' bankruptcy, and it suspended the contracts that made the Bank of Saint Ferdinand the government's cashier.[6] The whole situation grew even more complicated when the Director of the Bank of Saint Ferdinand, Joaquín Fagoaga, run away with securities and a large amount of money that provoked enormous losses to the bank and hurt its credit (Tortella and García Ruiz 2013, pp. 42–43; also Santillán 1966, pp. 256–312; Tedde de Lorca 2019, pp. 225–26). The reorganized bank took over the capital of its two predecessors, but failed to raise funding from the public, which forced yet another reorganization and capital reduction in 1851 (from 200 million *reales* to a 120 million) as questionable assets and loans were written off. The bank benefitted from the demand for Spanish grain and minerals stimulated by the Crimean War, and it managed to get its note circulation up to 120 million *reales* (Kindleberger 1984, p. 148).

The Bank of Barcelona was also severely impacted by the crisis for altogether different reasons: In February 1848, it was discovered that a significant amount of cotton had disappeared from the warehouse in which it kept deposited by its clients as collateral for loans (later it was discovered that it had been the clients themselves who had embezzled it). This provoked severe losses to the bank and also affected its credit and reputation. The bank was also impacted by the political crisis in neighboring France, when King Louis Philippe was deposed on February 1848, and the bank clients demanded conversion of their banknotes into specie, forcing the bank to diminish lending and discounts, and request a 'passive dividend' from its shareholders. The strong reputation of the bank allowed it to weather the crisis but henceforth it decided to reduce its operations using goods as collateral and manned its own warehouses (Tortella and García Ruiz 2013, p. 44; also Cabana 1978, pp. 31–33).

The government approved a new Banking Law in May 1848, inspired by the English Peel Act of 1844, that forbade the creation of new banks of issue, limited the maximum banknote circulation to one half of the disbursed capital, and in effect granted the Bank of Saint Ferdinand the

---

[6] The contract was restored in November of 1847 after the Salamanca government fell, and was replaced by a new cabinet.

monopoly of issue in Spain, with the exceptions of Barcelona and Cadiz, while it divided the bank into an issue department and a banking one. The new Law made life difficult for the Bank, which strongly resisted it and it lasted only two years. In December 1848, Santillán was appointed governor of the Bank, a position that he held until his death in 1863. With his military background and strong discipline, Santillán was unrelenting in his efforts to strengthening the Bank's financial position and resisting the government pressures for new advances (the government was so frustrated with his resistance that it dismissed him in April 1854, but reinstated him in August of that year). He was also instrumental in the drafting of two new bills that led to new banking laws in 1851 and 1856. The 1851 first banking law opened up the possibility of creating new banks, allowed for a higher banknote/capital ratio, and abolished the division of the bank into two departments. Under his leadership, the Bank became profitable again, and credit to the private sector exceeded credit to the state. His resistance to accommodate the government's demands led to the establishment of the *Caja de Depósitos* (the National Savings Bank) in 1852, which paid a 5% interest rate to depositors and was the custodian of judicial deposits. It was a success, and since it invested practically all of its funds in public debt, it soon became complimentary of the Bank of Saint Ferdinand and relived the government's pressure on the Bank for funds, thus allowing it to focus on the private sector and on floating debt. Ultimately, however, the model for the Caja was shaky because the price of public debt hinged on public confidence on the government; if that confidence was shaken the price would decline, thus placing the Caja in trouble and triggering a crisis. This is what happened later in 1864 (Tortella and García Ruiz 2013, pp. 45–46).

The experience of this banking crisis was yet another example of the consequences of the political games between politicians, bankers, bank shareholders, debtors, and depositors that influenced government regulations, which are at the heart of this book. This crisis showed the deep connection between business and politics at the root of banking in Spain. José Campo who founded the *Sociedad Valenciana de Fomento in* 1846 and José Salamanca from the Bank of Isabella II constantly mixed business and politics, and were masters at both. Salamanca was a deputy in the Spanish Congress from the city of Salamanca when he first arrived in Madrid in 1836, and he adroitly used his political contacts to open doors for him and "amassed a fortune in the stock exchange dealing in Spanish debt, and farming the salt state monopoly." His connections with the royal family were central to his businesses—including, as we have seen, a company to build a railroad between Madrid and Aranjuez where the royal family had a palace that they visited frequently. Campo for his part, who also promoted railroad companies, was ennobled with a marquisate and even became councilor in the Valencia Town Hall, and later major of the city. He was involved in infrastructure projects like railroad lines that connected the city with the tracks that linked Barcelona with Madrid and projects involving running water and gas illumination in Valencia. He founded the *Sociedad Valenciana de Fomento* to help finance those initiatives, using short-term promissory notes as means of payments, despite the fact that the *Sociedad* was not authorized to issue banknotes (Tortella and García Ruiz 2013, pp. 37, 40).

This was an example of an instance in which the autocratic centralized network that led to the establishment of the banking system advanced its interests and welfare. And it was a case in point, as we will see throughout the book, of how politics intruded into bank regulation, and how coalitions of politicians, bankers, and other interest groups generated policies that determined (in the words of Calomiris and Haber 2014) "who gets to be a banker, who has access to credit, and who pays for banks bailouts and rescues." In effect, that political deal established a rent-distribution system whereby the Queen and the government received a steady source of public finance; the bank insiders earnt rents and got loans for their own nonfinancial companies, and the shareholders earnt compensation in the form of above-normal returns in exchange for the real risk of expropriation.

In the end, the crony political bargains that we described, combined with the weak rule of law and the lack of constraints on the Crown's authority and discretion, which as we have seen retained the power to expropriate the banking system, together with the propensity from bank insiders to resort to inside lending, resulted in an unstable and fragile banking system that was prone to crisis. Hence, not surprisingly most of the new banks were short-lived: By the 1850s, the number of banks grew to nearly 60, and by the 1870s, it fell to less than 15. Since then, the number of banks grew slowly, but their size was relatively small.

Political instability continued in the second half of the 1850s. Following the toppling of a Conservative government in 1854, new elections produced a progressive parliament committed to putting an end to the corruption from the previous governments, do away to the right-wing

cronies that surrounded the Queen, as well as implement progressive policies, including a new disentailment law in 1855 to deal with Church and
secular properties that sought to raise new funds for the government.
Spain had adopted the decimal system in 1848 and a 1854 law allowed
coinage of gold, thus in effect shifting from silver to bimetallism.

A cornerstone of the new government policies was a new Railroad
Law, enacted also in 1855 that sought to build a railroad as a key to
foster economic development in the country. It included free entry of
capital goods, rolling stock, and fuel; it eliminated the restrictions from
the 1848 law that had limited concessions to ten years; and it required the
periodic revision of fares downward to prevent excessive profits. The Law
left the construction and management in private hands, but planning and
supervision were assigned to the state. It included subsidies, land grants,
and right of way privileges, as well as special guarantees to investors and
tariff remissions for imports used to build the railroad. In order to avoid
a repetition of the cronyism of previous attempts, the Law required that
all concessions had to be approved by the Spanish Parliament. Finally,
to promote and facilitate investment, the government approved two new
banking laws in 1856: the Bank of Issue Act and the Credit Company
Act. These laws were instigated by demands from the Bank of Saint
Ferdinand, which sought more fiduciary circulation, as well as permission
to increase its capital and open branches in the provinces. The Bank's
requests coalesced with demands by a coalition of provincial businessmen
and leftist politicians in Congress who resented the privileges of the Bank
of Saint Ferdinand and wanted more banks, and sought more competition
and more independent institutions. The outcome of the struggle between
these two groups was a compromise in which the Bank of Saint Ferdinand
was able to keep a privileged position and was renamed Bank of Spain
(*Banco de España*), but also established the principle of plurality of banks
of issue granting the Bank of Saint Ferdinand monopoly in Madrid as well
as in two other cities in which it opened branches: Valencia and Alicante.
The first law also established limits to the volume of banknotes issued by
each bank; granted the governor of the Bank of Spain supervisory powers;
considered banknote holders and current account creditors as depositors
with privileged creditor consideration to foster the growth of demand
deposits; established that banks had to be incorporated; and excluded
investment banking from the banks' purview. The Credit Company Act
dealt specifically with investment banks or credit companies (*sociedades de*

crédito) in response to request from foreign banks and investors (principally from France) who wanted to open branches in Spain, and it opened the door to investment companies or banks modeled on the French *Crédit Mobilier*. It allowed credit companies to make loans to the government, collect taxes, and practice commercial banking and industrial promotion. These laws opened a new era in Spanish banking that has been considered the period in which banks enjoyed the widest degree of freedom in their history. The laws also provided a strong incentive to invest in railroads. The 1856 Bank of Issue Act also included provisions to establish banks of issue in every city in the country that could be either branches of the Bank of Spain or a local bank with the right to issue banknotes. This led to the opening of new banks all over Spain. The impact of the new laws was almost immediate: The number of banks of issue more than doubled growing from 10 in 1857 to 21 in 1865 (including new banks in major cities such as Malaga, Seville, Saragossa, Valladolid, La Coruña, Bilbao, and Santander) (Tortella and García Ruiz 2013, pp. 48–58; Kindleberger 1984, pp. 147–48).

The Credit Company Act also led to the establishment of 'mercantile societies' (or credit companies). Indeed, the growth of credit companies was even more spectacular, growing more than fourfold from 6 to 35 (including the Spanish Industrial and Mercantile Company, *Sociedad Española Mercantil e Industrial*, SEMI; the *General de Crédito*; the Crédito Mobiliario Español; the *Sociedad Catalana General de Crédito*; and the *Sociedad Valenciana de Crédito y Fomento*). French bankers, like the Pereire brothers, the Rothschilds, and the Prost rushed to create them and to apply for railroad concessions. The Pereires set up *Crédito Immobiliario* and concentrated on the northern railroad, el Norte (which connected Madrid with Bayonne), the Rothschilds the *Sociedad Española Mercantil e Industrial* (SEMI), and Prost the *Compañía General de Crédito en España*. Neither the Rothschilds nor the Prost ever had their capital paid up (Kindleberger 1984, pp. 148–49). This growth, however, proved only temporary: By 1874, there was only 1 bank of issue and 13 credit companies.

The opening of new banks continued the pattern of bank bargains that was already well established in the country, marked by the insider interplay of politics and banking. For instance, Manuel A. Heredia, one of the most prominent businessmen in Spain, owner of the iron furnaces in Malaga and a relative by marriage of José Salamanca, the infamous finance minister to whom he had lent money in his early years in Madrid, was involved in the Bank of Malaga (*Banco de Malaga*); and the SEMI, the credit company founded by James de Rothschild, emerged in response to a bill submitted to Congress to establish an investment bank (the Credit Company Act) for the Pereire brothers, prominent bankers in France and owners of the *Crédit Mobilier*: The same day that the bill was submitted, another group of Spanish bankers put together by the Rothschilds and headed by a former minister of finance (Alejandro Mon) applied for a charter, and its board of directors included members of the nobility, recent finance ministers, and individuals who also served on the Bank of Spain's board. Furthermore, some of the banks had very close links with the credit companies: For instance, the *Compañia General Bilbaína de Crédito* was a subsidiary of *Banco de Bilbao*, and its founders were all members of the *Banco de Bilbao*'s board, and there were similar close links between the *Sociedad Catalana General de Crédito* and the bank, between the *Banco de Valladolid* and the *Crédito Castellano*, or the *Banco de Sevilla* and the *Crédito Comercial de Sevilla*. The banks, more tightly regulated by the new laws than the credit companies, benefitted from their participation in profitable business (particularly railroads), while the credit companies benefited from the banks' solvency and their loans.

Many of the new institutions, particularly the SEMI, continued providing funding to the government. Many of them also invested heavily in railroads (e.g., the SEMI invested heavily on railroads from Madrid to Saragossa and Alicante, and on the Alar del Rey-Santander line; the *Sociedad Valenciana de Crédito y Fomento* financed the Valencia-Jativa railway; and the *Sociedad Catalana General de Crédito* financed the Barcelona-Saragossa railroad, the Tarragona-Valencia line, and the Barcelona-France line; and the *Compañia General Bilbaína de Crédito* funded the Tudela-Bilbao line) and some in other ambitious ventures, such as mining (the General de Crédito created Spain's General Mining Company, the *Compañia General de Minas de España*, that invested in copper end lead mining) or water works (the *Sociedad Valenciana de Crédito y Fomento* undertook the water works in Valencia and led a project to improve the Valencia harbor). Very few, however, made industrial investments.

The railroad boom raised total kilometers of track in Spain from 332 km in 1854 to 5145 in 1866. It represented 1.55 billion pesetas in investment, or close to 90% of all investment during that period (in contrast 'only' 98 million were invested in Spanish manufacturing in the same period) (Harrison 1978, p. 48). Many of these railroad investments, however, proved disastrous for the banks and the credit companies. The blinded faith in the future of railroads, combined with the investment in public debt, became their undoing. The Rothschilds were ahead of the curve and their little faith in the Spanish economy led SEMI to divest itself of its railroad interest and speculate on public debt. But they were an exception, as most of the other institutions' assets remained invested in railroad securities and public debt. In the end, the SEMI was right and the investment in railroads proved to be a terrible mistake because of the costs (Spain is a very mountainous country and there were long distances between large population centers, which made the construction of these railroad lines much more expensive), and even more importantly, because the country was not developed enough to be able to take advantage of the new railroad lines, as there was little to carry on them. Furthermore, a significant proportion of the revenue generated from the construction was returned abroad (the law had not included linkages between rail and equipment manufacturer). Finally, the design favored interconnections with France, rather than interconnections among Spanish cities. When the railroad boom collapsed, they were completed but they had little passenger or freight traffic to haul. Hence, once the railroad lines were completed, revenues barely covered running costs, which not only did not allow credit companies to distribute benefits but also were often insufficient to even cover the interest on the bonds (Kindleberger 1984, p. 149).

When the collapse started in 1864 (partly as a result of the French crisis of that year), the Spanish government stopped paying railroad subsidies. This ended up being devastating for the credit companies who were not being reimbursed and often did not even receive interests on the bonds subscribed, but also on the banks of issue because, as we have seen, they had very close links with the credit companies and had lent massively to them. The Pereires' *Crédito Immobiliario* pulled out in 1864, followed by the Rothschilds in 1866. The situation was aggravated because the Bank of Spain not only did not act as a lender of last result, but withdrew liquidity from the market (Kindleberger 1984, p. 149).

The failure of these investments was compounded by the reckless decisions of the O'Donnell governments that provided generous railway subsidies and embarked in costly foreign adventures in Morocco and elsewhere, which dilapidated the revenues generated by the disentailment in land properties, as well as by other international factors such as the sharp decline in cotton prices that followed the end of the American Civil War in 1865 (which ruined many Spanish banks that had invested heavily in lands), as well as international tensions between Prussia and Austria that drove up interest rates and made access to international credit harder. All this happened in a context in which Spain's access to international credit and financial markets was already curtailed (the London and Amsterdam Stock Exchanges were closed to Spanish securities in retaliation for Bravo Murillo's government debt reform which had included a partial repudiation of debt), which forced the Bank of Spain to restrict banknote circulation to keep it within the legal limits, and other banks of issue to play a deflationary role, which aggravated the recession that started in 1864. Crop failures in 1866 and 1867 that caused hunger and a jump in the death rate further aggravated the crisis. The Alonso Martínez government tried to address the gaping budget deficits negotiating a loan with a syndicate of London bankers to get credit (in exchange for a charter of a bank), but it failed when London itself entered into a recession. As a result of these developments, the whole banking system started to collapse 'like a house of cards,' and a long line of bank failures ensued, particularly (and unsurprisingly) among those who had close links with commercial banks like the Banks of Valladolid, Seville, and the *Catalana*. In the end, 28 banks went out of business between 1865 and 1870, and the number of banks fell from fifty-seven to thirty during that period. This led to a period of capital withdrawals and domestic deflation, as interest rates increased for 6–12% between 1864 and 1866. The crisis also had political consequences: Alonso Martínez resigned in response to the crisis, and the death of his successor, Narváez, in April 1868, was followed by a revolution (the so-called Glorious Revolution and the *Sexenio Revolucionario*) led by a group of generals (Tortella and García Ruiz 2013, pp. 61–64; Kindleberger 1984, p. 149).

This banking crisis was another consequence of the bank bargains that had made the banking system so fragile. Once again, the political deals between the government and a group of bankers was shaped by the institutions that set the distribution of power in the political system. The crony political bargains that shaped bank risk-taking decisions and allocation of losses were a rent-distribution system that benefitted the government and the bank insiders, and it resulted in an instable and fragile banking system that was prone to crisis, and ultimately led to its collapse. Indeed, the new banking laws had created a system in which banks were strictly regulated and enjoyed local emitting monopolies, but also one in which credit companies were much freer. Many of them, with close links with the banks of issue, used this freedom to make very risky investment decisions that ended up causing their demise. The government's misguided decision to promote massive investment in railroads at a time in which the country was yet underdeveloped led them to establish a system that had the seeds of failure from its inception.

The revolutionary period that followed lasted six years (1868–1874). It led to the deposition of Queen Isabella II and constituted the most explicit attempt in Spanish history up to that point to establish economic liberalism in the country: It included decrees abolishing restrictions on the establishment of corporations, as well as inspection of credit companies and royal commissaries of banks of issue (with the exceptions of the Bank of Spain and the Bank of Barcelona, which were deemed too important, and hence were expected to render services to the government). The new government, led in financial and commercial matters by the Minister of Finance Laureano Figuerola, implemented a monetary reform that sought to align Spain with the Latin Monetary Union and introduced a new monetary unit, the *peseta*—equivalent of four *reales*. The new revolutionary government had to deal with the public debt, which was estimated at (an astonishing) 600 million pesetas, and the crisis of the railroad companies. Even wealthy people, who had invested in railroad securities and wanted the government to bail them out, supported the revolution (many prominent bankers, businessmen, and politicians had even petitioned the Queen in February 1866 for increased railway subsidies). The main government's creditor was the National Savings Bank (*Caja de Depósitos*), which was owed 310.8 million, and it was followed by banks, many of them foreign, with a combined total of 85.9 million, and short-term debt (Tortella 1977, p. 519). Since one of the main grievances that led to the revolution was the increasing fiscal pressure from the Queen, one of its consequences was that people stopped paying taxes, thus worsening the fiscal situation of the country. The government was compelled to request (again!) support from the bankers, who agreed to float a new 320 million loan on the condition that part of the loan would be destined to subsidize the railroad companies.

A new law in 1869 established general free incorporation for banks and credits companies. This law was exceedingly liberal: Limits to banknote issues were to be established by the bank's statutes; it set no legal limits to the issuing of shares, bonds, and other credit instruments; and it exempted companies established after the publication of the law from the inspection from the government. However, existing banks of issue retained their monopoly of issue in their towns until their charter expired, which in fact prevented the possibility of opening new banks. These economic policies favored cotton imports and iron output (Figuerola had enacted a new laissez-faire tariff law that lowered import duties), and railroads (they received subsidies and also benefitted from increasing traffic and new laws that facilitated mergers), which led to a period of sustained growth that averaged 7.8% between 1869 and 1873. Banks also benefitted from increasing demand for credit. The implementation of the disentailment (*desamortización*) program that had allowed for the massive transfer of land from the Church, nobility, and towns into private hands (mostly middle- and upper-middle class) created a demand for capital in agriculture that led to the emergence of new mortgage banks, like the Bank of Castile (*Banco de Castilla*) founded in 1871 by prominent bankers and politicians with international connections (it had links with the French *Paribas*) that focused initially in agricultural credit; the Spanish Popular Bank of Barcelona (*Banco Popular Español de Barcelona*) that specialized in lending to municipalities; and the Mortgage Bank (*Banco Hipotecario*) that provided advances to the government on the promissory notes from the buyers of nationalized land, and was granted a monopoly on the right to issue mortgage bonds.

These bold reforms and the desperate steps to deal with the country's intractable fiscal challenges (including the selling of copper and sulfur mines to international investors) proved to be insufficient. The liberal government was unable to secure funding from international lenders, and the domestic banking system was in shambles: Only 25 incorporated banks remained in 1874 (compared to 60 in 1864) and most of them were in dire financial straits. So desperate was the situation, that against all its principles, the government ended up granting the Bank of Spain in 1868 the monopoly of banknote issue in exchange for a loan, which deprived other banks of issue of that privilege (most provincial banks ended up as branches of the Bank of Spain, exchanging their shares for those of the Bank at par value, which was quite advantageous to most of them; the five who refused, including Barcelona, Santander, and Bilbao, were reconverted as commercial and deposit banks). The government efforts were further hindered by political turmoil, as it confronted a war in Cuba and a new Carlist War on the Basque Country and Catalonia, which forced increased expenditure on an already bankrupt treasury. Ultimately, the experiment with democracy and liberalism failed in nineteenth-century Spain, and the Bourbons returned to the throne in the hands of Alfonso XII, son of Isabella II (Tortella and García Ruiz 2013, pp. 70–76).

The restoration of the Bourbon monarchy in 1874 brought about a new and long political regime that ended in 1923 with Primo de Rivera's assumption of dictatorial powers. The new regime was characterized by bipartisan rule, in which two parties (the Conservative Party led by Cánovas and the Liberal Party led by Sagasta) alternated in power. A new Constitution (1876) enshrined the new regime that survived the death of the King in 1885. His wife María Cristina of Habsburg-Lorraine served as regent (she gave birth to a posthumous son who became Alfonso XIII and was crowned in 1902).

Despite the very precarious fiscal inheritance that it received from the republic (e.g., after ten years of crises, the number of banking corporations in the country was reduced to 16, nine of which were non-issuing banks), the Restoration governments benefited from the end of the Carlist and Cuba Wars, which alleviated the depleted treasury's coffers. The new government also maintained the monopoly of issue vested upon the Bank of Spain, which now had new powers to expand circulation up to 500 million pesetas and to increase its capital by 50%, and took advantage of the Bank's new firepower to access steady credit. Demands for money from the public and the government led to new decrees and laws that increased the maximum volume of circulation, which led to a significant growth of the money supply. It was at this time that the Bank, taking advantage of its status as the official banker of the state (even though it was still privately owned, although the government appointed its governor), became a truly national bank. Political stability and economic growth led to the emergence of new banks, notably the Hispanic Colonial Bank (*Banco Hispano Colonial*), founded by Antonio López y López who had started his business career in Cuba and had close links with the Barcelona business elite and became Marquis of Comillas. Once again, another bank was created based on the special relationship that its founder had with the Spanish state: He had been asked in 1876 by the government to coordinate a 100 million loan to defeat an uprising in Cuba, and he founded the bank to manage the operations of the loan. In exchange, he received the management of Cuba's custom duties. When the war in Cuba ended in 1878, the Bank obtained a new contract with the government in which he kept Cuba's custom duties and also become an agent to manage the colonial finances.

This was the period of steady development for banks in Bilbao and Vizcaya (benefitting from the emergence of iron and steel industries in the city) and Madrid, but less so in other parts of Spain, notably Catalonia, which was hurt by the predominance of small and median enterprises that self-financed themselves, and by the mistakes of bank managers that speculated in the stock market; or in Santander hurt by the decline of the city as exporter of grain and flour. Moreover, while the Rothschild branch entered in slow decline as it focused less on government finance, the *Urquijo* became very successful lending to the government and operating in the stock exchange (Estanislao Urquijo's support for the King had brought him the marquisate of Urquijo) and with investments in metallurgy tobacco, mining, shipbuilding, iron and steel companies. The improved fiscal position of the Spanish government made lending to the government less attractive, and many of these banks capitalized from the implementation of the 1885 Commerce Code, which discouraged long-term lending by commercial banks, but allowed merchant banks to invest in companies by participating in their capital. Henceforth, merchant banks took advantage of the development of the Spanish economy by investing in industry, textiles, mining, electricity, utilities, transportation, and public work companies. This was also a time of expansion for savings banks, which grew from 9 in 184 to 43 in 1900, and turned them, in addition to their charitable work, into a major source of financial funds because they collected more money than they could absorb and hence decided to invest in public bonds and private securities (Tortella and García Ruiz 2013, pp. 83–86).

Conclusion

As we have seen throughout the chapter, up to the beginning of the twentieth century, Spanish banking was characterized by steady progress frequently marred by periodic crises closely associated with political and/or economic crises, and by the need to support the state's funding needs. The origins of the Spanish financial system led to the development of a financial system that was characterized by limited credit growth; specialization; and the centrality of banks. Banks were also close to power and provided a source of income to the government. This banking system was the result of political choices that made it vulnerable. Indeed, political institutions structured the incentives of actors to form coalitions that shaped laws, regulations, and policies in their favor. Therefore, it is not surprising that bank insolvency and illiquidity crises followed.

Spanish banks also played a different role in the country's development than that of other European countries. They did not follow Gerschenkron's model (1962), as banks did not substitute for entrepreneurship (as in Germany), and were not sufficiently large for economic development. On the contrary, they were often more interested in profit generation, as shown during the railroad boom, which did not led to much development. As a result, development was set back. Indeed, Spain lacked the appropriate institutional framework and socioeconomic mix to speed up development, and the political bargains that we have described throughout the chapter led largely to the development of an inadequate legal and regulatory framework that gave way to an extractive model (Acemoglu and Robinson 2012). While Spain had a dual economy with a small number of commercial cities in the periphery and Madrid, the rest of the country was largely agricultural based. And the railroads failed to close that gap. On the contrary, their focus on connectivity with France (with a neocolonialist flavor) hindered efforts to breach that divide and close the gap. And foreign banks largely pursued their own interests and did not help speed up development either (Kindleberger 1984, p. 150).

As we will see in the next chapter, the growth, modernization, and diversification of the economy that followed in the second half of the twentieth century was accompanied by banking growth, a process that continued for most of the twentieth century. Throughout most of the twentieth century, the largest joint-stock banks practiced mixed banking promoting actively industrial firms and public utilities, controlling industrial groups, and acting like holding companies, while they also dealt with commercial banking and attracted deposits, and made short-term loans and discounting. This lasted through the end of Franco's dictatorship, with the banking crises of the late 1970s and early 1980s, which led to the consolidation and liberalization of the sector, as well as divestment from the industrial sector.

REFERENCES

Acemoglu, Daron, and James Robinson. *Why Nations Fail: The Origins of Power, Prosperity and Poverty*. New York: Random House, 2012.

Cabana, Francesc. *Historia del Banc de Barcelona 1844–1920*. Barcelona: Edicions 62, 1978.

Calomiris, Charles W., and Stephen H. Haber. *Fragile by Design: The Political Origins of Banking Crises & Scarce Credit*. Princeton: Princeton University Press, 2014.

Ehremberg, Richard. *Capital and Finance in the Age of the Renaissance: A Study of the Fuggers*. New York: Harcourt Brace, 1928.

Gerschenkron, Alexander. *Economic Backwardness in Historical Perspective*. Cambridge, MA: Harvard University Press, 1962.

Harrison, Joseph. *An Economic History of Modern Spain*. New York: Holmes & Meir, 1978.

Kindleberger, Charles. *A Financial History of Western Europe*. New York: Routledge, 1984.

Piketty, Thomas. *Capital e Ideologia*. Bilbao: Deusto, 2019.

Santillán, Ramón. *Memorias, 1808–1856*. Madrid: Tecnos and Banco de España, 1966.

Tedde de Lorca, Pedro. "La Evolución del Sistema Bancario Español en el Siglo XX." In *Guía de Archivos Históricos de la Banca en España*, edited by María de Inclán Sánchez, Elena Serrano García, and Ana Calleja Fernández. Madrid: División de Archivos y gestión Documental del Banco de España, 2019.

Tortella, Gabriel. *Banking, Railroads and Industry in Spain*. New York: Arno Press, 1977.

Tortella, Gabriel, and José Luís García Ruiz. *Spanish Money and Banking: A History*. New York: Palgrave, 2013.

Vicens Vives, Jaume. *An Economic History of Spain*. Princeton, NJ: Princeton University Press, 1969.

CHAPTER 3

Spanish Banking in the Twentieth Century

INTRODUCTION

As discussed in the previous chapter, after decades of development, the modern Spanish banking system started in 1856 with the approval of the Banks Law and the Credit Banks Law. By the end of the nineteenth-century banks' businesses had taken off, but their level of development and modernization was still lagging behind their leading European counterparts. As we have seen, as a result of *political bargains* they had been hampered for decades by little and inadequate regulation, and most of them had not been able to go beyond simple non-specialist deposit-takers. Hence, in 1900 the financial intermediation ratio (e.g., the ratio between total assets issued by financial institutions and the national wealth), which indicates the level of development of the banking system, was 39% compared to the European average of 104% (with 100% being considered the indication of an advanced financial sector) (Martín-Aceña 1985, pp. 134–39).

The opening of numerous entities had been the most important development in the banking sector in the early years of the twentieth century. While there were a few large institutions like the *Hispano Americano*, the *Bilbao*, or the Spanish Credit Bank (*Banco Español de Crédito*) which was the result of the reorganization of *Crédito Immobiliario* in 1902, the majority of the banks continued to be small and their activities were largely circumscribed to the town or city in which they were based. These banks were typical credit and discount firms, but they were also playing a

© The Author(s) 2020
S. Royo, *Why Banks Fail*,
https://doi.org/10.1057/978-1-137-53228-2_3

(limited) role through their investments in mining, industry, or export-led agriculture, to promote the economy of the town/city/region in which they were based. Moreover, the growth of the banking sector during the first decade of the twentieth century was not linear as many institutions were forced to close: Between 1900 and 1914, twenty institutions folded.

This chapter examines the evolution of the Spanish banking system in the twentieth century. Between 1900 and 1975, the banking sector was transformed and modernized, growing from an underdeveloped structure into a comparatively modern sector. Still, most of that period was marked by intense state intervention in the sector and the maintenance of the status quo. Prior to the 1960s, the banking sector's institutional framework was composed of the following elements: the status quo of the 1940s and 1950s which limited the emergence of new banks and the opening of new branches; strong regulation of interest rates that capped deposits and minimum lending rates to protect banks' margins; and a financial system highly compartmentalized that sought to divide the banking system into commercial and investment/industrial banks. Yet the structure of the banking sector underwent substantial changes as the number of credit institutions multiplied, and some of them gained in size and became 'mixed banks,' combining retail and investment banking. The degree of financial intermediation also rose, and many credit institutions diversified and specialized in different credit segments. Despite progress toward increasing competition, those decades also witnessed the increasing concentration in the banking sector, a process that intensified in the period between the two world wars and after the Spanish Civil War, with the emergence and consolidation of the eight large banks: *Bilbao, Central, Español de Crédito, Hispano, Vizcaya, Urquijo, Santander* and *Popular*. Restrictive entry rules and tight regulation of credit activities, combined with the emergence of economies of scale, contributed to the increase in the level of financial concentration. Those institutions dominated the financial landscape until recent decades.

The process of financial liberalization and deregulation that started in the 1960s contributed to further growth in the banking sector. Yet, by the onset of Democracy in the 1970s Spanish banks were still relatively inefficient and lacked the ability to compete abroad. Many of them were too small and lacked adequate management structures. Finally, the regulatory framework based on rudimentary prudential regulation and underdeveloped supervision that limited competition was still inadequate for the times. While this framework was able to prevent systemic crises until the 1970s (the relatively small banking crises that arose in the twentieth century prior to that decade were settled discreetly by the banks and/or the Bank of Spain), it proved unsustainable. The economic crisis of the 1970s and 1980s brought those deficiencies to the fore bringing the expansion phase to a halt and overturning the map of the Spanish banking system. Indeed, the severe banking crisis that run from 1977 to 1985 had a tremendous impact on the banking sector and led to far-reaching changes to the regulatory and supervisory frameworks, accounting standards, and prudential regulations.

As in previous decades, throughout the Twentieth century progress in the development of the banking system was often marred by crises, typically closely associated with the country's economic and/or political crises (and even a devastating Civil War). Furthermore, successive governments continued establishing institutional and regulatory frameworks that favored both the government and other privileged actors' access to finance at the expense of an environment conductive to a stable banking system. Indeed, political institutions continued to structure the incentives of bankers, and political and economic actors continued forming coalitions that shaped regulations and policies in their favor This institutional framework was the result of political choices that made the banking system vulnerable, because prudent lending practices continued being influenced by the desires of the groups that were in control of the government and other leading institutions, who often channeled credit to groups that were considered politically crucial. Therefore, it is not surprising that the banking system was still fragile and crises prone.

The First Decades of the Twentieth Century

The 'Disaster of 1898' in which Spain lose its last colonies (Cuba, Puerto Rico, and Philippines) was major blow to Spain. It led to calls for the regeneration of the country and the emergence of nationalisms (including regional ones in Catalonia and the Basque Country, fueled by fears for the loss of the colonial markets). The country, however, benefitted from capital repatriated from the former colonies (between 1892 and 1902, it was estimated that remittances amounted to 1000 million pesetas) that led to the emergence of a number of new modern banks: Between 1899 and 1914, 50 new banks were created, principally in the Basque Country (like the Bank of Vizcaya-*Banco de Vizcaya*) and Madrid (like the Hispanic American Bank-*Banco Hispano Americano*).[1]

After a short period of turmoil and uncertainty that included bank runs (e.g., in Bilbao), the closing of some institutions (like Credit of the Mining Union, *Crédito de la Unión Minera* affected by the halt in iron exports); drop in profits; the suspension of some banks' shares (Bilbao and Vizcaya); and strains in liquidity, which were addressed by the Bank of Spain which made extraordinary credit lines available; the First World War turned out to be very beneficial to the Spanish banking sector. The country's neutrality placed it in an ideal situation to serve the needs of the warring nations for goods and services. This led to a sharp increase in exports, balance of payment surpluses, the rise of the *peseta*, and inflows of gold and foreign currency. Spanish exporters saw their profits soar and banks saw money flowing into their accounts. This spurred a process of import substitution that led to the emergence of new companies (and banks). Not only the number of banks increased, but also deposits (which reached 1653 million pesetas in 1920), lending and investment (reaching almost one billion pesetas in 1918), profits, credit, and paid-up capital went up. Many banks took advantage of the boom to expand and open branches across the country. This led to a significant increase in the intermediation coefficient (the ratio of bank deposits to GDP) that went up from 3.6% in 1913 to 10% in 1919; as well as in the composition of the money supply: Private sector bank money represented 12% of the total in 1913, and almost 40% in 1920. Another development was the consolidation of mixed banks, as banks stepped up their investment in industry underwriting third-party debt issues or owning a fixed stake in industrial firms to control them. Banks were trying to address a chronic shortcoming of Spanish capitalism, namely that lack of sufficient national savings outside of the large banks to make possible the develop the nation's industry. Banks seeing an opportunity to increase their profits beyond solely commercial transactions stepped in this vacuum and invested in these industrial firms with a view to controlling them.

The end of the war in November 1918, however, led to a new economic crisis caused by the widespread paralysis of industry and trade. On the banking sector, it put a sudden stop to the period of growth. Indeed, a year after the end of the war many institutions were

---

[1] Yet again, 13 of these banks were liquidated by 1914.

experiencing liquidity and solvency problems. While Spanish banks had expanded during the boom, they had failed to seek specialization or differentiation, and while they had tried to cover the whole range of bank transactions, many of them lacked sufficient equity capital and assumed positions of excessive risk. Moreover, many banks did not have the appropriate management structures, and their managers often lacked sufficient training to manage risk or experience in the business/industrial world. Finally, they often had too many people in their board of directors (part of the bank bargains game!) many of whom lacked detailed understanding of the banking business and just focused on dividends (a persistent problems in future decades, as well as we will see later in the book). The economic crisis made it much harder for debtors to meet their obligations with the banks at a time in which banks were experiencing losses on their private and public debt portfolio caused, in the latter case, by the declined market value of their debt securities. While the impact of the crisis was felt throughout the country, it was particularly severe in Catalonia where the Bank of Barcelona closed its doors as a result of its imprudent investment decisions and excessive speculation in foreign currency markets, particularly in German marks. The Tarrasa Bank was also severely impacted by the crises and also closed for similar reasons (Martín-Aceña 2012, pp. 104–9).

During the post-war period, Spanish banks benefited from the expansion of the Spanish economy promoted by the favorable international climate and boosted by strong public and private investment that led to notable growth in national income and the transformation of the Spanish economy as agriculture and mining's share of the country's GDP declined, while industry and services increased: By 1935, the participation of the agricultural sector in the country's GDP represented 23% (41.2% of the active population), while the industrial one represented 24.3% (19.3% of the population) (Tedde de Lorca 2019, p. 13).

Spanish banks were also able to take advantage of the 1917 Law for the Protection of National Industry that recommended the creation of a state bank to provide long-term credit to industry, which eventually led to the creation by a consortium of banks, with the support of the state, of the Bank for Industrial Credit (*Banco de Crédito Industrial*) in 1920; and Fernández Villaverde's stabilization plan that allowed the Bank of Spain to accept public bonds as collateral with low interest rates and with a favorable tax treatment, which made them very attractive to banks. Over the next two decades, the Bank of Spain would grow and consolidate its role as a true central bank, including becoming a 'lender of last resort,'[2] a process that culminated with the Banking law of 1921, developed by the Minister of Finance Francesc Cambó, which sought to bring some order to the financial system at a time of crisis, by increasing the state's control over the Bank of Spain and over private banks. This Law placed private banks in the hands of the newly created High Council for Banking (the *Consejo Superior Bancario*, or CSB), which was dominated by the larger banks, thus establishing a system of self-regulation that set the tone for the next few decades. The CSB set working rules for banks inscribed in a new Register of Banks and Bankers, compiled statistics, and established accounting systems. In exchange, participating banks (foreign ones were excluded) enjoyed benefits in their taxes and operations with the Bank of Spain (Tortella and García Ruiz 2013, pp. 3, 37–39).

As a result of intensifying and unprecedented social unrest the King appointed General Primo de Rivera as president in 1923 and the new government suspended the 1876 Constitution. The government, led on the economic side by the nationalist Minister of Finance Calvo Sotelo, implemented nationalist, protectionist, and interventionist policies, and embarked in an ambitious public works program (including waterworks, road construction, electrification, and the renewal of rail infrastructure) that contributed to the growth of the economy. However, it failed in the face of great opposition from landowners, among others, to implement a tax reform that sought to tax income. Calvo Sotelo was very enthusiastic about monopolies and established an oil monopoly (CAMPSA) to produce, refine, and distribute petrol; and a telecommunications one (Telefónica), which were enthusiastically supported by the banks. It also supported public banks like the existing Mortgage Bank, which received a boost from the public works, and the Bank for Industrial Credit. In 1925 it launched the new Bank of Credit for the Local Administration (*Banco de Crédito Local*) to lend to local and municipal entities, and the National Farming Credit Service (the *Servicio Nacional de Crédito Agrícola*, the embryo of the future Farming Credit Bank-*Banco de Crédito Agrícola* created in 1926). In 1928, it established the Spanish Bank of

---

[2] It first acted as 'lender of last resort' in 1914 when it rescued the Hispanic American Bank, which had been a victim of the Mexican Revolution. However, until the banking crisis of 1931, its interventions were selective, for instance it did not rescue the Bank of Barcelona in 1920, and they were largely driven by orders from the government to intervene.

Foreign Trade (*Banco Exterior de España*), which held a monopoly of export credit insurance. All these official banks had private capital until their nationalization in 1962, and the private sector was given the task of managing them, while the authorities reserved the right to appoint the governor. For instance, the management of the Bank of Credit was assigned by the Bank of Catalonia, whose director and founder, Eduard Recasens, had been a major proponent of its establishment and an enthusiastic supporter of the government's policies; and the *Exterior* was managed by the Peninsular and American National Credit (the *Crédito Nacional Peninsular y Americano*), a consortium led by the Bank of Catalonia and the Central Bank. These official banks were very specialized and they were created to serve the specific needs of particular industries or economic activities. They did generally not take deposits from private individuals, but they received funding directly from the Treasury and took advantage of privileges granted by the state like the right to issue special securities that were treated as public-backed debt. The lending from the official banks increased overtime, reaching 33% of the total banking system in 1925, and 45 in 1930. Finally, the government also regulated the savings and loans (*cajas de ahorro*), which had originated in 1838 linked to beneficial institutions and pawnshops and over time had evolved into non-for-profit credit and deposit entities that sought to foster savings across all segments of society particularly low-income people and small investors and reinvested their benefits in social and cultural projects in the areas in which they were established, with a 1926 decree that placed them under the Ministry of Labor (before they had been considered 'charities' and they had been lightly supervised by the Ministry of Interior) and seeking funds it required a mandatory investment ratio whereby 50% of their deposits (by 1920 they held 20% of deposits) had to be invested in government securities (Martín-Aceña 2012, pp. 114–15; Tortella and García Ruiz 2013, pp. 100–5; Tedde de Lorca 2019, pp. 17–20).

In the twenties, the Spanish financial system grew, as banks' assets increased from 7657 millions *pesetas* in 1923 to 12,245 millions in 1930. They continued purchasing public debt in large quantities to cover the government's deficit and finance infrastructure projects, benefitting from the good returns and high liquidity (it could be used automatically as collateral at the Bank of Spain and banks could immediately monetize it, thus serving as a very helpful stabilizing mechanisms during crises) of these loans. This period also witnessed the consolidation in the banking sector as five banks (*Hispano, Español de Crédito, Central, Bilbao* and

Vizcaya) came to control 50% of paid-up capital and 70% of deposits in the system. The crisis of 1924–1926 had a limited effect as six small institutions folded, but the Bank of Spain (still reluctant to play the role of lender of last resort), under pressure from the authorities, was forced to support *Central* which was facing serious liquidity problems (Martín-Aceña 2012, pp. 111–14; Tedde de Lorca 2019, p. 16).

The expansion of the 1920s ended in 1930 with the onset of the Great Depression. The weakness of the Spanish economy, which was severely affected by the deteriorating international economic situation and had an impact on various sectors including banking, contributed to the fall of the Primo de Rivera's dictatorship in January 1930. The volume of loans, credit, and deposits suffered from the economic and financial crisis. The municipal elections of 1931, which resulted in the defeat of the parties that supported the monarchy, led to the departure of the King and the proclamation of the Second Republic. In response, most wealthy Spaniards transferred their capital to foreign banks, and deposits decreased by 20%. While some Spanish banks were severely affected by the capital flight, notably the Central Bank, the crisis did not lead to a series of bank failures. There were three main reasons for this: First, the Bank of Spain, acting as lender of last resort and taking advantage of an increase in the fiduciary circulation ceiling that had been approved by the government, was instrumental in that outcome as it lent freely to other banks. Moreover, banks were also able to take advantage of an instrument that gave them access to credit with public debt as collateral. Finally, Spanish banks also benefitted from the nature of the structure of their assets: As noted before, they were able to use public debt securities and collateral, and thus they could automatically monetize this public debt and immediately draw upon cash reserves. These factors were crucial to avoid the collapse of the banking system and a repetition of previous banking crises. In the end, only the Bank of Catalonia, which had been closely associated with the Primo de Rivera dictatorship, collapsed. However, successive Republican governments were consumed with trying to prevent the fall the exchange rate of the peseta, which led to a policy of high interest rates as the instrument to sustaining the peseta exchange rate fully endorsed by the Bank of Spain, which hampered economic recovery at a time in which most other European countries were implementing loose monetary policies. The appreciation of the peseta and the high interest rates contributed to increasing social unrest, which together with political polarization, increasing violence, growing divisions over the autonomy of some regions, and the controversial land reform, ultimately led to the Civil War of 1936. Notably, however, this crisis did not lead to a systemic failure in the banking system, and Spanish banks were able to sail throughout the Great Depression with limited damage for the reasons stated above. Only banks that had been closely associated with the Primo de Rivera dictatorship, like the Bank of Catalonia, the Spanish Bank of Foreign Trade, and the Central Bank, closed down or were severely impacted (Tortella and García Ruiz 2013, pp. 108–14; Martín-Aceña 2012, p. 113; Tedde de Lorca 2019, p. 17).

The brutal Civil War that raged the country between 1936 and 1939 had profound consequences at all levels, divided the country into two opposing sides, and caused massive destruction. From an economic standpoint, it led to the breakdown of the monetary union, as each side used its own banknotes: the Republican peseta and the nationalist peseta. The war also led to massive state intervention in the economy, depleted the Bank of Spain's gold reserves, increased massively public debt, and printing of cash that caused rampant inflation and the depreciation of the peseta. Banks could not conduct their business normally and the war deeply affected their investment activities.

The end of the war on April 1, 1939 led to the restoration of the monetary union (Republican pesetas lost their legal tender), the rebuilding of the financial system and the unification of the two Banks of Spain that had operated during the war, one of each zone, into a new Bank of Spain (Tedde de Lorca 2019, pp. 23–24). One of the most important initiatives took place in October 1939 when the Franco government approved a Ministerial Order establishing a *numerus clausus* preventing the creation of new credit institutions and the opening of new branches without prior government approval. This crucial decision with lasting impact in the banking system sought to preserve the status quo. The government followed that decision with the approval of the Unfreezing Law (*Ley de Desbloqueo*) in December 1939, which sought to restore normal financial transactions (Republican transactions during the war had been considered *sub-judice*); and an additional decree that prohibited the creation of new banking institutions, the opening of new agencies and branches, the change of premises or the nature and legal status of banks, and the signing of agreements for transfer or mergers. This decision sought to prevent the financial crisis of the past, but it also signaled the mistrust that the new nationalist government had of the private sector (Martín-Aceña 2012, pp. 120–21).

The Franco dictatorship accentuated the nationalistic and protectionism of previous governments and sought to establish an autarkic economic system. From a banking perspective, the Franco government, divided between the fascist who pushed for the nationalization of the banks and the monarchist who were more pragmatic, followed a strong interventionist policy and severely restricted the possibility to open new banks as well as restricting the operations of existing ones. It approved new decrees on banking secrecy and limiting the distribution of dividends to avoid public outrage and prevent crowding out public debt bonds. The more pragmatic sectors of the new regime carried the day and the 1946 Banking Law (called the Benjumea Law after the Minister of Finance) did not nationalize the Bank of Spain or the private ones, although it gave the state strong regulatory powers over the banking sector and ended the Bank's privilege of issuing banknotes. The new law was a continuation of the past: It strengthened the regulation of banks' operations, including interest rates; reinforced the control of the Ministry of Finance over the Bank of Spain, which was now prevented from dealing directly in commercial banking operations; maintained the oligopolistic structure of the banking sector; required foreign banks to submit guarantees to the Bank of Spain in the form of deposits and subjected them to the principle of reciprocity; and separated the internal and external components of monetary policy, creating, as part of the Ministry of Commerce, the Spanish Institute of Foreign Currency (the *Instituto Español de Moneda Extranjera* or IEME). More significantly, in order to avoid crises and make capital increases unnecessary, the Law forced banks to build reserves quickly placing legal restrictions on dividend payments: When banks had net profits of more than 4% of their paid-up capital plus reserves, they had to deduct at least 10% to set aside reserves of up to half of the bank's capital. This provision, together with the fact that lending and borrowing interest rates were also controlled by the authorities, sought to ensure the solvency of these institutions (Martín-Aceña 2012, pp. 126–27; Tedde de Lorca 2019, p. 24).

This Law was understood as a tacit agreement between the new regime and the economic powers to cooperate while respecting each other. While the Benjumea Law did not impose mandatory investment coefficients in public funds (that would happen later in 1959 with a new banking law), the government expected the banks to continue with the strong provision of public funds tradition, and the president of the CSB sent a circular letter to banks in 1947 'recommending' that they invest 45% of their deposits in public funds (for local banks the requirement was 50%) under a 'veil threat' that the Bank of Spain would take banks' contributions into account whenever they would get in need of financial assistance.[3] Not surprisingly the share of public funds in the hands of banks increased rapidly. This had a significant inflationary impact as banks took full advantage of the right to pledge those bonds in the Bank of Spain (Tortella and García Ruiz 2013, pp. 115–20).

This period also witnessed further consolidation of the banking sector through mergers and acquisitions: In 1950, there were 306 institutions on the Register of Banks and Bankers, and in 1960 the number was 261, and the fourteen national banks accounted for almost 80% of the branches and agencies (the biggest seven banks: *Español de Crédito, Central, Hispano, Vizcaya, Bilbao, Santander* and *Popular* had 71% of the total). The government massive interventionism, however, did not prevent the growth of the financial system: banks' assets increased from 65,234 million pesetas in 1947 to 648,471 million pesetas in 1962; and credit and loans to business rose from 13,409 million pesetas to 104,430 million during the same period. Spanish banks largely lacked any true specialization, focusing both on retail and investment banking, and building a portfolio of securities that included both private and public debt. National banks in particular had a significant investment portfolio and close relations with the manufacturing companies that they invested in (Martín-Aceña 2012, pp. 125–28; Tedde de Lorca 2019, p. 25).

After the civil war, nine banks (*Banesto, Bilbao, Castellón, Central, Guipuzcoano, Hispano, International de Industria y Comercio, Valencia* and *Vizcaya*) signed a private pact (the *Arrangement to Moderate Banking Competition*) on maximum (passive) and minimum (active) interest rates and charges, to take effect on January 1, 1941. This private agreement, with the exception of the proviso that created a private court to penalize those who broke it, was informally approved by the Ministry of Finance. The agreement, however, did not have enough punitive instruments and a more comprehensive one (the *Interbank Agreement on Passive Interest Rates*) in which the responsibility to monitor compliance fell in the hands

---

[3] The Bank of Spain sent again a similar 'recommendation' in 1950 that included the 'threat' of a reduction of the Bank of Spain's rediscount limits to those banks that failed to invest 45% of their deposits in public funds.

of the deputy governor of the Bank of Spain, was signed on November 1949. This was renegotiated and signed in June 1952 as the *Regulatory Arrangement of Banking Conditions*.[4]

The economic failure of the autarkic approach led to a policy shift marked by a more balanced approach to economic growth, which led to yearly average per capita growth of 5% in the 1950s. A surprising development, however, was that accelerated growth was not accompanied of higher inflation. On the contrary, average inflation averaged 5.5% in the 1950s (compared to 12.2% in the 1940s). Some scholars, as we will examine later, have attributed the high inflation to the oligopolistic structure of the banking system, which allowed them to ensure high financial margins by charging higher prices for credit, and thus raising inflation. According to this perspective, banks took advantage of cross-shareholding and cross-directorship between banks and between banks and other companies and industrial firms; and market sharing price agreements (cartelization) (Pérez 1997). Others have contended, however, that cross-shareholding and cross-directorship were not so relevant because only the top management was involved in future decisions, and that the country also benefitted from these arrangements, which were instrumental to provide credit to companies and thus contributed to the development of the country. Moreover, the amount of credit from the banks to their subsidiaries was relatively small (around 4% of total credit in national banks). Others have pointed out to the comparatively limited profitability of banks prior to 1970 (their performance throughout the 1950s and 1960s was similar to that of other industrial companies) despite the oligopolistic structure of the banking sector (Tedde de Lorca 2019, pp. 30–31). Finally, the cartel agreements were not consistently applied, were often broken (e.g., some accounts received far more than the maximum agreed interest rates), and there was real competition among banks. Indeed, the failure of those agreements led to the adoption in February 1960 of the *Bank Terms Convention*, which included 27 rules including maximum interest rates for deposits and maturities; minimum interest rates for loans and discounts; a process to determine the responsibility of those who breached the agreement as well as the penalties.

---

[4] Luís Saenz de Ibarra, deputy governor of the Bank of Spain between 1947 and 1956, was instrumental in the negotiation and implementation of these agreements, and he played a key role as intermediary between the banks and the government/regulators (Tortella and García Ruiz 2013, pp. 124–25).

These authors attribute the inflation to the financial repression and the economic policies of the 1940s and 1950s (Tortella and García Ruiz 2013, pp. 122–25).

During the Franco regime, banks continued serving the economic needs of the government, and the 'banking games' included the following components: Oligopolistic banking (*numerus clausus* limited the number of banks and ensured the status quo); control of interest rates; external capital controls; investment channeled through the government; compulsory purchase of debt; and reliance on public banking. Banks were also forced to give out loans to meet the government policy goals (e.g., they had to give loans to improve farms, or to allow poor tenants to purchase their houses), and they were compelled to male mandatory investments that were channeled toward the acquisition of public funds. But banks also benefitted from these arrangements: not only they were able to maintain their hegemony in the financial system (by 1960 commercial banks controlled 73.2% of combined bank deposits[5]), but also their securities for agricultural credit could be discounted at the Bank of Spain, thus serving as an automatic pledge for public funds (and thus contributing to inflationary pressures). They largely had the support of the Ministry of Finance and the Bank of Spain, who were able to withstand the pressures from other sectors of the government controlled by the fascists (*falangistas*) who sought to erode their power (Tortella and García Ruiz 2013, pp. 125–26).

The shift in economic policy toward a more liberal model that left behind autarchy was enshrined by the 1959 *Stabilization and Liberalization Plan* which consolidated the victory of the most liberal and technocratic elements of the regime versus the *falangistas*. The Plan sought balanced budgets, positive balance of payments, and price stabilization through deficit limits and tax increases. The Plan also sought the liberalization of the Spanish economy through progressive lower tariffs; the promotion of foreign direct investment; ceilings on bank loans; the convertibility of the peseta; limits to monopolistic practices; and the progressive opening of markets. Banks, happy with a less interventionist approach and hoping that it would translate into less requests for loans and mandatory investments, supported it enthusiastically. This plan,

---

[5] According to a study from the School of Industrial Organization (EOI), the five largest banks (Banesto, Bilbao, Central, Hispano and Vizcaya) held 45% of the capital; 65% of demand deposits; 50% of industrial participations; and 72% of the branches in Madrid, and 68% in Barcelona (Tortella and García Ruiz 2013, p. 129).

which was implemented in the form of development plans (*Planes de Desarrollo*), led to a dramatic turnaround in the performance of the Spanish economy, and what is known as the 'Spanish economic miracle.' The country joined the industrialized world and left behind its endemic underdevelopment. Between 1959 and 1974, the economy grew an average of between 6 and 7%, enjoying the second highest growth rate in the world (only behind Japan) and became the ninth largest economy in the world (Tedde de Lorca 2019, p. 22). Spanish banks played a central role in this process, acting not only as financers but also as entrepreneurs, providing leadership, intermediation and funding that allowed the establishment of industries and companies that contributed to the economic development of the country.

In the financial realm, the liberalization was carried through a new banking law in 1962 (the Basic Law on Credit and Banking Organization-*Ley de Bases de Ordenación del Crédito y la Banca*, also known as the '*Navarro Rubio Law*,' after the Minister of Finance who spearheaded it), which represented a sharp departure from the past. This Law was developed in consultation with the financial institutions and other business representatives (these consultations had become an established component of the 'banking games' during the regime). The new Law identified bottlenecks and shortcoming in the Spanish banking system: inadequate instruments at the disposal of the Bank of Spain to conduct monetary policy; the shortage of credit that was holding back growth; the lack of specialization of Spanish banks; the absence of specialized institutions able to meet industry's funding needs; and finally, the lack of coordination between investment and saving processes. In order to address them, the new law sought to restructure the Bank of Spain to improve its operations; to boost bank lending; to limit the power of the banks over nonfinancial companies; and to eliminate the long period of the status quo by increasing the number of banks and branches in the system (thus eliminating the pre-existing *numerus clausus*). As a result of this law, the financial sector underwent a significant transformation: It led to the creation of new institutions; the further consolidation of the sector through mergers, liquidations and acquisitions; and the expansion of banking businesses. Moreover, the gradual deregulation introduced by the law increased the level of intermediation and led further diversification of operations and branch expansion (Martín-Aceña 2012, pp. 125–28; Tedde de Lorca 2019, pp. 24–25).

One of the most notable components of the new Law was distinction between retail banking (commercial banks) and wholesale banking (industrial banks). The new Law included regulations to create investment banks that were now required to have equity of at least one million pesetas at the time of their incorporation, and other banks were prohibited from holding more than 50% of their capital. The Law's intent to compel banks to choose their status between commercial and investment banking, however, had limited effect as mixed banks (which had the option to apply to be licensed as investment banks) survived even though their investment in new industrial stock was subject to authorization. In fact, the hegemony of the mixed banks continued through the end of the dictatorship and into the early 1980s. Yet, the market share of the largest six banks decreased over time (*Español de Crédito, Hispano, Central, Bilbao, Vizcaya* and *Santander*): While it grew from 64.55% of total deposits in 1934, to 70.63% in 1960, it decreased to 56.83 in 1975 as medium-size banks responded to the financing needs of small and medium enterprises, particularly in Catalonia (like Sabadell Bank, *Catalana* Bank and Atlantic Bank) (Tedde de Lorca 2019, pp. 27–29). Furthermore, the implementation of this law led to the nationalization of the Bank of Spain and the other public banks (with the exception of the Spanish Bank of Foreign Trade). The Bank of Spain would now depend on the Treasury and was charged with the regulation of credit; open market operations; rediscount to the private banks; the establishment of monetary policy instruments such as mandatory reserves, liquidity, and guarantee coefficients; the inspection of the private banks; and the centralization of reserves and foreign currencies (Tedde de Lorca 2019, p. 25). The government also established a new Risk Information Center (*Central de Información de Riesgos*) that operated as part of the Bank of Spain and was responsible for collecting quantitative information on loan risks above 5 million pesetas and revived the Registry of Unpaid Acceptances (*Registro de Aceptaciones Impagadas*), which collected information on unpaid bills.

The Navarro Rubio Law and the decrees that followed it strengthened controls over the banking sector and did not diminish the degree of state intervention: It ordered banks to specialize; established tight controls over interest rates (after 1966 the Ministry of Finance no longer established floors, only ceilings with the aim to lower the price of credit, reduce the financial costs for companies and lower inflation); established high standards and a strict process for the opening of new branches; imposed mandatory investments to support the goals established in the Development Plans; and forced them to provide loans to the public sector. Official banks for their part continued playing a central role in the financing of strategic public investments.

While interest rates were still set by the authorities and asset and liability operations remained subject to regulations and control, the 1962 Law paved the gave way for an opening, and these controls were relaxed in later years. Yet, obligatory coefficients for the banks and privileged credit circuits remained, and only in 1969 the authorities started to take small steps to deregulate interest rates, with the basic discount rate now being set by the Bank of Spain. However, despite the elaborate process (banks even had to publish their preferred rates in the legislative gazette) and the penalties established in the law to enforce discipline in interest rates (which were fixed by the Ministry of Finance), it failed to do so as bank continued resorting to new creative initiatives (for instance, offering tied-up incentives like insurance) to offer different interest rates and circumvent the interbank arrangements. Moreover, while the formal arrangements among banks to fix interest rates had ended in 1960, the presidents of the largest banks started to meet weekly to continue trying to curb passive interest rates above the legal limit in the so-called Big Seven Club (Banesto, Bilbao, Central, Hispano, Popular, Santander, and Vizcaya) and to coordinate their actions. Still, they had limited success and criticism of the power of the banks continued amidst accusations that many bank leaders were opposed to the regime, and despite the government's 'financial repression' concentration in the banking sector still persisted,[6] and some attempts for further consolidation were denied (e.g., when Central and Hispano tried to merge in 1965) (Tortella and García Ruiz 2013, pp. 128–33).

The 1962 Law also delegated the regulation of foreign banks to the government and applied the principle of reciprocity. The provisions regarding foreign banks, which were further articulated in a 1978 decree, allowed for the establishment of representative offices, subsidiaries, and branches of foreign banks. As a result, 24 branches of foreign banks entered the Spanish market between 1979 and 1981. Finally, the Law

---

[6] Despite the Navarro Rubio Law's provisions that forced banks to choose between becoming commercial banks or industrial banks had a limited effect: as late as 1966–1967, the six largest banks still exerted influence over 1000 companies through their presence in these companies' boards of directors; and the five largest banking groups (Banesto, Bilbao-Vizcaya, Central, Hispano-Urquijo, and Vizcaya) jointly controlled 68.7% of capital, 78.9% of deposits, and 80.9% of the branches (Martín-Aceña 2012, pp. 132–38).

created the Institute of Medium and Long Term Credit (*Instituto de Crédito a Medio y largo Plazo*, ICMLP), and nationalized three of the official banks: *Hipotecario*, *Crédito Industrial* and *Crédito Local*. The Spanish Bank of Foreign Trade was not nationalized, but it remained an official bank and was subjected to the same regulations as private sector banks.

The implementation of the 1962 Law which ended the 'status quo' led to the creation of new credit institutions: Between 1963 and 1975, 31 new institutions were added to the Registry, raising the total number to 132. The law also allowed for the expansion of new branches, through the so-called expansion plans prepared annually by the Bank of Spain to determine the growth capacity of each institution. As a result 2549, new offices were opened between 1963 and 1973. The final deregulation of the opening of new branches was introduced by decree in August 1974, which allowed credit institutions to open branches depending on their capacity ratios defined in relation to equity capital. This liberalization led to an accelerated expansion of branches: Between 1974 and 1978, 5658 new offices were opened, reaching a total of 11,095 (from 2775 in 1963!). This expansion became the main instrument to increase competition in the sector. Indeed, this was a period of rapid growth in the banking sector as noted by the growth in capital and reserves, as well as assets, which grew more than 15 fold between 1960 and 1975. However, an important development was the decrease in their participation on public fundsing: While public securities represented 70% of their portfolio in 1960, by 1975 their share has fallen to 66% (Martín-Aceña 2012, pp. 132–38).

The implosion of a major scandal, the so-called *Matesa scandal*, in which a private company that had received loans from a public bank to export textile machinery (in support of a law that subsidized exports of machinery with public credit), ended selling them to its own branches in other countries, exposed the weaknesses of the interventionist system and led to calls for further liberalization. The government responded with the 1971 Law of Organization and Framework for Official Credit which sought to align the financial operations of the official banks with the government's economic and social development targets and created the Official Credit Institute (*Instituto de Crédito Oficial*, ICO), which replaced the ICMLP and became the cornerstone for the management of official credit. The 1971 Law also sought to strengthen the control of state banks; simplified the mandatory investment coefficients; eliminated the option to rediscount or pledging in the Bank of Spain; and authorized foreign banks to operate in Spain as long as management remained in Spanish hands.

Spanish banks also were also impacted by the major changes that were taking place in other countries in the 1970s, in which deregulation, internationalization, and computerization had led to increasing economies of scale and the further development of universal banking. They started to adapt to the new environment and invested in computerization to improve their efficiencies (given their large workforce they faced very high labor costs) and services. But in order to improve competitiveness there was still a clear need for further deregulation of the financial system.

Pressures for increasing competition in the banking sector materialized in two new decrees approved by the government in 1974 that increased the degree of deregulation and pushed for greater uniformity. These decrees sought to increase more competition among banks by authorizing new institutions and branches; introducing more flexibility over interest rates; promoting more operational uniformity and treating saving banks/*cajas* and banks more uniformly. These decrees decreased market segmentation between industrial and commercial banks and between banks and savings banks (which could now perform the same operations as banks)[7]; established clearer criteria to create new banks and open new branches; introduced deposit certificates as a way to attract additional borrowing; deregulated long-term bank interest, liberalized interest rates with a maturity of more than 24 months; and made cash rations, collateral and investment coefficients more generally applicable. Spanish banks also faced pressures to open further its market to foreign banks: By 1977, there were only four foreign banks that controlled less than 1% of deposits; a new decree in 1978 authorized branches of foreign banks (Tortella and García Ruiz 2013, pp. 134–38).

The Spanish banking experience during the dictatorship conforms to Calomiris and Haber's (2014, pp. 41–53) 'autocratic regime' taxonomy. Indeed, Franco's regime built a 'centralized autocratic network of

---

[7] By 1974, savings banks, led by *La Caixa, Caja Madrid, Caja de Barcelona* and *Caja de Zaragoza* in that order, controlled approximately a third of the deposits. The 1977 decree eliminated the restrictions that did not allow savings bank to perform the same operations as banks, and in exchange it eliminated the privileges that savings banks enjoyed in the mortgage market. This decree in effect leveled the playing field between them.

alliances, in which bankers played an important role to finance expenditures and provide capital. This coalition undermined the stability of the banking system and its ability to provide credit. His governments used a multitude of policies to tax and extort banks, including mandatory investment rations; reserve requirements; maximum (passive) and minimum (active) interest rates and charges; transaction taxes; and the granting or revoking of privileges. The regime also owned and run public banks which it used to hold fiscal balances and to provide funding for government initiatives. Bankers were compensated for the risk of expropriation with higher rates of return on their capital above what they would have received in a competitive market. During the first decade of the regime, Franco restricted the number of bank charters in order to minimize competition among banks, thus allowing them to charge higher interest rates, which were set by the government. Indeed, despite the fact that it was not a perfect oligopoly, and that the private agreements to limit competition had not been effective, bankers still reaped oligopolistic rents. In exchange, the banks provided loans to his government and extended credit to companies associated with the leadership of the regime, thus forming a network of partnerships to share rents. Furthermore, bank insiders often held interest in nonfinancial enterprises and lent to those companies, and bank insiders borrowers faced preferential treatment at times of crises and often avoided the harsh enforcement of loans. It was a system that granted special privileges to insiders and restricted entry, while generating flow of rents to the government that gave the regime an incentive to maintain the status quo, and thereby reducing the risk of expropriation.

The outcome of this government-banker partnership was a semi-oligopolistic bank structure based on crony political bargains that produced a coalition of interests between the dictator, the minority shareholders, and the bank insiders. Franco's governments received a steady source of public finance; bank insiders earnt rents in the form of high salaries, and also through above normal returns from their participation in nonfinancial institutions that benefitted from special access to credit and preferential interest rates. Finally, minority shareholders earnt compensation in the form of stock returns, and depositors, who were trapped by the limited competition, still had access to relatively low risk means of savings with reasonable interest rates (as banks increasingly found ways to bypass the official rates, they benefitted from that), and their deposits were partially protected (at the end of the regime by the FGD). The government provided the regulatory framework that ensured the persistence of the system. The supply of credit, while still relatively scarce, improved over time. But the system was unstable because Franco largely lacked constraints on his authority and discretion, and even more importantly because of the propensity, as we have seen, by bank insiders to lend to their own nonfinancial enterprises and to use the banks to rescue their own enterprises at times of crises, which eventually led to the collapse of many banking institutions in the late 1970s–1980s.

Franco's political institutions created a rent-distribution system in which debtors, depositors, and taxpayers were sources of rent extraction from the banking system. These institutions generated a repressed banking system and a credit market that was not very efficient allocating credit, particularly during the first decades of the regime, in which competition in the credit market was limited and was not allocated broadly because bankers did not benefit as much from doing so. The regime also used to some degree inflation taxation to tax common citizens. The inflation tax devalued assets held in cash, thus taxing the public by making the cash in their pockets worthless. Bankers also shared the revenue of this tax because inflation erodes the value of checking account balances (see Calomiris and Hager 2014, pp. 41–58).

The political circumstances of the Franco regime defined the nature of the 'Game of Bank Bargains' which determined the type of banks that existed in Spain. In the end, the partnerships that we have described came at a cost to the country: not only they created a banking system that in which people had less access to credit that they would have otherwise, while they had to pay higher interest rates on loans and earn lower interest on deposits, but also a banking system that was unable to prevent the crisis of the 1970s–1980s.

THE 1977–1985 BANKING CRISIS

The death of the dictator on November 20, 1975 ushered a transition to democracy, a process that was accompanied by fiscal and financial reforms. It took place however during a period marked by a major international economic crisis caused by the second oil shock. Spain responded to these developments in a difficult environment marked by the oil crises that were ravaging Western countries. The Suárez democratic government elected in the June 1977 election responded to the economic crisis with the

October 1977 *Moncloa Pacts*, political agreements that brought together the main political parties and economic actors (Royo 2000).

In addition to the economic crisis, as mentioned before, Spanish banks were also impacted by international developments in the sector: technological innovation (particularly the introduction of ATMs that were transforming banking operations); improvements in telecommunications that threatened their traditional role of intermediation and forced them to consider their strategies and services that they offered to customers (many banks in other countries were responding by offering universal services); liberalization that now allowed banks to offer universal banking; and the emergence of new products, like securities. Furthermore, the country's application for membership to the European Economic Community (EEC) would also have a very important impact on Spanish banks, as the EEC was implementing a series of Directives to create a Single Market that would allow banks to operate throughout the Community.

The government and the Bank of Spain worked to address the problems in the banking sector caused by the economic crisis and they created the Deposit Insurance Fund (*Fondo de Garantía de Depositos*, FGD) in November 1977, with contributions from the banks (1.2 per 1000 deposits). Moreover, the Ministry of Finance, led by Fuentes Quintana (until his resignation in February 1978), who was very critical of the banks' oligopolistic practices, continued pushing for the financial liberalization of the sector. Banks renewed attempts (e.g., in 1975) to persist with the private arrangements of the previous decade to do away with passive interest rates above the maximum and to stop special 'conditions" to make their products more attractive and attract deposits, failed again. This practice became even harder once the Suárez government approved a ministerial order in 1977 that further liberalized interest rates for operations that exceeded the one-year limit. Yet banks' leaders, notably Aguirre Gonzalo chairman of *Banesto*, opposed this liberalization and continued trying to limit competition and as late as 1980 introduced an agreement to try to cap passive interest rates. The government, however, continued the liberalization process: In 1981, it liberalized all deposits with the exception of deposits of up to six months and special accounts, and it completed this liberalization process in 1987 when it abolished the maximum passive rates and removed caps on changes from active operations.

Still, Spanish banks were unable to escape the combined effects of the democratic transition, the domestic and international economic crisis, the liberalization and restructuring in the sector to address competitive and technological pressures, and the changes needed to prepare for the impending membership to the EEC. Indeed, the severe banking crisis that run from 1977 to 1985 had a huge impact on the banking sector bringing the banking expansion to a halt, and causing the disappearance of numerous institutions: Of the 110 banks in operation on December 31, 1977, the crisis ended up affecting 76 banks and banking firms (20 of which were created between 1963 and 1979). The crisis, which exposed their aggressive growth strategies as well as their management short-comings, made them vulnerable. It impacted approximately 27% of the deposits held at private banks, and it ended up costing 48,381 jobs in the sector. In addition, the crisis had a cost to the state estimated at 1216 billion *pesetas* (or 15% of GDP), and to the banking sector an additional 365 billion *pesetas* between contributions to the Deposit Guarantee Fund (FGD) and special operations (plus an additional 200 billion in intra-group restructuring costs). The FDG ended up assuming control of 29 banks, while the state expropriated 20 banks that were part of the *Rumasa* Group for reasons of public and social interest. Many other banks were also affected and experienced serious liquidity problems: They had to be absorbed by their parent institutions or other banks; or the Bank of Spain had to step into provide special credit support. Finally, the crisis also had a severe impact on credit unions (several of them were wound up and in many cases they were taken over by savings banks); and to a lesser degree on saving banks, as most of their problems were resolved discretely. Still the Savings Banks Deposit Guarantee Fund (*Fondo de Garantia de Depósitos de las Cajas de Ahorro*-FGDCA) had to grant 11.6 billion pesetas to four institutions (Martín-Aceña 2012, p. 138; Poveda 2012, p. 222; Tedde de Lorca 2019, p. 40).

The four most notable examples of failures were the Bank of Navarra, the Urquijo Bank, *Banca Catalana*, and *Rumasa*, but they were followed by a long list of other failures. The first major credit institution affected by the crisis was the Bank of Navarra, who had a long history of granting higher interest rates (*extratipos*) to depositors and had embarked in an aggressive expansion strategy absorbing *Caja de Crédito para la Vivienda*, the *Caja Continental* and the *Caja de Crédito y Ahorros*. It was inter-vened in 1978. In 1976, during an ordinary inspection by the Bank of Spain, the inspectors discovered severe financial irregularities. The bank, heavily indebted to other banks, had granted loans and guarantees above the legal limit to the Catalan holding *NIP*, which was presided by the same person, Juan Palomeras, who had used those loans and guarantees to take control over the bank. In response to this crisis, in 1978 the Bank of Spain and other private banks created a mixed entity, the Bank Corporation (*Corporación Bancaria*) to address the banking crisis in general but more specifically to acquire and clean up the Bank of Navarra.[8]

The Urquijo Bank, an industrial bank whose performance was closely connected to that of its industrial portfolio, was also overwhelmed by the industrial crisis and was bought in 1983 by the *Hispano*, with whom it had a long formal relationship (the *Hispano* held 12.3% of the *Urquijo's* stock at that time) to share the profits of industrial dividends (in 1944, the leaders of both banks, the marquises of Aledo and Urquijo, had signed an agreement, known as the *Jarillas Pact*, according to which all the industrial bank operations would be handled by the *Urquijo*). The purchase did not work well because the *Hispano* struggled absorbing the losses from the industrial portfolio, which forced the Bank of Spain to intervene, organizing a rescue operation and appointing a new chairman, Claudio Boada, in 1985. The bank ended up receiving 45 billion *pesetas* in assistance that returned when it sold the *Urquijo* sold to *Banca March* in 1988. The *Urquijo* was sold again in 2006 to Bank Sabadell.

Banca Catalana, which had been led by Jordi Pujol (who served as vice president of the bank and later was elected in 1980 as president of the *Generalit de Catalunya*, the regional government, a position in which he served for 23 years), was also insolvent as a result of its risky attempts to fund 'Catalan interests,' and its aggressive strategy to grow and gain market share. The bank funded all kind of Catalan related initiatives, including the transfer loans for one of Barcelona football club' stars, the Dutch Johan Cruyff. The bank went on a "shopping spree" absorbing the Commercial Expansion Bank in 1971; and at the Bank of Spain's request it took over the Bank of Gerona in 1975 and the Industrial Bank of the Mediterranean in 1979. In 1980, it acquired the Mercantile Bank of Manresa and the Bank of Barcelona, at the same time that it integrated the Industrial Bank of Catalonia. The Bank was also involved in

---

[8] Palomeras opposed that intervention and opted for the bankruptcy of the bank. He was the first banker detained and judged in Spain. He faced several crimes, including the falsification of the bank accounts, but was absolved by the Pamplona Audience and by the Supreme Court.

some questionable and irregular operations, like the real estate development projects in Montigalá-Batllòria and in Tabasa. By 1981, it was the tenth largest bank in Spain and the first one in Catalonia with 262 billion *pesetas* in deposits and 352 branches. The problems for the bank started in 1982 when a false announcements about its bankruptcy provoked a run on its capital. The Bank of Spain ended up appointing three administrators in 1982 who discovered a number of accounting irregularities that forced a capital increase funded by the FGD. The managers of the bank were accused in 1986 of misappropriation of funds, falsification of public documents, and attempts to modify prices fraudulently.[9] It was intervened by the Bank of Spain and later sold to the Vizcaya Bank in 1984.

Finally, the holding group *Rumasa*, which controlled several companies and banks including the Atlantic Bank, became insolvent in 1983 and was expropriated that year. The holding was founded by José María Ruiz Mateos, whose affiliation with the Catholic organization Opus Dei had provided him with privileged treatment from leaders of the Franco regime also affiliated with that organization. For a period of time, the group even received credit lines from the Bank of Spain when it purchased banks in difficulties, such as the *Banco de Siero*, the *Banco de Murcia* in 1967, and the *Banco del Noroeste* in 1974, "despite the fact that these acquisitions were made at absurd prices, higher than the value of the deposits of the banks acquired" (Poveda 2012, pp. 226, 235). The lending of the holding's banks had been concentrated on group companies, but the banks which had been acquired by crossed loans also lacked real capital. In a desperate quest to get new clients and capital the banks embarked

---

[9] The bank invested 195 million pesetas in the stock of the *Montigalá-Batlloria*'s real estate development project. The project located in Badalona and Santa Coloma de Gramenet planned to build 13,000 apartments to host 50,000 people, following the model of the Finish city, Tapiola. The opposition from the municipalities and the collapse of the bank made the project fail. Another questionable initiative promoted by Pujol was the purchase of 13 million in *Infraestructuras Sociedad Anónima*'s stock. This company held 58% of the firm *Túneles y Autopistas de Barcelona*, which won the official concession to build the tunnels that would go through the *Collserola* mountain to improve access to Barcelona. Different problems led to the stagnation of the project. Pujol also promoted the purchase of stock from the *Sociedad Anónima Grupo Alimenticio* in 1974, for 100 million pesetas; and another 100 million from the *Company Cervezas Ole Barcelona Sociedad Anónima* (CERBASA), which commercialized Moritz's beer. This firm closed even before the bank, amidst workers' protests and lockouts. See: "Banca Catalana repartió 516 millones en dividendos entre 1974 y 1976 año en que ya tenía un déficit de mil milllones," *El País*, August 26, 1986.

in campaign offering (unlawfully) high interest rates, that they were not in a position to pay. For a while, the managers, also benefiting from the fact that consolidation of account was not compulsory, were able to obstruct the ordinary inspection procedures. The economic crisis precipitated the situation. Rumasa's response to the economic crises was to double down and embark in an aggressive expansion strategy: Between 1978 and 1982, it almost doubled the size of the holding, disregarding all warning from the Bank of Spain. Ruiz Mateos refusal to allow the auditing of the holding group forced the Socialist government's hand and led to its expropriation. The government ended up expropriating all the shares of the group companies and financed the operation by means of an extraordinary debt issue. The group losses were estimated at 261.5 billions pesetas (around 1.5 billion euros) (see Tortella and García Ruiz 2013, pp. 142–45).

Causes of the 1977 Banking Crisis

The banking crisis was caused by a combination of economic, political, and banking factors (see Tortella and García Ruiz 2013, pp. 142–45; Poveda 2012, pp. 223–24). On the economic side, the first and second oil crises were devastating for the Spanish economy (which imports all its oil), causing a recession and having a huge inflationary impact (see Table 3.1). The crisis impacted particularly the country's industrial and tourism sectors, thus leading to a sharp increase in unemployment (at the end of 1975 the Spanish unemployment rate stood at 3.4%, then rose for ten years, peaking at 21.4% in 1985), followed by bankruptcies and closures. Governments across the world, including Spain, resorted to high interest rates to try to tame inflation, which impacted banks because it made the funding of their unproductive or speculative assets more expensive.

On the political side, the transition to democracy had generated uncertainties about the direction of the country and had led to capital flight, hurting banks. Furthermore, the political situation made it harder to introduce the necessary financial reforms and policies.

There were also institutional factors that were intrinsic to the banking sector. First, most banks suffered from high overheads costs from their oversized fixed assets and bloated organization structures. Second, increasing competition for deposits had led new banks to offer very high interest rates to attract new less solvent customers, hence hurting

Table 3.1 The Spanish economy (1980–1985)

Subject descriptor	1980	1981	1982	1983	1984	1985
GDP, current prices ($)	230.189	204.082	197.154	172.429	171.955	181.172
GDP per capita, current prices ($)	6112.8	5368.6	5158.7	4490.7	4460.6	4683.1
Output gap % of potential GDP	−1.416	−3.52	−3.989	−4.193	−4.437	−4.214
Inflation, average consumer prices (% change)	15.5	14.5	14.4	12.1	11.2	8.8
Unemployment rate (% labor force)	11	13.7	15.7	17.2	19.9	21.3
Government structural balance (national currency)	−4.905	−3.011	−11.939	−9.825	−10.861	−16.096
General government gross debt (% GDP)	16.577	20.017	25.141	30.378	37.075	42.059
Current account balance (% GDP)	−2.3	−2.5	−2.3	−1.4	1.2	1.1

Source IMF, *World Economic Outlook*, October 2019

their profitability. Eventually once these customers turned loss-making as a result of the crisis, it resulted in capital losses that threatened their survival. The problem was compounded by the poor management of many of these banks, who had made controversial decisions to increase their market share and had assumed too much risk. The high-risk investment of many banks (particularly the newer ones) in property and/or financial corporations haunted them once these investments collapsed as a result of the crisis. Third, their decision to intervene in the stock market to support their own shares also contributed to the depleted capital levels.

The break-up of the status quo was a decisive factor in the crisis (Poveda 2012, pp. 225–27). As we have seen, the 1962 Credit and Banking Law, which sought to encourage business development and long-term financing, had forced the breakdown of the status quo by splitting banks between commercial and industrial ones. Yet, while commercial banks were not allowed to hold shares or shareholding (this ban did not extend to existing banks), industrial banks had restrictions on their commercial activities, but were allowed to accept certificate of deposit and long-term deposits, issue short-term bonds with maturity of less than two years, and earn interest above the general schedule. These incentives attracted speculative investors and promoters to the banking sector that assumed significant risks: Between 1963 and 1965, twenty new licenses were granted. The Law preserved the restrictions on branching and reduced the scope of long-term business for non-industrial banks and established a series of incompatibilities that limited the number of posts that banks executives could hold in other companies as chair, director, or senior executive (a very common practice for mixed banks), but excluded existing banks.

The Law also imposed minimum capital requirements: Between 10 and 100 million pesetas depending on the population of the town or the city in which they were established, that proved to be clearly insufficient and ended up attracting speculative investors. The problem was compounded by the fact that the criteria for granting bank licenses only depended on "the interest for the national economy," a totally arbitrary criteria that did not include any assessment of either the viability of the projects or the suitability of the promoters. Not unexpectedly, many of the new banks that emerged lacked sufficient capital and/or strong management and embarked in expansion plans funded with high-cost borrowing that were often speculative and/or involved too much concentration of risk, making them very vulnerable to a crisis. Hence, it was not surprising that, according to the DGF, of the 51 banks affected by the crisis, 47 were new banks or bankers. Indeed, the crisis exposed the misguided way in which the 1962 Law had tried to dismantle the status quo, in the absence of adequate regulatory and supervisory instruments. In addition, some industrial banks were involved in industrial companies that suffered greatly as a result of the crisis, thus saddling the banks with overvalued assets, and trapping them into a situation in which they had to support the losses of their industrial firms at an increasing cost. In the end, the high concentration of risk in these money losing corporate groups that they were funding threatened their survival. And all this took place in a context of increasing competition and liberalization pressures that hurt many banks and led to their intervention. Finally, the DGF had also identified irregular and unlawful practices (Poveda 2012, p. 224).

The crisis also exposed the shortcomings in the regulatory framework and the supervisory mechanisms. Concentration of risk proved to be a serious weakness. Banking regulations included provisions to limit risk concentration, as well as insider and connected risk. The 1968 Law, for instance, barred banks from granting loans to their own executives without the prior approval from the Bank of Spain; and other regulations restricted crossed loans and established a limit on proprietary lending of 2.5% of own and borrowed funds on loans extended by mixed and commercial banks to single customers or groups of banks subsidiary companies (any loan above 2.5% had to be pre-approved by of the Bank of Spain). However, new and old banks resorted to questionable accounting practices and other ways (often illegal) to elude these limits. And while administrative approval was required for the purchases of one bank by another, these controls did not apply to purchases of banks by individuals or by other companies/financial corporations, which led to widespread transfer of banking licenses (as we discussed above the creation of the *Rumasa* banking group was a crash example of this). Finally, as a final step to break-up the banking status quo, successive regulations tried to expand the limit on branch numbers based on banks funds, and population of the town in which they were based. The restrictions were not very strict and, despite the economic crisis, there was a boom in the growth of branch offices: Between 1975 and 1977, 4600 branches were opened, and an additional 1000 per year opened over the next five years. While this increased competition, it also led to higher operating costs that grew on average from 2.5% of total assets to 3.5% between 1975 and 1980 (Poveda 2012, p. 227).

The break-up of the banking status quo was the outcome of the *games of bank bargains* that we discuss throughout the book: Existing banks had sufficient leverage to protect their interest and to ensure that the new law's ban on shares or shareholding for commercial banks would not apply to them. The dictatorship for its part, was desperate for funding to foster economic development and to provide long-term financing and it was wary to alienate such an important constituency and thus, compromised. At the same time, it pursued new sources of funding and was very interested in attracting new capital and new promoters. The coalition between the 'old' banks and bankers, the new capital, and the government's financial and economic elites led to a flawed reform that was compromised from the outset: It not only failed to force a true split between commercial and industrial banks, but the mixed banks continued with their dominance of the banking system, banks executives maintained their posts in other companies (using their banks to prop them up during the crisis), and finally, new groups of promoters and investors, many of them inappropriate for banking, who held unrealistic expectations (and often speculative goals) received banking licenses. The problem was compounded by the lack of supervisory criteria in the granting of licenses, that often depended "more on connections than on the suitability or viability of the project" (Poveda 2012, p. 226). The large existing banks, for their part, ended up creating industrial corporations and investment banking subsidiaries to take advantage of the advantages that the law granted to these institutions. The break-up of the status quo was a truly an exemplary 'insider game' with near-catastrophic impact on the Spanish banking system.

SPANISH BANKING IN THE 1980S AND 1990S

Spain joined the EEC in 1986. This development had profound consequences for the country and its banks. Not only it contributed to the transformation of the Spanish economy, but also of the country's society (Royo and Manuel 2003). EEC membership propelled the modernization of the Spanish economy and the country's economic structures. For the banking sector, the integration process took place in the midst of the technological transformation of banking that was accelerating disintermediation and increasing domestic competition. EEC accession, which would involve the country's integration in the Single Market, would also bring increasing foreign competition in the domestic market for Spanish banks. Finally, in response to the international banking crisis of the 1970s–1980s the Basel Committee on Banking Supervision (BSBS), which had been created in 1974 by the Bank for International Settlements as the global standard setter for the prudential regulation of banks and to provide a forum for regular cooperation on banking supervisory matters reached an agreement in 1988, Basel I, that established minimum capital requirements for banks. The liberalization process in Spain was consistent with that of other Western countries at the time. Following the 1984 increase of reserve requirements and mandatory investment in public debt, the Socialist government liberalized the opening of offices by commercial banks in 1985 and saving banks after 1988 (in fact establishing to the full operational equality between commercial and saving banks), accepted the full liberalization of rates and fees in 1987, and in 1989 it approved a timetable for the full liberalization of investment to eliminate the mandatory investment ratios by 1992. In 1990, it started a process to reduce the minimum cash reserve ratio, which continued in subsequent years and ended at 2%. Finally, the liberalization process also included foreign banks when the government set rules in 1988 and 1995 for the harmonization with domestic banks by the mid-1990s. This, however, did not lead to a significant growth of foreign banks in the country because it was a very expensive investment, thus allowing national banks to retain their hegemonic position in the Spanish market (Tortella and García Ruiz 2013, pp. 172–73).

In response to these developments, Spanish banks embarked in a process of consolidation that led to a number of mergers and acquisitions. The process had already started at a much smaller scale in the late 1970s as the larger banks cajoled to keep their supremacy in the banking size rankings: In 1997, *Banco Central* merged with the *Banco Ibérico*, a smaller institution, and *Banesto* acquired *Banco Coca* (in 1977) and *Banco Madrid* (in 1978). But those mergers were largely attempts to retain or gain leadership of the banking system in assets (and in the case of *Banesto* it proved to be a disaster when they discovered after acquiring those two banks that had very serious problems that resulted in significant losses, which were compounded by the losses of another subsidiary, *Banco Garriga Nogués*).[10] This process of consolidation took some time and it was somewhat contentious: While the leaders of banks like *Bilbao* and *Central*, seeking economies of scale and scope and afraid of potential takeovers of their banks by larger foreign entities supported it; several others (notably *Banesto*, *Popular* and *Vizcaya*) opposed it because they believed in specialization, were not convinced that size and efficiency were connected, and did not believe that they could exploit the economies of scale and economies of scope of the merged banks.

Nevertheless, the Bank of Spain, convinced that Spanish banks needed to be large to compete with their European counterparts and seeking to protect them from the potential colonization by foreign banks, pushed for mergers among the largest banks, and actively worked to facilitate them. The push for mergers was based on a number of arguments: to

---

[10] *Banesto* was also involved in other disastrous errors those years: the Bank Catalan for Development (*Banco Catalán de Desarrollo, Cadesbank*), which had to be restructured by the FGD; the Bank of Madrid (*Banco de Madrid, Bandri*) which had to be cleaned up by Banesto itself; and the Garriga Nogés Bank (*Banco Garriga Nogués, BGN*), which had been involved in questionable investment and in which clients connected with the bank chairman, Javier de la Rosa, concentrated 55.4% of the credits granted. No wonder why *Banesto*'s chairman Pablo Garnica was so reluctant to embark in other mergers (Tortella and García Ruiz 2013, pp. 154–55).

take advantage of economies of scale; to meet the financial needs of large borrowers; to achieve international status; to be able to compete with larger foreign competitors; and to rationalize the branch network (Poveda 2012, p. 249). The first major merge took place between *Bilbao* and *Vizcaya*. On September 1987, the chairman of *Bilbao*, José Angel Sanchez Asiaín, first proposed an understanding to his counterpart from *Vizcaya*, Pedro de Toledo. Sanchez Asiaín was able to overcome Toledo's initial opposition and the new megabank, the Bilbao Vizcaya Bank (the *Banco Bilbao Vizcaya*, BBV), was born in June 1988.

Yet other merger attempts like the *Bilbao-Banesto* and *Banesto-Central*, failed, and this despite the fact that the Bank of Spain had taken the initiative and supported those mergers. The crisis of *Banesto* was closely connected with the concentration process that took place those years. As noted above, the first round of acquisition of *Coca* and *Madrid* had resulted in significant losses for *Banesto*, that the bank had been able to disguise through a network of holding companies that had been created to park treasury stock and shareholdings. While *Banesto* was able to survive this first crisis, it was left vulnerable. On February 1986, the Bank of Spain's governor Mariano Rubio sent *Banesto*'s chairman, Pablo Garnica, a report indicating that *Banesto* had to mark down 95,000 million *pesetas* as additional losses. Bilbao, trying to seize the opportunity, proposed a friendly merger in 1987, which was rejected. It was followed by a hostile takeover attempt that also failed. These crises led to a dramatic (and very public) leadership crisis at *Banesto*. The Bank of Spain appointed José María López de Letona as chairman. He was in favor of Bilbao's bid against the opinion of the board, but when two investors, Juan Abelló and Mario Conde, purchased 2 million shares of the bank and became members of its executive committee, they engineered a 'coup' and Conde became chairman of *Banesto* in November 1987. Conde was supported by the bank's old-timers and maneuvered to have the *Bilbao*'s offer rejected. The Bank of Spain responded by forcing the new management to appoint three independent directors to monitor developments. Conde's appointment, however, did not end the bank's crisis: *Banesto* struggled to raise the 80 billion *pesetas* in provisions required by the Bank of Spain, and a year later it had lost market share while the default ratio had increased. In February 1989, Abelló decided to leave the bank and was followed that summer by three of the bank traditional families who had controlled the banks since its inception in 1902, in protest over the failed merge with *Central* and Conde's questionable accounting practices to inflate profits.

Conde responded with more reckless 'financial engineering' and aggressive credit-lending. He granted 2 billion pesetas in credits to companies related to his personal interests and created the Industrial Corporation, a holding company that sought to take advantage of tax credits established by the Socialist government in 1990, to bring together companies in which *Banesto* had equity participation, including prestigious ones like *Acerinox* (metal); *Agromán* (construction); *Asturiana de Zinc* (mining); *Carburos Metálicos* (chemical); *Petromed* (oil); Tudor (batteries); or *La Unión y el Fénix Español* (insurance). Conde's escape forward led him to create dubious subsidiaries to channel investments; divest from the Industrial Corporation (allegedly to reduce risk concentration and raise equity) and use its capital gains to cover the losses of its banking business; resort to questionable accounting practices like including the accounts of one of its subsidiary banks, the Portuguese *Totta e Açores*, into *Banesto*'s balance sheet to make them look better; embark in desperate search for capital (e.g., from investors like Corsair Inc.); or sell assets, like the Madrid Bank (which was sold to a Deutsche Bank subsidiary). In the end, all these maneuvers failed and by the end of 1993, the result of these practices and of the bank excessive credit-lending, led to a patrimonial hole of €3.6 billion (equivalent to roughly US$7.2 billion today). On 28 December 1993, Luis Carlos Croisser, President of the Stock Market National Commission (the *Comisión Nacional del Mercado de Valores*, CNMV) imposed a trading halt on the bank's stock, while Luis Ángel Rojo, the Governor of the Bank of Spain and its Executive Committee, determined that Conde and his team had gone too far in their attempt to cover their losses, and using the 1988 Law of Discipline and Intervention of Credit Institutions intervened *Banesto*. He appointed Alfredo Sáenz, a well-reputed banker and former chairman of Banco Bilbao Vizcaya, as temporary chairman, and removed *Banesto*'s entire board. At the time of the intervention, *Banesto*'s situation was extremely serious: The bank had important problems with asset quality and contingencies not covered; it faced a severe shortage of equity; it had very high risk concentration; and it faced uncertainties regarding its profits and loss accounts as well as its cash management. By the end of 1993, the requirements of provisions to balance the account had reached 605 billion *pesetas*. *Banesto* was subsequently restructured with funds from the FGD (at a loss of 193 billion *pesetas*) and the crisis was settled by a reduction and subsequent increase in capital. *Banesto* was auctioned and acquired in 1994 by *Banco Santander*, and following a lawsuit Conde was sentenced to 20 years in jail by the Supreme Court in 2002 (he served 11 years before being paroled) (see Tortella and García Ruiz 2013, pp. 154–61; Poveda 2012, pp. 250–51).

The problems (and conflicts) also extended to other banks. The Kuwait Investment Office (KIO) had been investing in *Banco Central*'s stock since 1986 and had joined other Spanish investors in *Cartera Central*. They were opposed to the merge with *Banesto* and had a public feud with Central's Chairman Alfonso Escámez who believed that the bank needed to be larger to face ECC's competition and hence was more supportive of the possible merger with *Banesto*. He was also attracted to Banesto's industrial portfolio as a way to increase *Central*'s participation in industry. The failure of that merge, and concerns about being absorbed by *Santander*, which had embarked in an aggressive growth strategy and had emerged as one of the country's top banks, pushed him (with the government's support) to merge with *Hispano* in the summer of 1991, and create the Central Hispanic American Bank (*Banco Central Hispanoamericano*, BCH). Escamez stepped down in 1992, and the BCH was led by José María Amusátegui (former chairman of the *Hispano*) and his team from the Hispano, who had close connections with the government and the financial authorities.

Moreover, in addition to *Banesto*, over a period of 17 years there were other crisis of smaller banks; three in the early 1990s, one in 1996 and another in 2003 that were settled through the regular resolution procedures. These institutions were declared insolvent and wounded up in an orderly manner (Poveda 2012, p. 252).

The new megabanks, BCH and BBV, competed for a few years but the consolidation process continued later in the 1990s led by an aggressive Santander Bank. Its new chairman, Emilio Botín III who had succeeded his father in 1986, was committed to shattering the status quo and bring what had been largely a regional bank (until 1946 when it absorbed the Mercantile Bank, which was larger, with the support of *Central*) to the top Spanish private financial institution. Botín's shake-up of the banking sector started with the introduction of a new bank account that payed higher interest (the '*supercuenta*') to attract more deposits. As we have seen, he took advantage of the *Banesto*'s intervention to purchase it in 1994, and he followed that up raising the ante and absorbing the BCH in 1999, creating the largest bank in Spain: the Santander Central Hispano Bank (the *Banco Santander Central Hispano*, or BSCH). The BBV, for its part, responded merging with *Argentaria*, a new institution that had emerged in 1991 from the consolidation of all former public banks (with the exception of ICO, which still exists), in 1999, and forming the *Bilbao Vizcaya Argentaria Bank* (Banco Bilbao Vizcaya Argentaria, BBVA), which would be led by Argentaria's chairman Francisco González.[11]

The role of the Bank of Spain throughout the banking crisis and the mergers cannot be sufficiently overstated. Led since 1992 by its prestigious Governor, Luís Angel Rojo, who had a long a brilliant history at the institution having served as the director of its research office and also as its deputy governor; it played a central role in the intervention and merging processes, which enhanced the reputation of the Bank. The *Banesto* crisis confirmed Rojo's belief that financial engineering and off-balance sheet operations could lead to problems caused by opacity and inadequate risk information and reaffirmed his commitment to financial stability. Following the Maastricht Treaty requirements, which set the framework for the launching of the Euro in 1999 and the establishment of the Eurosystem, the Spanish Congress approved a new law in 1994 enshrining the autonomy of the Bank of Spain, and establishing its mandate to seek price stability, as well as the prohibition to grant overdrafts or engage in any monetary financing to the public sector. Rojo was the architect of the creation (in the 9/1999 circular letter) of the new countercyclical provisions (the so-called dynamic provisions) that institutions had to accumulate during the boom years to use during periods of crises, a lasting legacy that played a central role (as we will examine on Chapter 5) in future decades (Tortella and García Ruiz 2013, pp. 159–61).

The consolidation of the sector was accompanied by the banks' expansion abroad. By the late 1990s, led by Spanish Multinational Corporations (MNCs) (such as *Endesa, Iberia, Iberdrola, Repsol* or *Telefónica*) and banks (BBVA and Santander), the country finally became a next exporter of capital and raised the stock of Foreign Direct Investment (FDI) from 1% of GDP in 1980 to 35% in 2004 (Guillén 2005, p. 11). Taking advantage of the combined domestic processes of EEC integration, liberalization and consolidation that took place in the country throughout the 1980s–1990s, as well as international developments (notably the liberalization and privatizations that took place in Latin American throughout

---

[11] González's appointment as chairman of Argentaria (and later of BBVA) was clouded in controversy, because he did not have a banking background and was a close personal friend of the Prime Minister, José María Aznar, since they went to school together.

the 1980s) Spanish firms embarked in a process of expansion abroad, in which banks played a central role. Banks sought to capitalize from privatizations, deregulation, and the high cost of capital in the countries of origin, as well as from the opportunity to serve their domestic clients who were investing in these countries abroad, to grow and diversify the risk in their domestic (Spanish) market. They started by opening subsidiaries and branches in Latin America, the UK, the United States and other European countries, a relatively cheaper and safe option. And they followed up with alliances and acquisitions.

The first mover was Santander, which had been able to gain market share and financial muscle with the launching of the '*supercuenta*.' Santander benefited from its own branches as well as BCH's operations in Latin America, which led to the establishment of *Santander Chile* and *Santander Río*. In 1988, it signed an alliance with the Royal Bank of Scotland and followed with acquisitions in Mexico (*Banca Serfín* in 1991), Brazil (*Banespa* in 2001), and the UK (*British Abbey National Bank* in 2004). BBVA, which did not want to be left behind, and took control of other banks in Peru (*Continental* in 1995); Mexico (*Probursa* in 1995, *Banca Cremi y Oriente* in 1996 and *Hipotecaria Nacional* in 2004); Colombia (*Ganadero* in 1996 and *Granahorrar* in 2004); Argentina (*Francés* in 1996); Venezuela (*Provincial* in 1997); Chile (*BHIF* in 1998 and Forum in 2007); Brazil (*Excel Económico* in 1998); and Puerto Rico (*Poncebank* in 1998). BBVA most ambitious acquisition was *Bancomer* in Mexico in 2001 (the purchase of all its capital was completed in 2004) (Tortella and García Ruiz 2013, pp. 163–65). This internationalization strategy allowed Santander and BBVA to grow, gain market share other countries, and to balance their risk in the Spanish market, which would prove prescient as we will see later in the book during the 2000s banking crisis.

The 1992–1993 European economic crisis that followed speculator attacks on the European Monetary System (EMS) seriously impacted Spain, pushing unemployment over 20%, and led to subsequent devaluations of the peseta (as well as other European currencies). The crisis exposed the shortcomings and loss of competitiveness of the Spanish economy caused by the unbalanced policies and growth of the 1980s. The larger Spanish banks, however, capitalized from their investment abroad and were able to survive the crisis relatively unscathed. They mobilized resources from abroad to address the demands from their domestic clients and took advantage of the record low interest rates set by the European Central Bank (ECB) (which were too low for Spain) to expand credit to the private sector, which grew from 63.1% of GDP in 1975 to 169% in 2008, with some deleterious consequences as we will examine later in the book.[12]

During the last decade of the twentieth century, the banking sector experienced additional changes marked by the process of 'disintermediation' that had been accelerated by technological changes like the Internet, which allowed clients to operate directly in the financial markets bypassing traditional intermediaries. This forced Spanish banks to invest heavily in information and communication technologies. Yet, branches were still more important in Spain than in other countries because of low population density in some areas, the high specialization of many branches, and the Conservative nature of banking clients who still relied in branches to fulfill their financial needs. Indeed, the strong density of branch networks persisted as a distinctive feature of the Spanish banking system, and in fact there was an increase in the number of branches, particularly among savings and loans (*cajas*) that expanded their branch network aggressively across the country, as we will see later in the book.

The 1977–1985 crisis also had a significant impact on prudential and accounting regulations. It led to new resolution mechanisms and procedures for banks that focused on matters of financial health and solvency. On the resolution regime, the establishment of the DGF in 1977 was followed by the creation of the Bank Corporation (*Corporación Bancaria*) in 1978, jointly owned by the Banks of Spain and the banks. It was charged with the task to purchase the shares of banks in problems to restore them to health and then sell them to other institutions as soon as feasible. Henceforth, the *Corporación Bancaria* took control and assumed management of some problem institutions. Its limited financial resources (it was only allocated capital of 500 million *pesetas*) led to the granting of legal personality to the DGF in 1980, which was now authorized to purchase assets, hold shares of banks in trouble and participate in their

---

[12] The growing demand for credit, boosted by record-low interest rates, led to a sharp increase in the country's external financing needs: in 1998, Spain did not need outside funding and was able to meet credit demand with domestic bank deposits alone, yet by 2007 the external financial needs exceeded 9% of GDP (Malo de Molina 2012, p. 205; see Tortella and García Ruiz 2013, p. 167).

capital increases (thus, rendering the *Corporación Bancaria* unnecessary). Contributions to the DGF were adjusted over time. The crisis also led to the review of procedures for sanctions and intervention in credit institutions in the 1988 Banking Discipline and Intervention Law, which closed loopholes, and defined infringements, sanctions and procedural rules. It also strengthened the power of the Bank of Spain to take control of those banks or replace their management in cases in which it was not possible to discern their account situation, or in exceptional circumstances that affected their stability, liquidity or solvency. Furthermore, the crisis also led to the establishment of Conservative accounting rules, introducing the accrual accounting principle, and the progressive adaptation of Spanish accounting rules to international accounting standards. On the solvency side, new rules (equivalent of a risk-weighted ratio) based on the assumed risk level of assets and off-balance sheet exposures were introduced in 1985. Risk categories were also introduced as well as solvency requirements. The reform process culminated when Spain had to adapt its banking legislation to the EEC's *acquis communautaire*, a process that included a seven-year transition period. The final addition to the regulatory system, as noted above, which sought to strengthen banks for future contingencies was the introduction of dynamic provisioning in 1999 to cover potential losses both on transactions with clear signs of impairment and also on apparently healthy transactions (Poveda 2012, pp. 232–53). These provisions, as we will see later in the book, played a positive role in the initial stages of the 2007 financial crisis.

CONCLUSION

The 1977 banking crisis was considered the most serious one in Europe: Overall the cost of the crisis was estimated at 1.5 trillion pesetas. As we have seen, Spanish banks were largely unprepared to face it: The sector was too fragmented with too many banks; banks lacked the appropriate dimension because the Franco regime had been opposed to the consolidation of the sector (for instance, as noted before, it rejected the merge of *Hispano* and *Central*); banks had been unable to take advantage of the economies of scale that emerged when the sector was liberalized by the Navarro Rubio Act; and in many cases the qualifications (and ethics) of the banks' managers were questionable at best (a problem that reoccurred again in the 2007 crisis, particularly with many savings and loans/*cajas*, as we will see later in the book). These sectoral problems were compounded by the economic and political challenges described above. In the end, the combination of these factors made the crisis inevitable.

In general, smaller banks suffered the most, while the larger ones, who were often instrumental in helping rescuing some of the troubled banks, were able to capitalize from the crisis and purchase some of their smaller competitors at relative low prices once they had been recapitalized by the FGD. These banks (like *Santander, Banesto, Central, Hispano, Bilbao* and *Vizcaya*) emerged stronger from the crisis, as they used it as an opportunity to reduce their operating costs, professionalize and modernize their management and operations, and improve their competitiveness. Indeed, by the mid-1980s the Spanish banking system was comparable to that of other European countries as measured by the degree of 'bancarization' (financial assets/GDP); the capitalization of the banks; the number of employees, which was around the EEC average; and the intermediation margins. However, the number of offices was still comparatively high at 850 per million of population (the highest in the developed world); and their branch structure was still too large. Moreover, the size of the Spanish financial institutions was still comparatively low. This was a historical challenge that had prevented the development of larger banks, and it was the result of the low level of internationalization (international financial activity was less than 1% of the world total in the late 1980s), as well as insufficient competition (official interest rates had ensure very healthy margins), rather than low concentration (the market share of the largest institutions was comparable to that of its European counterparts). Finally, another outcome of the crisis was the fact that it broke the traditional link between banks and industrial companies, which had dragged some of them down, and forced banks to look for more profitable alternatives (see Tortella and García Ruiz 2013, pp. 145–48).

By the beginning of the twenty-first century, the Spanish banking system was characterized by: the centrality of the banking system in the Spanish economy; strong credit growth; specialization in retail banking; focus on the property market; a dense branch network (by 2007 Spain still had the highest branch and ATM density per head of population[13]); increasing concentration; high market power compared to its European

---

[13] This unique feature of Spanish banking has been explained by several factors: low population density; specialization in retail banking; limited financial education (people still prefer going to a bank office or using an ATM, than using the Internet); and low penetration of the broadband in telecommunications (Tortella and García Ruiz 2013, p. 174).

neighbors; high profitability; rapid internationalization; low default rates; and solvency at the EU-15 levels (Vives 2012, pp. 392–98). These characteristics coupled with the changes to the regulatory and supervisory frameworks that followed the crisis seemed to place Spanish banks in a strong position to address the challenges of the new millennium. This proved a mirage, as we will examine later in the book.

The Franco dictatorship provided an opportunity to examine how authoritarian political leaders formed coalitions with other groups to create a banking system. The Spanish government regulated bank entry tightly to increase rates of return sufficiently to compensate bank insiders and shareholders for the risk of expropriation. Political bargains gave big banks rents in the form of market power and lax prudential regulation in exchange for their commitment to share those rents with favored constituents. The transition to democracy in the 1970s led to the establishment of a new democratic regime, which led to new political bargains regarding the banking system.

The examination of the Spanish case shows that domestic social, political, and economic factors are crucial to understand coalition formations and policy choices. These coalitions are not neutral, and they largely influence the stability and resilience of the banking system and its ability to provide credit (see Calomiris and Haber's 2014). However, the Spanish transition to democracy, which coincided with a global economic crisis, shows the persistence and stickiness of these coalitions and bargains, as it did not result in an instantaneous reorganization of the banking system. On the contrary, the new democratic government attempts to dislodge the existing oligopolistic banking structure largely failed, and the election of a new Socialist government in 1982 (which had called for the nationalization of the banking sector) did not led to a new coalition or new policies (Pérez 1997). The new Socialist government appointed the same central banker reformers to key economic policy-making positions, who sustained the accommodative partnerships with the banks, allowing the Big Seven banks oligopoly to persist well into the mid-1980s while protecting them from strong competition. All this planted the seeds for future crises.

In the next part of the book, we turn our attention to baking in Spain throughout the first two decades of the twentieth century to explain the banking crises that started in 2008.

REFERENCES

Calomiris, Charles W., and Stephen H. Haber. *Fragile by Design: The Political Origins of Banking Crises & Scarce Credit*. Princeton: Princeton University Press, 2014.

Guillén, Mauro. *The Rise of Spanish Multinationals*. Cambridge: Cambridge University Press, 2005.

Malo de Molina, José Luís. "The Macroeconomic Basis of the Recent Development of the Spanish Banking System." In *The Spanish Financial System: Growth and Development Since 1900*, edited by José Luís Malo de Molina, and Pablo Martin-Aceña. New York: Palgrave, 2012.

Martín-Aceña, Pablo. "Desarrollo y Modernización del Sistema Financiero, 1844–1935." In *La Modernización Económica de España*, edited by Nicolás Sánchez Albornoz. Madrid: Alianza Editorial, 1985.

Martín-Aceña, Pablo. "The Spanish Banking System from 1900 to 1975." In *The Spanish Financial System: Growth and Development since 1900*, edited by José Luís Malo de Molina and Pablo Martin-Aceña. New York: Palgrave, 2012.

Pérez, Sofia A. *Banking on Privilege*. New York: Cornell University Press, 1997.

Poveda, Raimundo. "Banking Supervision and Regulation Over the Past 40 Years." In *The Spanish Financial System: Growth and Development since 1900*, edited by José Luís Malo de Molina and Pablo Martin-Aceña. New York: Palgrave, 2012.

Royo, Sebastián. *From Social Democracy to Neoliberalism*. New York: St. Martin's Press, 2000.

Royo, Sebastián, and Paul C. Manuel (Eds.). *Spain and Portugal in the European Union*. Portland: Frank Cass, 2003.

Tedde de Lorca, Pedro. "La Evolución del Sistema Bancario Español en el Siglo XX." In *Guía de Archivos Históricos de la Banca en España*, edited by María de Inclán Sánchez, Elena Serrano García, and Ana Calleja Fernández. Madrid: División de Archivos y gestión Documental del Banco de España, 2019.

Tortella, Gabriel, and José Luís García Ruiz. *Spanish Money and Banking: A History*. New York: Palgrave, 2013.

Vives, Xavier. "The Spanish Financial Industry at the Start of the 21st Century: Current Situation and Future Challenges." In *The Spanish Financial System: Growth and Development since 1900*, edited by José Luís Malo de Molina and Pablo Martín-Aceña. New York: Palgrave, 2012.

CHAPTER 4

From Boom to Bust: The Economic Crisis in Spain 2008–2013

INTRODUCTION

The economic crisis that hit the country in 2008 cannot be separated from the subsequent financial crisis. In order to contextualize the banking games that inform the next chapters and to understand the overall consequences that the economic crisis had on Spanish banks and *cajas*, this chapter examines the economic crisis and analyzes its causes and consequences. Yet again, the performance of the Spanish banking system was deeply connected to the performance of the Spanish economy, and progress in the banking sector was marred by the performance of the Spanish economy at large. The economic crisis that started in 2008, part of the *great recession* that engulfed most countries, had profound consequences for the Spanish banking system. This book will show that the financial crises were the result of a political bargain in which incentives and a lax regulatory framework favored developers, property owners, and bankers, thus confirming a central tenant: the crucial importance of domestic political institutions, the rules of the game, and the role of domestic players operating within those institutions.

From: Sebastián Royo, "After Austerity: Lessons from the Spanish Experience," in *Towards a Resilient Eurozone: Economic, Monetary and Fiscal Policies*, ed. John Ryan (New York: Peter Lang, 2015).

The 2008 economic crisis, while not fully unexpected, came as a relative surprise given the strong performance of the Spanish economy during the first years of the twentieth century. The overall pattern of Spanish economic history has been described, crudely, as a graph shaped like an upside-down version of the letter 'V'. That is, the graph rises—bumpily at times, through 600 years under the Romans, 700 years under or partly under the Moors, and a century of empire-building—to the peak of Spanish power in the sixteenth century. After that, the history of the nation goes downhill until the 1970s. A vast empire was gradually lost, leaving Spain poor and powerless. And there was much political instability: Spain suffered forty-three *coup d'états* between 1814 and 1923, a horrendous civil war between 1936 and 1939, followed by thirty-six years of dictatorship under *Generalísimo* Franco.[1]

After Franco's death in 1975, the graph turned upward again. King Juan Carlos, Franco's heir, oversaw the return of democracy to the country. A negotiated transition period, which has been labeled as a model for other countries, paved the way for the elaboration of a new Constitution, followed by the first free elections in almost forty years. These developments were followed by the progressive return of Spain to the international arena—where they have been relatively isolated during the dictatorship. The following decade also witnessed the Socialist Party being elected to actual power in 1982, bringing a new aura of modernity to the country. The 1980s also witnessed Spain's integration into NATO (1982) and the European Community (1986). The following two and a half decades were a period of phenomenal growth and modernization.

Indeed, before the global crisis that hit Spain in the spring of 2008 the country had become one of Europe's most successful economies.[2] While other European countries had been stuck in the mud, Spain performed much better at reforming its welfare systems and labor markets, as well as improving flexibility and lowering unemployment. Over the decade and a half that preceded the 2008 global financial crisis, the Spanish economy seemed to had been able to break with the historical pattern of boom and bust, and the country's economic performance was nothing short of

---

[1] See "After the Fiesta," *The Economist*, 25 April–1 May 1992, p. 60.

[2] This chapter draws upon S. Royo, *From Social Democracy to Neoliberalism* (New York: St. Martin's Press, 2000); S. Royo, *Varieties of Capitalism in Spain* (New York: Palgrave, 2008); and S. Royo, *Lessons from the Economic Crises in Spain* (New York: Palgrave, 2013).

remarkable. Yet all this came to a halt when the global financial crisis hit Spain in 2008. As a result, Spain is suffering one of the worst crises since the 1940s (Royo 2013).

Following the transition to democracy and the country's European integration, Spain was, prior to the 2008 crisis, a model country. But then the (debt fueled) dream was shattered and the country's economy imploded after 2008. How did this happen? Policy choices and the structure of decision making; the role of organized interest; the structure of the state; and institutional degeneration all played an important role in explaining the severity of the economic crisis in Spain; as did the country's membership under an incomplete monetary union. The country had to face a triple crisis: financial, fiscal, and competitiveness. This chapter seeks to provide an overview of the country's evolution since the transition to democracy, and to explain its economic collapse after 2008 (see Royo 2000, 2008, 2013).

The first section of the chapter outlines the main features of the Spanish growth model, and the challenges that it faced. Section two describes the scale of the shock it underwent from 2008 onward and analyzes the triple crisis in financial, fiscal, and competitiveness performance. The chapter concludes with brief lessons from the Spanish experience.

The Miraculous Decade[3]

European integration was instrumental in the modernization of the country. Indeed, before the global crisis that hit Spain in the spring of 2008 the country had become one of Europe's most successful economies (see Table 4.1). Propped up by low interest rates and immigration, Spain was (in 2008) in its fourteenth year of uninterrupted growth and it was benefiting from the longest cycle of continuing expansion of the Spanish economy in modern history (only Ireland in the Euro zone has a better record), which contributed to the narrowing of per capita GDP with the EU. Indeed, in 20 years per capita income grew 20 points, one point per year, to reach close to 90% of the EU15 average. With the EU25, Spain already reached the average in 2008. The country grew on average 1.4 percentage points more than the EU since 1996.

[3] This section Borrows from S. Royo, *Varieties of Capitalism in Spain* (New York: Palgrave, 2008).

Table 4.1 The boom years (2000–2008)

SPAIN	Units	Scale	2000	2001	2002	2003	2004	2005	2006	2007	2008
GDP constant prices	National currency	Billions	546.9	566.8	582.2	600.2	619.8	642.2	668.0	691.8	697.7
GDP constant prices	Annual percent change		5.1	3.7	2.7	3.1	3.3	3.6	4.0	3.6	0.9
GDP per capita, constant prices	National currency	Units	13.6	13.9	14.1	14.3	14.5	14.8	15.2	15.4	15.3
Output gap in percent of potential GDP	Percent of potential GDP		1.9	1.5	0.3	0.1	0.5	1.4	2.9	3.9	3.1
GDP based on purchasing-power-parity (PPP) share of world total	Percent		2.2	2.2	2.2	2.2	2.1	2.1	2.1	2.1	2.0
Inflation, average consumer prices	Annual percent change		3.5	2.8	3.6	3.1	3.1	3.4	3.6	2.8	4.1
Unemployment rate	Percent of total labor force		13.9	10.6	11.5	11.5	11.0	9.2	8.5	8.3	11.3
Employment	Persons	Millions	16.4	16.9	17.3	17.9	18.5	19.3	20.0	20.6	20.5
General government balance	National currency	Billions	−6.2	−4.4	−3.3	−1.6	−2.9	8.8	19.9	23.3	−41.9
General government balance	Percent of GDP		−1.0	−0.6	−0.5	−0.2	−0.3	1.0	2.0	2.2	−3.8
Current account balance	Percent of GDP		−4.0	−3.9	−3.3	−3.5	−5.3	−7.4	−9.0	−10.0	−9.6

Source International Monetary Fund, *World Economic Outlook Database*, October 2009

Unemployment fell from 20% in the mid-1990s to 7.95% in the first half of 2007 (the lowest level since 1978), as Spain became the second country in the EU (after Germany with a much larger economy) creating the most jobs (an average of 600,000 per year over that decade). In 2006, the Spanish economy grew a spectacular 3.9%, and 3.8% in 2007. As we have seen, economic growth contributed to per capita income growth and employment. Indeed, the performance of the labor market was spectacular: Between 1997 and 2007, 33% of all the total employment created in the EU-15 was created in Spain. In 2006, the active population increased by 3.5%, the highest in the EU (led by new immigrants and the incorporation of women in the labor market, which increased from 59% in 1995 to 72% in 2006); and 772,000 new jobs were created.

The economic success extended to Spanish companies, which expanded beyond their traditional frontiers (Guillén 2005). In 2006, they spent a total of €140 billion on domestic and overseas acquisitions, putting the country third behind the UK and France in the EU. Of this, €80 billion were to buy companies abroad (compared with the €65 billion spent by German companies). In 2006, Spanish Foreign Direct Investment (FDI) abroad increased 113%, reaching €71.5 billion (or the equivalent of 7.3% of GDP, compared with 3.7% in 2005).[4] In 2006 *Iberdrola*, an electricity supplier purchased *Scottish Power* for $22.5 billion to create Europe's third largest utility; *Banco Santander*, Spain's largest bank, purchased Britain's *Abbey National Bank* for $24 billion, *Ferrovial*, a family construction group, concluded a takeover of the *British BAA* (which operates the three main airports of the UK) for £10 billion; and *Telefonica* bought *O2*, the UK mobile phone company. Indeed, 2006 was a banner year for Spanish firms: 72% of them increased their production and 75.1% their profits, 55.4% hired new employees, and 77.6% increased their investments.[5]

The country's transformation was not only economic but also social. The Spanish became more optimistic and self-confident (i.e., a Harris poll showed that Spaniards were more confident of their economic future than their European and American counterparts, and a poll by the *Center for Sociological Analysis* showed that 80% were satisfied or very satisfied with

---

[4] Emilio Ontiveros, "Redimensionamiento Transfronterizo," *El País*, July 15, 2007.

[5] Deloitte's "Barometro de Empresas," from "Un año de grandes resultados," *El País*, Sunday, January 14, 2006.

their economic situation). Spain became 'different' again and according to public opinion polls it had become the most popular country to work for Europeans.[6] Between 2000 and 2007, some 5 million immigrants (645,000 in 2004 and 500,000 in 2006) settled in Spain (8.7% of the population compared with 3.7% in the EU15), making the country the biggest recipient of immigrants in the EU (they represented 10% of the contributors to the Social Security system). This was a radical departure for a country that used to be a net exporter of people, and more so because it was able to absorb these immigrants without falling prey (at least so far) to the social tensions that have plagued other European countries (although there have been isolated incidents of racial violence) (see Calavita 2005).[7] These immigrants contributed significantly to the economic success of the country in that decade because they boosted the aggregate performance of the economy: They raised the supply of labor, increased demand as they spent money, moderated wages, and put downward pressure on inflation, boosted output, allowed the labor market to avoid labor shortages, contributed to consumption, and increased more flexibility in the economy with their mobility and willingness to take on low-paid jobs in sectors such as construction and agriculture, in which the Spanish were no longer interested.[8]

Indeed, an important factor in the per capita convergence surge with the EU after 2000 was the substantive revision of the Spanish GDP data as a result of changes in the National Accounts from 1995 to 2000. These changes represented an increase in GPD per capita of 4% in real terms (the equivalent of Slovakia's GDP). This dramatic change was the result of the significant growth of the Spanish population since 1998 as a result of the surge in immigration (for instance in 2003 population grew 2.1%). The key factor in this acceleration of convergence, given the negative behavior of productivity (if productivity had grown at the EU

[6] According to the *Financial Times*, 17% of those polled selected Spain as the country where they would prefer to work ahead of the UK (15%) and France (11%). See "España vuelve a ser diferente," *El País*, February 19, 2007, and *Financial Times*, February 19, 2007.

[7] Calativa provides a detailed analysis of the immigration experience in Spain and exposes the tensions associated with this development. She also highlights the shortcomings of governments' actions in regard to integration, and the impact of lack of integration on exclusion, criminalization, and radicalization. See 2005.

[8] "Immigrants Boost British and Spanish Economies," *Financial Times*, Tuesday, February 20, 2007, p. 3.

average Spain would have surpassed in 2007 the EU per capita average by 3 points), was the important increase in the participation rate, which was the result of the reduction in unemployment, and the increase in the activity rate (the proportion of people of working age who have a job or are actively seeking one) that followed the incorporation of female workers into the labor market and immigration growth. Indeed between 2000 and 2004, the immigrant population has multiplied by threefold.

As a matter of fact, most of the 772,000 new jobs created in Spain in 2006 went to immigrants (about 60%). Their motivation to work hard also opened the way for productivity improvements (which in 2006 experienced the largest increase since 1997, with a 0.8% raise). It is estimated that the contribution of immigrants to GDP had been of 0.8 percentage points in the four years to 2007.[9] Immigration represented more than 50% of employment growth, and 78.6% of the demographic growth (as a result Spain led the demographic growth of the European countries between 1995 and 2005 with a demographic advance of 10.7% compared with the EU15 average of 4.8%).[10] They also contributed to the huge increase in employment, which was one of the key reasons for the impressive economic expansion. Indeed, between 1988 and 2006, employment contributed 3 percentage points to the 3.5% annual rise in Spain's potential GDP (see Table 4.1).[11]

The Basis for Success

What made this transformation possible? The modernization of the Spanish economy in the two and half decades prior to 2008 had been intimately connected to the country's integration in the European Union. Indeed, European integration was a catalyst for the final conversion of the Spanish economy into a modern Western-type economy. Yet, membership was not the only reason for this development. The economic liberalization, trade integration, and modernization of the Spanish economy started in the 1950s and 1960s and Spain became increasingly prosperous over the two decades prior to EU accession. However, one of the

---

[9] Guillermo de la Dehesa, "La Próxima Recesión," *El País*, January 21, 2007.

[10] "La Economía española creció en la última década gracias a la aportación de los inmigrantes," *El País*, Monday, August 28, 2006.

[11] See Martin Wolf, "Pain Will Follow Years of Economic Gain," *Financial Times*, March 29, 2007.

key consequences of its entry into Europe was that it consolidated and deepened that development processes, and it accelerated the modernization of the country's economy. Indeed, EU membership facilitated the micro- and macroeconomic reforms that successive Spanish governments undertook throughout the 1980s and 1990s. Spain also benefited extensively from European funds those two decades: approximately 150 billion euros from agricultural, regional development, training, and cohesion programs.

Moreover, European Monetary Union (EMU) membership was also very positive for the country: it contributed to macroeconomic stability, it imposed fiscal discipline and central bank independence, and it lowered dramatically the cost of capital. One of the key benefits was the dramatic reduction in short-term and long-term nominal interest rates: from 13.3% and 11.7% in 1992, to 3.0% and 4.7% in 1999, and 2.2% and 3.4% in 2005. The lower costs of capital led to an important surge in investment from families (in housing and consumer goods) and businesses (in employment and capital goods). Indeed, EMU membership (and the Stability Pact) provided the country with unprecedented stability because it forced successive governments to implement responsible economic policies, which led to greater credibility and the improvement of the ratings of Spain's public debt (and consequently to lower financing costs).

Another important factor to account for the country's economic success was the remarkable economic policy stability that followed the economic crisis of 1992–1993. Indeed, there were few economic policy shifts throughout the 1990s and early 2000s, and this despite changes in government. Between 1993 and 2009, there were only two Ministers of Finance, Pedro Solbes (from 1993 to 1996, and from 2004 to 2009) and Rodrigo Rato (from 1996 to 2004); and the country only had three Prime Ministers (Felipe González, José María Aznar, and José Luís Rodríguez Zapatero). This pattern was further reinforced by the ideological cohesiveness of the political parties in government and the strong control that party leaders exercise over the members of the cabinet and the parliament deputies.

In addition, this stability was reinforced by the shared (and rare) agreement among Conservative and Socialist leaders regarding fiscal consolidation (the balance budget objective was established by law by the Popular Party), as well as the need to hold firm in the application of restrictive fiscal policies and the achievement of budgetary surpluses: As a result, a 7% budget deficit in 1993 became a 2.2% surplus in 2007, and public debt decreased from 68% of GDP in 1998 to 36.2% in 2007.

Finally, other factors that contributed to this success included the limited corruption and the fact that politics were fairly clean and relatively open; that Spain had a flexible economy; and the success of Spanish multinationals: There were eight firms in the *Financial Times* list of the world's largest Multinationals in 2000, and 14 in 2008.[12]

The Challenges

However, this economic success was marred by some glaring deficiencies that came to the fore in 2008 when the global financial crisis hit the country, because it was largely a "miracle" based on bricks and mortar.[13] The foundations of economic growth were fragile because the country had low productivity growth (productivity contributed only 0.5 percentage points to potential GDP between 1998 and 2006) and deteriorating external competitiveness.[14] Over the decade that preceded the 2008 crisis Spain did not address its fundamental challenge, its declining productivity, which only grew an average of 0.3% during that decade (0.7% in 2006), one whole point below the EU average, placing Spain at the bottom of the EU and ahead of only Italy and Greece (the productivity of a Spanish worker was the equivalent of 75% of a US one). The most productive activities (energy, industry, and financial services) contributed only 11% of GDP growth.[15]

---

[12] According to 2007 data from the *World Bank Governance Indicators* (http://info. worldbank.org/governance/wgi/sc_chart.asp), Spain was ranked in the 75–100th country's percentile ranks in control of corruption, government effectiveness, regulatory quality, rule of law, and voice and accountability.

[13] According to Martinez-Mongay and Maza Lasierra, "The outstanding economic performance of Spain in EMU would be the result of a series of lucky shocks, including a large and persistent credit impulse and strong immigration, underpinned by some right policy choices. In the absence of new positive shocks, the resilience of the Spanish economy to the financial crisis might be weaker than that exhibited in the early 2000s. The credit impulse has ended, fiscal consolidation has stopped, and the competitiveness gains of the nineties have gone long ago." See Martinez-Mongay and Maza Lasierra (2009).

[14] "Fears of Recession as Spain Basks in Economic Bonanza," *Financial Times*, Thursday, June 8, 2006.

[15] "Los expertos piden cambios en la política de I+D," *El País*, Monday, December 18, 2006.

Moreover, growth was largely based on low-intensity economic sectors, such as services and construction, which were not exposed to international competition. In 2006, most of the new jobs were created in low-productivity sectors such as construction (33%), services associated with housing such as sales and rentals (15%), and tourism and domestic service (30%). These sectors represented 75% of all the new jobs created in Spain in 2006 (new manufacturing jobs, in contrast, represented only 5%). Furthermore, the labor temporary rate reached 33.3% in 2007, and inflation was a recurrent problem (it closed 2006 with a 2.7% increase, but the average for that year was 3.6%), thus the inflation differential with the EU (almost 1 point) had not decreased, which reduced the competitiveness of Spanish products abroad (and consequently Spanish companies were losing market share abroad).[16] Competitiveness was further hindered by a deep process of economic deindustrialization, low value added and complexity of exports, and low insertion in global value chains.

In addition, family indebtedness reached a record 115% of disposable income in 2006, and the construction and housing sectors accounted for 18.5% of GDP (twice the Eurozone average). House prices rose by 150% since 1998, and the average price of a square meter of residential property went up from 700 Euros in 1997 to 2000 at the end of 2006, even though the housing stock had doubled. Many wondered whether this bubble was sustainable.[17] The crisis that started in 2008 confirmed the worst fears, and the implosion of the housing bubble fueled corruption and bad practices in the cajas sector of the financial system.

Moreover, between 40 and 60% of the benefits of the largest Spanish companies came from abroad. Yet, in the years prior to the crisis this figure had decreased by approximately 10 percentage points, and there had been a decline in direct foreign investment of all types in the country, falling from a peak of 38.3 billion euros in 2000 to 16.6 billion euros in 2005.[18] The current account deficit reached 8.9% of GDP in 2006 and over 10% in 2007, which made Spain the country with the largest deficit in absolute terms (86,026 million euros), behind only the United

[16] Angel Laborda, "El comercio en 2006," *El País*, Sunday, March 11, 2007, p. 20.

[17] Wolfgang Munchau, "Spain, Ireland and Threats to the Property Boom," *Financial Times*, Monday, March 19, 2007; "Spain Shudders as Ill Winds Batter US Mortgages," *Financial Times*, Wednesday, March 21, 2007.

[18] "Spanish Muscle Abroad Contrast with Weakling Status among Investors," *Financial Times*, December 11, 2006.

States; imports were 25% higher than exports and Spanish companies were losing market share in the world. Hence, the trade deficit reached 9.5% in 2008.[19]

While there was overall consensus that the country needed to improve its education system and invest in research and development to lift productivity, as well as modernize the public sector, and make the labor market more stable (i.e., reduce the temporary rate) and flexible, the government did not take the necessary actions to address these problems. Spain spent only half of what the Organization of European Co-operation and Development (OECD) countries spent on average on education; it lagged most of Europe on investment in Research and Development (R&D); and it was ranked 29th by the UNCTAD as an attractive location for research and development. Finally, other observers noted that Spain was failing to do more to integrate its immigrant population, and social divisions were beginning to emerge.[20]

By the summer of 2008, the effects of the global crisis were evident in Spain, and between 2008 and 2013 the country suffered one of the worst recession in modern history. This collapse was not wholly unexpected. The global liquidity freeze and the surge in commodities, food, and energy prices brought to the fore the unbalances in the Spanish economy: the record current account deficit, persisting inflation, low productivity growth, dwindling competitiveness, increasing unitary labor costs, excess consumption, and low savings, had all set the ground for the current devastating economic crisis (see Royo 2013).

AFTER THE FIESTA: THE GLOBAL CRISIS HITS SPAIN

As we have seen, the imbalances in the Spanish economy came to the fore in 2007–2008 when the real estate market bubble burst and the international financial crisis hit Spain (see Table 4.2). In just a few months the 'debt-fired dream of endless consumption' turned into a nightmare. By the summer of 2013, Spain faced the worst economic recession in half a century. According to government statistics, 2009 was the worst year since there has been reliable data: GDP fell 3.7%, unemployment

---

[19] "La Comisión Europea advierte a España de los riesgos de su baja competitividad," *El País*, February 4, 2007.

[20] "Zapatero Accentuates Positives in Economy, but Spain Has Other Problems," *Financial Times*, April 16, 2007, p. 4.

Table 4.2   The economic crisis: 2008–2013

Subject descriptor	Units	2007	2008	2009	2010	2011	2012	2013
Gross domestic product, constant prices	% change	3.5	0.9	−3.8	−0.3	0.4	−1.4	−1.6
Output gap in percent of potential GDP	% potential GDP	3.8	2.3	−2.8	−3.4	−3.2	−4.5	−5.4
Total investment	% GDP	31.0	29.1	24.0	22.8	21.5	19.6	18.1
Inflation, average consumer prices	% change	2.8	4.1	−0.2	2.0	3.1	2.4	1.9
Unemployment rate	% total labor force	8.3	11.3	18	20.1	21.7	25	27
General government structural balance	% potential GDP	−1.1	−5.4	−9.5	−8.0	−7.8	−5.7	−4.5
General government net debt	% GDP	26.7	30.8	42.5	49.8	57.5	71.9	79.1
Current account balance	% GDP	−1.0	−9.6	−4.8	−4.5	−3.7	−1.1	1.1

Source International Monetary Fund, *World Economic Outlook Database*, April 2014

reached over four million people (eventually reaching over 27% in 2012, with more than 6 million people unemployed), and the public deficit reached a record 11.4% of GDP (up from 3.4% in 2008). Consumer confidence was shattered, the implosion of the housing sector reached historic proportions, and the manufacturing sector was also suffering.

Initially, the Zapatero government was reluctant to recognize the crisis, which was becoming evident as early as the summer of 2007, because of electoral considerations: The country had a general election in March 2008. And after the election, the Zapatero government was afraid to admit that it had not been entirely truthful during the campaign. By 2007, there was increasing evidence that the model based on construction was already showing symptoms of exhaustion. Yet, the Spanish government not only refused to recognize that the international crisis was affecting the country, but also that in Spain the crises would be aggravated by the very high levels of private indebtedness. As late as August 17, 2007, Finance Minister Solbes predicted that 'the crisis would have a relative small effect' in the Spanish economy.

When it became impossible to deny what was evident, the government's initial reluctance to recognize and address the crisis was replaced by frenetic activism. The Zapatero government introduced a succession of plans and measures to try to confront the economic crisis, and specifically to address the surge of unemployment (Royo 2013) (see Fig. 4.1).

The sharp deterioration of the labor market was caused by the economic crisis and the collapse of the real estate sector, and it was aggravated by a demographic growth pattern based on migratory inflows of labor: In 2007, there were 3.1 million immigrants in the country, of which 2.7 million were employed and 374,000 unemployed. In 2008, the number of immigrants increased by almost 400,000–3.5 million (representing 55% of the growth in the active population), but 580,000 of them were unemployed (and 2.9 million employed), an increase of 200,000. In the construction sector alone, unemployment increased 170% between the summer of 2007 and 2008. Meanwhile, the manufacturing and service sectors (also battered by the global crisis, lower consumption, and lack of international competitiveness) proved unable to incorporate these workers.

The pace of deterioration caught policy-makers by surprise. The Zapatero government prepared budgets for 2008 and 2009 that were utterly unrealistic in the face of rapidly changing economic circumstances

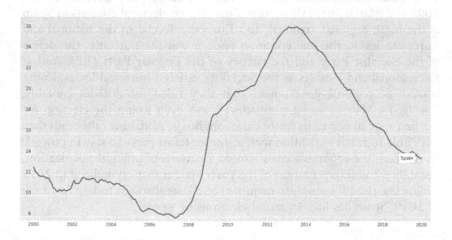

Fig. 4.1 Unemployment rate (2000–2020) (*Source* OECD Labor market statistics)

(as did all other advanced countries, the G-20 agreed on a plan for fiscal stimulus that would later prove relatively ineffective and dangerous for Spain as it increased the country's debt). As a result, things continued to worsen over the new four years. The most significant decline was in consumer confidence, which was hammered by the financial convulsions, the dramatic increase in unemployment, and the scarcity of credit. As a result, household consumption, which represented 56% of GDP, fell 1% in the last quarter of 2009 for the first time in the last 15 years. According to the Bank of Spain, this decline in household consumption was even more important in contributing to the recession than the deceleration of residential investment, which had fallen 20%, driven down by worsening financial conditions, uncertainties, and the drop in residential prices. The government actions had limited effect stemming this hemorrhage, and their efficacy was inadequate.

Finally, the impact of the global economic crisis was felt well beyond the economic and financial realms. The crisis also had severe political consequences. Spain followed in the path of many other European countries (including Ireland, Portugal, Greece, and France) that saw their governments suffer the wrath of their voters and have been voted out of office. The Socialist Party (PSOE) was re-elected in a general election on March 9, 2008. Soon thereafter, economic conditions deteriorated sharply and the government's popularity declined rapidly. Between March 2008 and March 2012, there were a number of electoral contests in Spain at the local, regional, national, and European levels. At the national and European levels, the one common pattern was the outcome: the defeat of the Socialist Party and the victory of the Popular Party (PP). And at the regional and local levels the Socialists suffered historical losses, losing control of regional government that they ruled for decades (notably, Castilla-La Mancha and Extremadura), and even losing the election for the first time in one of its historical strongholds, Andalusia (although they were able to reach a coalition with a smaller leftist party to stay in power). In the end, the economic crisis ignited a pattern of political polarization, instability, and fragmentation of the party system that crystallized in 2015 (following the PP's absolute majority) and lasts through today: As of fall of 2019, Spain has had 4 general elections in 4 years.

THE TRIPLE CRISIS

The Fiscal Crisis

One of the most common misinterpretations regarding the crisis in Southern Europe was attributing it to mismanaged public finances. Many policy-makers across Europe, especially in the creditor countries (crucially Germany), still insist today (2019) that the crisis was caused by irresponsible public borrowing, and this, in turn, led to misguided solutions. In fact, with very few exceptions, notably Greece, that interpretation is incorrect. In Spain, the crisis did not originate with mismanaged public finances. On the contrary, as late as 2011, Spain's debt ratio was still well below the average for countries that adopted the euro as a common currency: While Spain stood at less than 60% of GDP, Greece stood at 160.8%, Italy at 120%, Portugal at 106.8%, Ireland at 105%, Belgium at 98.5%, and France at 86%. On the contrary, prior to 2007, Spain seemed to be in an enviable fiscal position, even when compared with Germany.[21] Spain ran a budget surplus in 2005, 2006, and 2007. It was only when the crisis hit the country and the real estate market collapsed that the fiscal position deteriorated markedly and the country experienced huge deficits.

The problem in Spain was the giant inflow of capital from the rest of Europe; the consequence was rapid growth and significant inflation. In fact, the fiscal deficit was a result, not a cause, of Spain's problems: When the global financial crisis hit Spain and the real estate bubble burst, unemployment soared, and the budget went into deep deficit, caused partly by depressed revenues and partly by emergency spending to limit human costs. The government responded to the crisis with a massive €8 billion public works stimulus. This decision, combined with a dramatic fall in revenue, blew a hole in government accounts resulting in a large deficit.

Furthermore, the conditions for the crisis in Spain were created by the excessive lending and borrowing of the private sector rather than the government. In other words, the problem was private debt and not public debt. Spain experienced a problem of ever-growing private sector indebtedness, which was compounded by the reckless investments and loans of banks (including the overleveraged ones), both integral components of

---

[21] See Martin Wolf's blog: "What Was Spain Supposed to Have Done?" June 25, 2012, http://blogs.ft.com/martin-wolf-exchange/2012/06/25/what-was-spain-supposed-to-have-done.

the bank bargains that we examine later in the book, that were aggravated by competitiveness and current account imbalances. In Spain, the private sector debt (households and nonfinancial corporations) was 227.3% of GDP at the end of 2010; total debt increased from 337% of GDP in 2008 to 363% in mid-2011.

Yet, although Spain entered the crisis in a relatively sound fiscal position, that position was not solid enough to withstand the effects of the crisis, especially being a member of a dysfunctional monetary union with no lender of last resort. The country's fiscal position deteriorated sharply—collapsing by more than 13% of GDP in just two years. Looking at the deficit figures with the benefit of hindsight, it could be argued that Spain's structural or cyclically adjusted deficit was much higher than its actual deficit. The fast pace of economic growth before the crisis inflated government revenues and lowered social expenditures in a way that masked the vulnerability hidden in Spanish fiscal accounts. The problem is that it is very difficult to know the structural position of a country. The only way in which Spain could have prevented the deficit disaster that followed would have been to run massive fiscal surpluses of 10% or higher during the years prior to the crisis in order to generate a positive net asset position of at least 20% of GDP.[22] This, for obvious reasons, would not have been politically feasible.

The Loss of Competitiveness

There is another way to look at the crisis. Many economists argue that the underlying problem in the euro area was the exchange rate system itself, namely, the fact that European countries locked themselves into an initial exchange rate. This decision meant, in fact, that they believed that their economies would converge in productivity (which would mean that the Spaniards would, in effect, become more like the Germans). If convergence was not possible, the alternative would be for people to move to higher productivity countries, thereby increasing their productivity levels by working in factories and companies there (or to create a full fiscal union to provide for permanent transfers, as argued by OCA theory). Time has shown that both expectations were unrealistic and, in fact, the opposite happened. The gap between German and Spanish (including other

---

[22] From Martin Wolf's blog: "What Was Spain Supposed to Have Done?" June 25, 2012.

peripheral country) productivity increased, rather than decreased, over the past decade and, as a result, Germany developed a large surplus on its current account; while Spain and the other periphery countries had large current account deficits that were financed by capital inflows.[23] In this regard, one could argue that the incentives introduced by EMU worked exactly in the wrong way. Capital inflows in the south made the structural reforms that would have been required to promote convergence less necessary, thus increasing divergence in productivity levels.

In addition, adoption of the euro as a common currency fostered a false sense of security among private investors. During the years of euphoria following the launching of Europe's economic and monetary union and prior to the onset of the financial crisis, private capital flowed freely into Spain and, as a result as we have seen, the country ran current account deficits of close to 10% of GDP. In turn, these deficits helped finance large excesses of spending over income in the private sector. The result did not have to be negative. These capital inflows could have helped Spain (and the other peripheral countries) invest, become more productive, and "catch up" with Germany. Unfortunately, in the case of Spain, they largely led to a massive bubble in the real estate market, consumption, and unsustainable levels of borrowing. The bursting of that bubble contracted the country's real economy and it brought down the banks that gambled on loans to real estate developers and construction companies.

At the same time, as noted above, the economic boom also generated large losses in external competitiveness that Spain failed to address. Successive Spanish governments also missed the opportunity to reform institutions in their labor and product markets. As a result, costs and prices increased, which in turn led to a loss of competitiveness and large trade deficits. This unsustainable situation came to the fore when the financial shocks that followed the collapse of Lehman Brothers in the fall of 2007 brought "sudden stops" in lending across the world, leading to a collapse in private borrowing and spending, and a wave of fiscal crisis.

The Financial Crises

A third problem had to do with the banks. As we will see in much greater detail later in the book, this problem was slow to develop. Between

[23] Simon Johnson's blog: "The End of the Euro: What's Austerity Got to Do With It?" June 21, 2012.

2008 and 2010, the Spanish financial system, despite all its problems, was still one of the least affected by the crisis in Europe. During that period, of the 40 financial institutions that received direct assistance from Brussels, none was from Spain. In December 2010, Moody's ranked the Spanish banking system as the third strongest of the Eurozone, only behind Finland and France, above the Netherlands and Germany, and well ahead of Portugal, Ireland, and Greece. Finally, Santander and BBVA had shown new strength with profits of €4.4 billion and €2.8 billion, respectively, during the first half of 2010. Spanish regulators had put in place regulatory and supervisory frameworks, which initially shielded the Spanish financial system from the direct effects of the global financial crisis. Indeed, the Bank of Spain had imposed a regulatory framework that required higher provisioning, which provided cushions to Spanish banks to initially absorb the losses caused by the onset of the global financial crisis. And there were no toxic assets in bank´s balance sheets.

Nevertheless, this success proved short-lived. In the summer of 2012, Spanish financial institutions seemed to be on the brink of collapse and the crisis of the sector forced the European Union in June (2012) to devise an emergency €100 billion rescue plan for the Spanish banking sector (see Chapter 1). When the crisis intensified, the financial system was not able to escape its dramatic effects. By September 2012, the problem with toxic real estate assets forced the government to intervene and nationalize eight financial institutions. Altogether, by May 9, 2012, the reorganization of the banking sector involved €115 billion in public resources, including guarantees.

As we will examine later in the book, there are a number of factors that help account for the deteriorating performance of the Spanish banks after 2009. The first was the direct effect of the economic crisis. The deterioration in economic conditions had a severe impact on the bank balance sheets. The deep recession and record-high unemployment triggered successive waves of loan losses in the Spanish mortgage market coupled with a rising share of nonperforming loans. Like many other countries such as the United States, Spain had a huge property bubble that burst. Land prices increased 500% in Spain between 1997 and 2007, the largest increase among the OECD countries. As a result of the collapse of the real estate sector had a profound effect in banks: Five years after the crisis started, the quality of Spanish banking assets continued to plummet. The Bank of Spain classified €180 billion euros as troubled assets at the end of 2011, and banks were sitting on €656 billion of mortgages of which 2.8% were classified as nonperforming.

A second factor was concern over the country's sovereign debt. As mentioned before, the crisis in Spain did not originate with mismanaged public finances. The crisis has largely been a problem of ever-growing private sector debt, compounded by reckless bank investments and loans, particularly from the *cajas*, as well as aggravated by competitiveness and current account imbalances. To place the problem in perspective, the gross debt of household increased dramatically in the decade prior to the crisis, and by 2009 it was 20 percentage points higher than the Eurozone average (86% of GDP versus 66%). But the austerity policies implemented since May 2010 aggravated the fiscal position of the country. The ratio of Spain's debt to its economy was 36% before the crisis and reached 84% in 2013. In sum, Spain fell into the "doom loop" that had already afflicted Greece or Portugal and led to their bailout. The sustainability of the Spanish government debt was affecting Spanish banks (including BBVA and Santander) because they had been some of the biggest buyers of government debt in the wake of the ECB long-term refinancing operation liquidity infusions (the percentage of government bond owned by domestic banks reached 30% in mid-2012). Again, the doom loop was a result of EMU weakness, namely the lack of a banking union with a centralized EU funded mechanism to bail out banks.

Spanish banks were also suffering the consequences of their dependence on wholesale funding for liquidity since the crisis started, and, in particular, their dependence on international wholesale financing, as 40% of their balance depends on funding from international markets, particularly from the ECB. Borrowing from the ECB reached €82 billion in 2012, and Spanish banks had increased their ECB borrowings by more than six times since June 2011, to the highest level in absolute terms among Euro area banking systems as of April 2012.

The crisis also exposed weaknesses in the policy and regulatory framework, part of the banking bargains that we will examine later in the book. The most evident sign of failure was the fact that the country had already adopted five financial reforms in three years and had implemented three rounds of bank mergers. The results of these reforms were questionable at best. The fact that Spain had five reforms in less than three years, instead of one that really fixed the problem, says it all. They had been perceived largely as "too little and too late," and they failed to sway investors' confidence in the Spanish financial sector.

Finally, the financial crisis can also be blamed on the actions (and inactions) of the Bank of Spain, one of the key actors involved in the bank bargains. At the beginning of the crisis, the Bank of Spain's policies were all praised and were taken as model by other countries. Time, however, tempered that praise and the Bank of Spain was criticized for its actions and decisions (or lack thereof) during the crisis. Spanish central bankers chose the path of least resistance: alerting about the risks but failing to act decisively.

CONCLUSION

The economic crisis that started in Spain 2008 was largely a problem of ever-growing private sector debt, aggravated by competitiveness and current account imbalances, and compounded by reckless bank investments and loans, particularly from the *cajas*, which by over-lending freely to property developers and mortgages contributed to a real estate property bubble. This outcome was a result of the political bargains at the heart of the game of bank bargains focus of this book. The bubble contributed to hide the fundamental structural problems of the Spanish economy outlined in the previous sections and had an effect in policy choices because no government was willing to burst the bubble and risk suffering the wreath of voters. Furthermore, cheap credit also had inflationary effects that contributed to competitiveness losses and record balance of payment deficits. Therefore, three dimensions of the crisis (financial, fiscal, and competitiveness) are interlinked in their origins. The crisis exposed the underbelly of the financial sector and showed that many banks (particularly the *cajas*) were not just suffering liquidity problems but risked insolvency, which led to the EU financial bailout of June 2012. The bailout had onerous conditions attached and it limited national economic autonomy (see Dellepiane and Hardiman 2011). Finally, the financial and fiscal crises were made worse by the incomplete institutional structure of EMU and by bad policy choices at the EU level (excess austerity and refusal to act as a lender of last resort for sovereigns by the ECB) (Royo 2013).

In the end, the crisis exposed the weaknesses of the country's economic model. Indeed, despite the previous two decades' significant progress and achievements, the Spanish economy still faced serious competitive and fiscal challenges. Unfortunately, the economic success the country prior to the crisis fostered a sense of complacency, which allowed for a delay in the adoption of the necessary structural reforms. And this was not a surprise as the Spanish economy was living on borrowed time, despite all the significant progress, and the country still had considerable ground to cover, given the existing income and productivity differentials, to catch up with the richer EU countries and to improve the competitiveness of its economy (see Royo 2013).

The sudden collapse of the Spanish economy came as a shock. Yet, in retrospect it should not have been such a surprise. The policies choices and political bargains taken during the previous decades led to an unsustainable bubble in private sector borrowing that was bound to burst. Moreover, as we will examine on Chapter 7, the institutional degeneration that led to systemic corruption and contributed to the implosion of parts of the financial sector made the crisis almost unavoidable.

As we have seen, much of Spain growth during the 2000s was based on the domestic sector and particularly on an unsustainable reliance on construction. As we will later in the book, this outcome was part of a political bargain in which tax incentives and a lax regulatory framework favored developers, property owners, and bankers (particularly *cajas*). The particular regulation of the *cajas* proved fatally flawed, as it provided incentives that favored local and regional government actors' access to finance at the expense of an environment that would have provided a stable and efficient banking system. On the contrary, it led to a form of crony capitalism Spanish style, in which they invested massively in the construction sector in search of rapid growth and larger market share. These decisions proved fatal once the real estate bubble burst, and they led to the nationalization of several *cajas*, including *Bankia*, and the financial bailout from the European Union.

Membership in the European single currency was not the panacea that everyone expected to be, thus confirming the crucial importance of domestic political institutions and how domestic players operate within those institutions. In Spain, the adoption of the euro led to a sharp reduction in real interest rates that contributed to the credit boom and the real estate bubble. However, it also altered economic governance decisions. Successive Spanish governments largely ignored the implications of EMU membership and failed to implement the necessary structural reforms to ensure the sustainability of fiscal policies and to control unitary labor costs. These decisions led to a continuing erosion of competitiveness (and a record current account deficit), and a huge fiscal deficit when the country was hit by the global financial crisis.

Indeed, the experience of the country shows that EU and EMU membership had not led to the implementation of the structural reforms necessary to address these challenges. On the contrary, EMU contributed to the economic boom, thus facilitating the postponement of necessary economic reforms. This challenge however is not a problem of European institutions, but of national policies. The process of economic reforms has to be a domestic process led by domestic actors willing to carry them out.

The Spanish case serves as an important reminder that in the context of a monetary union, countries only control fiscal policies and relative labor costs. Spain proved to be weak at both. It failed to develop an appropriate adjustment strategy to succeed within the single currency, and it ignored the imperative that domestic policy choices have to be consistent with the international constraints imposed by euro membership. On the contrary, in Spain domestic policies and the imperatives of participating in a multi-national currency union stood in uneasy relationship to one another. The crisis was the tipping point that brought this inconsistency to the fore, which led to the worst economic crisis in Spanish modern history (before COVID-19). Next we turn to the elements of domestic bargains that underline the financial crisis.

REFERENCES

Calavita, Kitty. *Immigrants at the Margins*. New York: Cambridge University Press, 2005.

Dellepiane, Sebastián, and Niamh Hardiman. "Governing the Irish Economy." UCD Geary Institute Discussion Series Chapters, Geary WP2011/03, University College Dublin, Dublin, February 2011.

Guillén, Mauro. *The Rise of Spanish Multinationals*. Cambridge: Cambridge University Press, 2005.

Martinez-Mongay, Carlos, and Luís Angel Maza Lasierra. "Competitiveness and Growth in EMU: The Role of the External Sector in the Adjustment of the Spanish Economy." Economic Papers, No. 355, October 2009.

Royo, Sebastián. *From Social Democracy to Neoliberalism*. New York: St. Martin's Press, 2000.

Royo, Sebastián. *Varieties of Capitalism in Spain*. New York: Palgrave, 2008.

Royo, Sebastián. *Lessons from the Economic Crises in Spain*. New York: Palgrave, 2013.

Royo, Sebastián. "After Austerity: Lessons from the Spanish Experience." In *Towards a Resilient Eurozone: Economic, Monetary and Fiscal Policies*, edited by John Ryan. New York: Peter Lang, 2015.

The Global Financial Crisis and the Spanish Banking System: Explaining Its Initial Success (2007–2010)

INTRODUCTION

Contrary to their counterparts all over the world, during the first three years of the global financial crisis (2007–2010), most Spanish banks appeared to have escaped the worst direct effects of the crisis (Royo 2013). The timing of the problems (the effect of the crisis was less intense in Spain in its early stages) and the nature of the problem (Spanish securitization levels were moderately high by EU level, and this was largely undertaken by the *cajas*). Indeed, as noted on Chapter 1, when looking at the performance of the Spanish financial system, it is very important to distinguish between the large banks and the *cajas*; the large banks performed relatively well while the *cajas* suffered from a somewhat traditional financial crisis. Much of what the *cajas* were doing was lending to Non-Financial Corporations (NFCs) in construction, rather than just mortgage lending. Yet, a key element that separates this crisis from a traditional one, particularly on the *cajas'* side, is the role of market-based banking (MBB) liabilities (Carballo Cruz 2011). This distinction

From: "Why the Spanish Financial System Survived the First Stage of the Global Crisis?" *Governance* 26, no. 4 (October 2013): 631–56.

is unique in comparative terms; in no other country were small banks uniquely hit and large banks left largely unscathed.[1]

This chapter shows that despite all the challenges facing the Spanish financial sector, it was one of the least affected during the first phase of the crisis. Of the 40 financial institutions that requested direct assistance from Brussels prior to mid-2010, none were from Spain, and in December 2010, Moody's ranked the Spanish banking system as the third strongest of the Eurozone, only behind Finland and France, above the Netherlands and Germany. Finally, Santander and BBVA showed new strength with profits of 4.4 billion and 2.8 billion euros, respectively, during the first half of 2010.

This chapter seeks to examine the performance of Spanish financial institutions during the first three years of the crisis and to explain their initial success. This relative success, however, proved to be short-lived. Indeed, while Spanish banks faced fewer problems in the wake of the Lehman Brothers collapse compared to those countries with banking systems that were more market-based, eventually the crisis caught up with them when more traditional problems with lending and government debt emerged after 2008. Once again, the financial crisis was the result of a political bargains. In this case, it was the result of bargains in which incentives and a lax regulatory framework favored developers, property owners, and bankers, thus confirming a central tenant of this book: the crucial importance of domestic political institutions, the rules of the game, and the role of domestic players operating within those institutions We will examine this turn of events in greater detail on Chapter 6.

The Initial Impact of the Crises on the Spanish Financial System

Initially, one of the few encouraging outcomes of the global financial crisis in Spain was the positive performance of the Spanish financial sector during first stages of the crisis (2007–2010), as they seemed to have escaped the crisis's worst direct effects. Indeed, despite all the challenges facing the Spanish financial sector, it was one of the least affected by the

---

[1] In France, the investment banks connected to the mutuals were also hit very badly (notably Natixis but also Calyon). In the UK, Northern Rock (a building society) was also devastated and a few other building societies were affected badly. Even in Germany, a mortgage lender, HRE, was severely hit, and IKB was the first one to go under in 2007.

crisis in Europe. As noted before, of the 40 financial institutions that had requested direct assistance from Brussels, none was from Spain (as of June of 2010: a dozen were from Germany, five from the UK, six from the Benelux, and four from Ireland). Furthermore, the amounts that were committed to support those institutions were astronomical: 3300 billion or 28% of the EU's GDP, in capital injections (315 billion), the purchase of damages assets (103 billion) and guarantees (2900 billion). In contrast, the support provided to Spanish institutions (loans at 7.7% interest rate) would only reach 30 billion (just 1% of the total).[2] And in December 2010 Moody's ranked the Spanish banking system as the third strongest of the Eurozone, only behind Finland and France, above the Netherlands and Germany, and well ahead of Portugal, Ireland, and Greece. Finally, Santander and BBVA, showed new strength with profits of €4.4 billion and €2.8 billion, respectively, during the first half of 2010.

Indeed, initially (as of fall 2010), the crisis had a relative mitigated impact on Spanish banks (see Barrón 2012). Spain still had two of the most credit worthy banks in the world, *Banco Santander* and *BBVA*, and the crisis did not affect their solid foundations in terms of exposure to MBB (although they were severely affected by the wider economic crisis in Spain). On the contrary, they continued to generate profits in the aftermath of the international financial crisis. Of the top five banks in the world in terms of 2008 pre-tax profits, two—Santander and BBVA— were Spanish (the other three, Chinese) with BBVA making 2.8 billion euros and Santander 4.5 billion of net profit. Moreover, while their main competitors in other countries were severely affected, and in many cases had to be rescued by their governments, in Spain, Santander and BBVA were in need of neither state assistance nor liquidity injections. In contrast, Spanish banks emerged as rescuers in foreign markets; Santander expanded in the United States, UK, and Germany, while BBVA and *Banco Sabadell* (another private bank) in the United States. They also successfully raised funds from the flotation of their subsidiaries (i.e., Santander did it with its Brazilian subsidiary in one of the world's largest public offerings in 2009).[3] By November 2009, only one financial institution had collapsed due to solvency problems after overextending itself in loans

---

[2] "¿Por qué Berlín ataca a España?" *El País*, June 20, 1010.

[3] See Victor Mallet, "Prudence Pays Off for Big Banks." In "Investing in Spain," special report, *Financial Times*, October 2, 2009, p. 3.

to local projects and had to be taken over by the Bank of Spain, *Caja Castilla la Mancha* (CCM). But this only represented 1% of total assets.

Concerns over the strength and solvency of the financial sector led Spanish authorities to push for the publication of the "stress tests" that the main European banks underwent in 2010. When the results of these tests were made public on Saturday, July 23, 2010, only five of the 17 *cajas* failed: *Banca Cívica, Banca Espiga, Caja Catalunya's group, Unimm,* and *CajaSur.* It is important to note, however, that although five of the seven institutions that failed the tests were Spanish savings banks, it was partly because all Spanish lenders were subjected to the tests. Indeed, contrary to other European countries, Spain subjected 95% of its banking system to these stress tests (the average in the EU was 50%). The overall good result for the Spanish financial system confirmed the benefits of the countercyclical capital regime, which proved to be the best rescue plan for the Spanish banks because it allowed many of these institutions to pass the stress tests. The 27 banks and *cajas* that were included in the tests still held 19,796 billion euros in these kinds of provisions, available in case of a further crisis. The initial response of the markets to these results was very positive and helped to push down the country's risk premium. Given what we learnt later, those assumptions proved overoptimistic.[4] We look next at the factors that made banks (<u>not cajas,</u> we examine them on later) so resilient.

What Made Spain Different?

Institutional Setting

The structure of the banking sector conditioned the response of Spanish financial institutions to the crisis. Spanish banks had vibrant and efficient branch networks that helped them to develop strong relationship with their customers. They proved that despite the widespread perception that electronic banking is the future, the combination of large branches with the latest technologies is compatible with cost efficiencies, and they were crucial to attract deposits and strengthen their balance sheets.

An additional factor that explains the performance of the sector was the existence of a well-coordinated regulatory framework, in which the Bank

---

[4] Subsequent events seemed to show that they were not facing up to the reality of the situation, the assumed fall in property prices proved to be overoptimistic.

of Spain (BoS) played a crucial role as a supervisory body, but worked closely with the consumer protection agency, and the finance ministry.[5] For instance, the Bank of Spain discouraged lenders from adopting risky 'off balance sheet' accounting methods and from acquiring billions of Euros of repackaged US subprime mortgages and other toxic assets.

At the same time, the Bank's onsite supervisors, who are posted in the head offices of commercial banks and *cajas*, were diligent overseeing risk decisions. Moreover, the Bank of Spain was instrumental in forcing financial institutions to focus on quality of capital, limiting their leverage, and the debt to equity ratio. For the Bank of Spain, enforcement was key and they focus on principles, not merely the legalistic approach that has been more typical of the Anglo-Saxon countries. Finally, the reports from the Bank of Spain's were instrumental in pushing for the necessary reforms. In sum, the Bank of Spain played a central role of management and supervision. Not surprisingly, those financial systems with close and proactive supervision, like Spain, were able initially to weather the crisis better than those without. [The role of the Bank of Spain, however, was questioned and criticized later, as we will examine in Chapter 6.]

Regulatory Framework

The single most important factor to account for the performance of the Spanish financial system during the initial stages of crisis had been the implementation prior to the crisis of a "dynamic provisioning system," which established a countercyclical capital regime for banks. As a result, banks were forced to provision for latent portfolio losses, defined as those likely to occur but which may be undetected by conventional accounting. This regime prevented the provision of dividend increases at times of growth that might undermine banks' solvency in the long term, and it also prevented off-balance sheets activities.

Indeed, there was consensus at the time that the stern regulations of the Bank of Spain played a key role in this outcome. Indeed, the Bank of Spain had been so traumatized by previous crises that it prevented the banks from indulging into reckless leverage and off-balance sheet accounting that became so prevalent in other countries. One of the key factors that help account for the ability of the Spanish banks to come

---

[5] Emilio Botín, "Why Banks Must Adopt a 'Back-to-Basics Approach'," *Financial Times*, special section on the *Future of Finance*, Monday, October 19, 2009, p. 9.

through the initial states of the crisis in much better shape than most of their counterparts was the creation of a countercyclical capital regime for banks: the so-called Dynamic provisioning.

a. Starting in 2000, the Bank of Spain forced banks to make provisions for latent portfolio losses, defined as those likely to occur but which may be undetected by conventional accounting. This method allowed for the creation of a buffer in the form of a reserve deducted from capital in good times and released at times of downturn. The way this was calculated was by comparing long-term credit growth in the economy with the current rate of credit growth. This approach provided a better estimate of profitability and solvency over time. This approach left Spanish banks with larger capital buffers than banks in most other countries.

b. At the same time, the Central Bank also prevented the provision of dividend increases at times of growth that might undermine banks' solvency in the long term.

c. Finally, the BoS prevented off-balance sheets activities. As we have seen, the Central Bank made it so expensive for financial institutions to establish off-balance sheet vehicles, which have sunk banks elsewhere that Spanish banks stayed away from such toxic assets.[6] They focused on retail and commercial banking and largely ignored the risky attractions of opaque investment banking operations.

This mandate provided Spanish banks with a countercyclical mechanism that was essential to overcome the initial stages of the crisis. For instance, *Banco Santander* built more than 6 billion euros of generic loan loss provisions. This strategy was praised globally since the crisis started and may have sheltered Spanish banks initially from the worst effects of the crisis. As a result, the Spanish Central Bank gained international prestige, and these policies were replicated in other countries throughout the world. [According to estimates from *Iberian Equities*, had the US Fed imposed similar requirements on US banks between 2003 and 2007 they would have built a cushion of 109,000 million euros, an amount large enough to cover the losses of US commercial banks (37,000 million).]

---

[6] J. Plender, "Respinning the Web," *Financial Times*, Monday, June 22, 2009, p. 5. It is important to note, however, that the Spanish model is not complaint with global accounting standards.

The overall good result of the Spanish financial system confirmed the benefits of the countercyclical capital regime, which proved to be (at least for a time) the best rescue plan for the Spanish banks.

Historical Learning Process

Spanish bankers learned from the crisis of the 1970s, 1980s, and early 1990s, and they adopted measures to prevent the repetition of these crises. The idea of an anti-cyclical fund was an old one and had been a source of constant confrontations with the banks. As we examined in Chapter 3, between the banking crises of the 1970s until the bankruptcy of *Banesto* in 1993, the Bank of Spain witnessed the disappearance of 40 financial institutions. This tragic experience convinced the leaders of the Bank of Spain of the need of establishing special anti-crisis preventive measures. Mariano Rubio, the Bank's governor between 1984 and 1992, initiated the introduction of these measures through the establishment in 1988 of the *Law of Discipline and Intervention in Credit Institutions*, which was the basis of the banking legislation at the time of the crisis. The economic crisis of the early 1990s and the collapse of *Banesto* in 1993, which the Bank of Spain was almost powerless to prevent, illustrated the need to deepen the preventive measures. This crisis led Luís Angel Rojo, Governor of the Bank at that time, to establish such provisions in an internal document (a 'circular'); and his successor, Jaime Caruana (200–2006), started to implement them in 2000.[7]

Spanish governments were also committed to make sure these crises would not happen again and they strengthened the role of the regulator, giving it the authority to tell banks what they can and cannot do; supported more regulation and higher capital requirements; and forced the restructuring of the sector through mergers. Finally, it is important to note that the Spanish financial sector was less affected and influenced by the London–New York competition, which had led to a deregulatory race in other countries.

[7] "Los bancos y cajas disponen aún de un 'colchón' extra de 20,000 millones," *El País*, Monday, July 26, 2010, p. 15.

Strategic Decisions

The initial strong performance of Spanish banks was also explained by their focus on retail and commercial banking. They largely ignored the risky attractions of opaque investment banking operations. An additional important strategic factor, particularly for the larger banks, has been their strategic decision to diversify internationally, primarily in Latin America, which allowed them to compensate for the poor performance in the Spanish market, raise capital in local markets and diversify their risk and revenue sources. Indeed, the internationalization strategy that started during the 1990s led them first toward Latin America and later toward Europe, the United States and others parts of the world. This strategy was partially prompted by the saturation of the Spanish market, which had an estimated 30% excess capacity. For instance, Brazil represented in 2010 approximately 18% of Santander's global profits, while Santander's activities outside Spain represented about 50% of the institution's entire business, and it was estimated that within 8–10 years it would reach 75%.[8]

The crisis confirmed the benefits of this decision. Emilio Botín, then chairman of *Banco Santander*, had stated that "the crisis has strengthened our determination to grow and reinforce our international diversification, which is the path to guarantee a sustainable growth and a moderation of risks."[9] In October 2010, Santander launched an initial public offering (IPO) for its Brazilian subsidiary, Santander Brazil; and during the crisis it bought the 75% of Sovereign Bank that it did not control, as well as Bradford & Bingley, Alliance & Leicester in the UK.

For its part, BBVA acquired Guaranty Financial in August 2010, a Texas bank that had been subject to intervention by government authorities. It already owned Valley Bank, having integrated its Mexican subsidiary *Bancomer* as well as Laredo National, Texas Regional and State National. Its biggest acquisition was completed in 2007, when it paid €6.655 billion for Compass Bancshares, which enabled BBVA to enter the list of the top 20 banks in the United States.

---

[8] See "Spain's Big Banks Continue Their Global Expansion," in http://www.wharton.universia.net/index.cfm?fa=viewArticle&id=1786&language=english.

[9] "Retrato del Poder," *El País*, October 20, 2010.

Cultural Arguments

Individual and group psychology also influences economic and financial outcomes. In this regard, Spanish bankers were usually considered boring and Conservative bankers. They were viewed as less aggressive and ideological than their US counterparts, more egalitarian and less hierarchical, and more risk-averse. Moreover, in Spain compensation schemes were generally based on long-term performance and linked to group-wide profitability, and they did not provide incentives for taking excessive risk.

An additional contributing factor was the banking culture cultivated throughout the banks: At the time, banks claimed that they focused on their responsibility to customers, employees, and shareholders, not just on quarterly financial reports. *Banco Santander*, for instance, considered its banking culture (based on "ample resources, commercial dynamism, prudence in risk management, dedication and discipline for our human teams, and unity within the top management bodies...[The secret] is an orthodox, persistent and efficient management that always pursues a single objective: profitability for investors and shareholders"[10]) as one of the key reasons for its historical success. Emilio Botín, former chairman of the bank, always emphasized in his public comments the strong "culture of risk management" of the bank.[11]

Finally, at the time it was belived that Spanish bankers were generally more proficient at managing credit risk: Appropriate risk management was a crucial factor to account for their success. As opposed to banks

---

[10] Pablo Matín Aceña, *Universia Business Review-Actualidad Económica. 150 Aniversario Banco de Santander.* November 5, 2007, p. 23.

[11] See Emilio Botín, "Sage Advise," *Financial Times*, October 16, 2009. In this article, Mr. Botín states "banks must recruit the most talented and ethical people from society. An important lesson of the last two years is just how much people matter. The human factor in leadership, strategy, management, and execution has been the differentiator between good and bad banks. To work for a bank, whether as cashier or chairman, is to assume a huge responsibility: You are safekeeping people's savings, their security, and their opportunities. Young people entering banking must understand this, and not enter the industry simply because it offers the potential for large financial reward. That requires ethics in character, in culture and in training." See also Emilio Botín "La Experiencia Internacional del Santander-Central Hispano," in *ICE*, April–May 2002, n. 799, p. 122. Santander stresses this culture in their corporate training as well: The "Top 200" senior executives from the group's global network convene at the Training Center at least once a year to exchange ideas on strategy, risk, technological changes and the main corporate themes of the day, and they are also charged with disseminating corporate culture and values throughout the group. See "Santander fired by education," *Financial Times*, March 20, 2006.

from other countries (and the *cajas*) which lost sight of the fundamentals of good risk management, ignored the trade-off between risk and profit, or forgot the difference between credit risk and market risk; Spanish banks did not stray far from these basic principles or their core businesses. Spanish banks, like *Santander*, kept their risk management separate from business areas and with independent reporting lines; had risk policies decided at board level; and their risk management was focused on knowing the customer.[12] But, it is also important to acknowledge that they were aggressive expanding internationally, and that Spanish banks have also been affected by scandals.

The Financialization of the Spanish Financial System

Prior to the liberalization of the late 1980s and 1990s, the Spanish financial system (like those of Greece and Italy) was a typical credit-based Mediterranean system, characterized by extensive interventionism, state control over the banking system, and underdeveloped capital markets (Pérez and Westrup 2010). The role of financial institutions was to provide funding to contribute to the process of economic development and industrialization (Pérez 1997; Lukauskas 1997), and bank deposits were turned into low-interest credit for industrial enterprises and the government (Deeg and Pérez 2000). As we examined on Chapter 3, the oligopolistic nature of the sector generated significant costs for Spanish firms outside the finance sector (Pérez 1997).

The liberalization of the sector started in the second half of the 1980s, driven by the country's accession to the European Community and the subsequent European Single Market program. In Spain's case, the process of financial liberalization was also part of the Bank of Spain's effort to achieve effective disinflation (Pérez and Westrup 2010). The European integration process, and particularly the Single Market Program and the European Monetary Union, was a driving factor in subsequent developments. While financial regulations remained the responsibility of the national governments, in reality over those two decades there was a harmonization process throughout all the member states. This process led to the liberalization, modernization, consolidation, and opening up of the system.

[12] Emilio Botín, "Why Banks Must Adopt a 'Back-to-Basics Approach'," *Financial Times*, special section on the *Future of Finance*, Monday, October 19, 2009, p. 9.

Yet, the degree of financialization of the sector was relatively low prior to the crisis. Indeed, the majority of banks' *assets* were loans to customers, and a significant part of banks' assets involved Spanish government securities, which were considered among the safest possible asset investments. It is on the liabilities side of the balance sheet where market-based banking was most present in the Spanish banking system and, as we will see contributed significantly to the crisis as it unfolded in Spain. As in Italy and Greece, on the *liabilities* side Spanish banks had a broad and stable funding base. However, while funding from retail customers constituted a large share of their total liabilities, Spanish banks still had a significant dependence on wholesale interbank funding. The regulatory framework, as we see below, imposed important restrictions on the extent to which there could be securitization on the liabilities side; and banks sponsored by Asset-backed commercial chapter (ABCP) were close to zero (see Royo 2013).

At the outset of the crisis, the governor of the Bank of Spain characterized Spanish banks' assets and liabilities the following way:

> On the liabilities side...they have focused on the longer maturities. This has given them greater peace of mind with regard to the structure of their liabilities...Spanish banks' recourse to wholesale markets has taken the form of various instruments, including asset-backed securities. However, Spanish securitization bears little resemblance to that which is currently proving highly problematic. First, because of the high quality of the securitized portfolios. And second, because securitization was not conceived of as a business in itself to transfer risk...It should not be forgotten that Spanish banks have traditionally maintained a solid deposit base. This is largely explained by the particular banking model followed in Spain. Activity is focused on retail business....
>
> On the assets side, Spanish banks have not invested in the complex and opaque products that have turned out to be highly problematic or, as people usually now say, "toxic". Spanish banks' assets, even taking into account the rise in doubtful assets, are largely of good quality. (Fernández Ordóñez 2008)

Indeed, Spanish banks had "traditional" assets and liabilities. However, the funding gap was high, thus market-based banking was largely concentrated on the liabilities side of the balance sheet, with a significant element of bank lending prior to the crisis funded on wholesale markets (see Fig. 5.1). In this speech, Governor Ordoñez underestimated the vulnera-

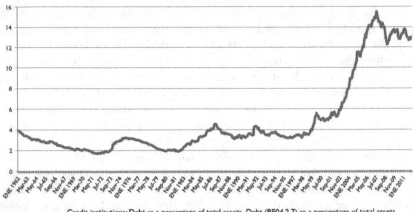

Credit institutions: Debt as a percentage of total assets. Debt (BE04.2.7) as a percentage of total assets (BE04.2.1). Monthly: January 1962 to January 2012.
Data source: Bank of Spain

Fig. 5.1   Spanish credit institutions: wholesale funding

bilities of MBB liabilities; he downplayed the risks associated with Spanish banks' interbank borrowing (McGuire and von Peter 2009). As we will examine later, he would come to regret it. In this regard, Spain is similar to Ireland, where Northern Rock's mortgage assets were similarly seen as high quality (and, it appears, unlike in Spain, will have low default rates) and securitizations were long term. However, by September 2007, Northern Rock needed liquidity support from the Bank of England because the securitization market dried up.

Once the crisis started, this became a major source of vulnerability because banks were no longer able to access funding on wholesale markets. Governor Fernández Ordóñez recognized as much later on when he stated that 'the high dependence of the Spanish economy on external financing, as a result of many years of spending reliant on foreign savings, made it particularly sensitive to any turbulence that might undermine confidence and obstruct the access of banks to international financial markets.'[13] This is exactly what happened when the wholesale financial markets on which Spanish banks depended for liquidity dried up starting in particular in May 2010 after the bailout of Greece. In addition, stage

[13] "Dublin Bail-Out Spooks Investors in Spain," *Financial Times*, November 27–28, 2010.

one and two of the crisis (global banking crisis, credit crunch, and the recessionary impact on the real economy) led to the bursting of the real estate market bubble, which led to significant problems on the assets side of bank balance sheets because of the property loans.

The relatively low degree of financialization of Spanish banks (as with Italian and Greek banks) can be explained by the slow evolution of the national financial system and Spanish banks' reluctance to change a business model that had been consistently successful for decades. At the same time, as we have seen, Conservative regulatory framework and supervisory practices have played a crucial role in limiting the risk exposure and excessive leverage of a significant part of the Spanish financial system prior to the crisis.

Yet, as noted before, the lending was largely concentrated in the construction/real estate sector, thus creating a very traditional bubble. The relatively high customer funding gap meant that market-based liabilities "turbocharged" traditional lending. Further difficulties were created because Spanish banks' market-based liabilities were also disproportionately sourced in more fragile markets, especially securitization and the inter-bank market.

Spain had very high levels of securitized lending (i.e., banks transforming mortgages into asset-based securities [ABS]), which contributed to the property market bubble. Moreover, while Spanish banks were relatively unexposed to market-based assets because gains in the domestic real estate market and through retail expansion abroad were profitable enough; market-based assets were still relevant and Spanish banks had write-down linked to securitized assets. Indeed, Santander and BBVA were already making substantial losses in 2007 and 2008 before the real estate collapsed.[14] While these write-downs were well below the levels of the UK or the United States, they were still significant.

Indeed, between 2000 and the summer of 2007 both the securitization market and the covered bond market expanded significantly in Spain. The outstanding amount of ABSs and covered bonds issued by Spanish institutions rose from 18 billion euros and 12 billion in 2000 to 298 billion euros and 168 billion in the second quarter of 2007. Furthermore, the outstanding amount of other fixed-income securities issued by financial institutions also expanded rapidly, albeit to a lesser extent. The growth

---

[14] According to Reuters Factbox Santander: 4.8 billion euros in 2007 and 8.3 billion in 2008. BBVA had 2.7 billion in 2007 and 4.2 billion in 2008.

of these markets in Spain was much faster than in other European countries; between 2004 and the second quarter of 2007, the weight of the Spanish market for ABSs in Europe was 14%, the second largest after the UK. In the covered bond market, Spain had a share of 21% during the same period, again the second largest in Europe, although in this case the leading market was Germany. In terms of outstanding amounts, as of June 2007 Spain had a share of around 13% and was ranked second (after the UK) in Europe in the securitization market and third (after Germany and Denmark) in the covered bond market. Yet, it is important to highlight that in Spain the originators of ABSs did not issue these securities as a means to transfer the credit risk of their loan portfolios to other investors, as it happened in other countries. On the contrary, these securities were issued for funding purposes. Between 2000 and 2007, the net flow of loans was persistently above the flow of deposits, and Spanish financial institutions relied on the bond markets as a way of funding this gap. Financial institutions relied not only on the securitization market, but also on other fixed-income markets for the same purpose, including notably the covered bond market (Blanco 2011, pp. 17–18).

While that model was successful for a while, in the end, one of the worse consequences of the crisis for Spanish banks and *cajas* was the evaporation of market liquidity. Its impact was particularly severe for ABSs because market liquidity is a very important characteristic of those assets. Shortening the maturity of these products was not possible with ABSs because the maturity of these bonds was the same as the maturity of the underlying assets. At the same time, another consequence of the financial crisis was the deterioration of the credit quality of financial institutions' portfolios. The impact of this factor was stronger in the case of ABSs owing to the lower level of protection in comparison with covered bonds (i.e., covered bonds have dual protection against losses, the issuer, and the underlying portfolio, but ABSs do not) (Blanco 2011, p. 19).

THE INTENSIFICATION OF THE CRISIS

Yet, Spanish banks did not survive unscathed this first stage of the crises. Indeed, Spanish banks and *cajas* were unable to escape the effects of the crisis. The weak underbelly was principally the *cajas*—unlisted in the stock market, regionally based, and often politicized savings banks—accounting for half of the financial sector's assets. During the boom years, they had successfully captured market shares from the banks, investing heavily in real estate (lending both to consumers and companies). In the end, as we will examine in detail in Chapter 6, the collapse of that sector was caused in large parts by the payment difficulties of debtors and mortgage holders.[15] In addition, *cajas*, highly dependent on international wholesale financing, were also forced to turn to the government and the European Central Bank for liquidity when wholesale markets froze, thus confirming the vulnerability associated with MBB liabilities.

The crisis also affected bank profitability, particularly in the Spanish market where their profits decreased 10% in 2009. Fortunately for the largest Spanish banks, their geographical diversification helped them to counterbalance their domestic losses. Indeed, as we have seen, the largest Spanish banks were active with foreign acquisitions during the previous two decades in the United States, Mexico, Brazil, and the UK, among other countries. For instance, about a fifth of Santander's profits came from Brazil and another fifth from the UK in 2008. This limited their exposure to the weak economy of the home country and allowed Spain's largest banks to fund themselves locally in countries where they had substantive investments, particularly throughout Latin America and the UK

And the loan delinquency rate continued to rise, reaching 5.35% in August 2010 and, for the first time during the crisis, the delinquent rate was higher for the banks (5.35%) than for the *cajas* (5.31%).[16] At the same time, the problems of the sector were reflected in the difficulties that they experienced raising funds in international markets. The European Central Bank (ECB) played a crucial role to compensate for this situation; the debt of Spanish financial institutions with the ECB reached 81.88 billion euros in March 2010. This unprecedented level of borrowing from

[15] A recent study from Vicente Cuñat and Luis Garicano shows that the main difference between banks and *cajas* was not so much the latter's political nature, but the lower level of professionalization of their managers: only 31% of their presidents has postgraduate studies, half of them had banking experience and half of them have occupied political positions before becoming presidents. According to them, this development means that *cajas* could have saved 12,000 million euros have they had better prepared and qualified managers and without a political past. *Cajas* with a political president have on average 0.93 points more of delinquent loans than those who do not, if the president does not have post-graduate studies 0.98 more, with no financial experience 0.93 more, and 2.84 more for those who meet the three conditions. See "La Politizacion eleva la morosidad de las cajas en 12.000 millones," *El País*, October 31, 2010.

[16] This data has been questioned, as many doubted whether the *cajas* were still recognizing the reality of their loan problems.

the ECB was a consequence of the struggle that Spanish financial institutions had in raising funds from the international capital markets and provided further evidence of the acute tensions in the Spanish banking system.

The results of the *cajas* continued to deteriorate during the first quarter of 2010; *Caja Madrid* announced a fall of almost 80% in revenues and *La Caixa*, 11.4%. The larger banks, however, continued growing, albeit many with lower profits. By May 2010, 34 of the 45 *cajas* were involved in merger and restructuring discussions; it was expected at the time that would lead to the creation of 11 new *cajas*. It also became apparent that simply merging or combining weaker institutions would not fully address the sector's woes. The Bank of Spain also pushed for significant restructuring that would deliver efficiency savings and cost cuts.

And the crisis extended to the banks as well. *Banif*, the private banking subsidiary of Banco Santander, suffered the collapse of Lehman because it had invested more the 500 million euros from its investors in Lehman's bonds. *Banif*, under threats of litigation from its customers, decided to compensate them partially, with the overall result that they lost 40 million last year. *Bankinter, BBVA, Bancaja, Altae* (*Caja Madrid*) and *Banco Sabadell* also sold these products to their investors and lost money, as well. In addition, *Banco Santander* and BBVA were also immersed in the *Madoff* fraud case: customers of Optimal, *Banco Santander's* Swiss-based hedge funds management, lost up to €2.3 billion through investments with Madoff; and BBVA announced that it may have lost up to €300 million. *Santander* had to announce a plan to compensate its clients. Private equity firms were also unable to float, refinance, or sell some of the leveraged companies that they bought during the boom years. Banks, who were also very dependent on international wholesale financing, were also forced to turn to the government and the European Central Bank for liquidity when wholesale markets froze.

Their biggest threat in at that stage of the crisis (prior to 2010) was their heavy exposure to the Spanish property crash, and their dependence on international wholesale financing (40% of their balance depends from funding from international markets). Prior to the crisis, they were so busy making money out of lending to the domestic housing market boom that the shifting market conditions were a surprise. Indeed, the situation deteriorated rapidly and it affected both banks and savings banks, but particularly the latter. The delinquent rate increased to 5% in 2009. In addition, *CCM* recognized a delinquent rate of 17.5% (in accounts that were considered technically irreproachable), which illustrates that there was still some legal margin to present better results and make-up results through the purchasing of properties, or the extension of mortgages, among other schemes.

Indeed, the exchange of debt for unpaid real estate assets at fictitious prices had become a somewhat typical instrument to hide their real delinquent rates for many institutions. According to some estimates, commercial banks exchanged debt for real estate assets valued at 10,000 million euros; and according to UBS if *Cajas de Ahorros* (saving banks) had done the same, their delinquent rate would be 5.6%-and not the stated 4.6%. *Cajas* also tried to preserve the accounting value of these assets through operations like *Aliancia*, a society established by eight saving banks to manage 200 million euros in real estate assets, with the acknowledged objective to take these assets out of their balance accounts and "not to lose money." Given this degree of 'manipulation,' it is not surprising that many analysts estimate that the delinquent rate is only half of what it would be is the accounts would be more transparent.[17]

Moreover, the very high level of unemployment (it reached 20% in 2010),[18] and the collapse of the real estate market, which constituted 60% of the banking loans (i.e., loans to families, to enterprises in the real state sector, or direct real estate assets), also increased the delinquent rate. This dire situation was compounded by the decrease in revenues caused by the collapse of interest rates and the lower chapter of business, which means that revenues decreased and also that banks had to take more provisions for portfolio losses. It is therefore not surprising that the governor of the Bank of Spain called for the urgent "restructuring" of the

---

[17] Vicente Cuñat and Luís Garicano, "¿Para cuando la reestructuración del sistema financiero español?" *El País*, September 13, 2009.

[18] In Spain, there has been a crucial link between unemployment and the health of the banking sector. The degree of 'bankarization' (the proportion of active and passive financial assets from all economic actors with banking intermediaries) of the Spanish economy is one of the highest of the OECD; and the source of business funding outside of the three main financial institutions (banks, *cajas* and credit cooperatives) is the lowest in the OECD. The increase in unemployment has been expected to cause further damage to financial institutions because it will lead to the further deterioration in the quality of their assets and their capacity to absorb additional risks (more than half the aggregated assets of the Spanish banking system is linked directly or indirectly to real estate assets: loans to families, to companies in that sector or direct real estate assets).

sector to address the overcapacity of the system, and tackle the irrational operations of savings banks, which in the absence of stockholders gained market share through unfairly competitive loan conditions.

In June 2009, the Spanish government created the *Fondo de Restructuración Ordenación Bancaria* (the Fund for the Orderly Restructuring of Banks, FROB), which theoretically paved the way for the restructuring of the Spanish financial sector. It created instruments to allow the Bank of Spain to reject the viability plans of financial institutions and to modify them, dismiss the managers, impose new management, and restructure institutions. The Bank of Spain also announced in July of 2009 that the 9 billion euros from the FROB could be leveraged up to 99 billion, with the aim to force a series of merges of *cajas*. Unfortunately, it took months for the situation to change significantly: For the first few months after the package was approved, no one requested any funds, and the only mergers confirmed were regional in nature and included relatively small institutions (i.e., in the first phase of consolidation, they included partnerships among: *Caja Duero* with *Caja España* and *Caja Burgos* in Castile-León; *Caixa Catalunya* with *Manresa and Tarragona*; and *Tarrasa, Sabadell, Girona and Manlleu* in Catalonia; *Unicaja* initiated the process to absorb *CajaSur* and *Caja Jaén* in Andalusia; and *Caja Navarra* and *Caja Canarias*, from Navarra and the Canaries, created a holding to share business and their risk and capital policies). One of the main reasons for the inaction was concerns about the role that Brussels would play in the merger processes (i.e., the former Competency Commissioner, Neelie Kroes, had demanded in November 2009 that the government define clearly the role and functioning of the FROB), and also there were questions on whether each merger would have to be approved by the EU to make sure that it would not violate European laws.[19]

Despite deteriorating conditions, the larger *cajas* refused at the beginning to take part in this process, and there was the widespread suspicion (based on what happened with CCM's accounts) that savings banks continued to manipulate their accounts, which did not inspire any confidence. In this regard, one of the main obstacles to the restructuring of the sector was the attempt to avoid direct conflict with the regional-autonomous governments that controlled many of these institutions (for

---

[19] Iñigo de Barrón, "El supervisor mete presión a las cajas," *El País*, October 18, 2009.

instance, Galicia appealed to the Constitutional Court against the FROB). The fact that their revenues were better than expected the first half of 2009 also lessened the pressure on these institutions.

However, as expected, the Bank of Spain used its most powerful instrument to promote the consolidation of the sector by forcing institutions to increase their provisions and reduce their benefits when they presented their end-of-the-year balances. The average decrease in 2009 was 50%; and the next two years were even more difficult: credit fell over 6% in 2010 and 4% in 2011; margins also declined as a result of the decrease of the Euríbor; international regulations were also requiring more capital; and the delinquent rate, particularly the one associated with real estate properties, continued to grow in parallel with the need to increase provisions. Therefore, there were clear signs that the worst was not over in 2009 because banks and *cajas* had underestimated the extent of their bad loan problems. As late as the end of 2009, the general consensus was that these institutions were still burdened with billons of Euros of bad loans extended to property developers, construction companies, and homeowners during the housing boom that peaked in 2007.

At that time, there were ten *cajas* with delinquent coverage lower than 40%, which represented 29% of the sector's assets; and 50 saving banks had coverage rates below 50% (including some of the leading ones, such as *Caja Madrid, Bancaja, CajaSur, Caja España, Caixa Tarragona, Caixa Girona, CAM*, and *Caja Canarias*). With these perspectives, there was no doubt that mergers were unavoidable in that sector to reduce branches and employment (there was consensus that they are overstaffed: Banks had almost the same market share but use 20,000 employees and 9000 branches less). According to analysts between 25 and 30% of the branches were not needed (or around 10,000 branches and 35,000 employees).

Moreover, the crisis of the Spanish financial sector was also manifested in the difficulties that Spanish companies and individuals were having in obtaining credit: By the end of 2009, four out five small and medium enterprises (SMEs) that had requested credit from the sector (banks, *cajas*, and credit cooperatives) had problems to obtain it, 17% of the requests were denied (and it was already 12% at the end of 2008), and for 59% of them credit was reduced, as well as the time to pay it back. That "credit dry" was the result of the dependency of the Spanish banks on international markets for funding (in the following three years they had to renew 400,000 million euros); the weight of the credit to real estate related activities in banks' balance accounts (as we have seen, they represented 20%) and the negative perspectives for the sector in the short term; and finally, the fact that although the Spanish financial sector initially faced the crisis with capital ratios significantly higher than those of their competitors, they were finally confronting the progressive erosion in the value of their assets, and they had to increase their provisions, and the prospects for reduced business activity.[20] The lack of credit, in turn, made the economic recovery process slower and more uncertain.

By the end of 2009, there was little doubt that a round of mergers was imminent. Deteriorating economic conditions pushed some *cajas* to the brink and gave further ammunition to regulators to demand changes. According to the Bank of Spain, the coverage against delinquent loans (defined as those in which at least a quota has not been paid in 90 days) directly linked with the real state sector, which represent 9.8% of the total (or 42.8 billion euros), increased to 41.4%. The Bank also identified 59.7 billion in other sub-standard loans (operations that show some weakness). In total, the Bank estimates that 165.5 billion euros were exposed to the constructions sector and are considered problematic loans.[21] And the delinquent rate continued growing: By April of 2010, it reached 5.387%, the highest level in 15 years, with the *cajas* suffering the highest levels at 5.518% (banks' were at 5.40% and credit cooperatives at 4.089%). As a result, the Bank of Spain continued exercising pressure on the *cajas* to restructure the sector, and its Governor, Miguel Angel Fernandez Ordonez (MAFO), led the charge to push them to merge and consolidate, while demanding greater transparency on the impact of the crisis on the balance sheet of these institutions.[22] *Citi* reported that the Spanish

[20] Alfonso García Mora, "El gran ausente," *El País*, November 22, 2009.

[21] "El Banco de España insta a acelerar el cierre de oficinas," *El País*, March 26, 2010; and "la banca acumula 60,000 millones en activos inmobiliarios por los morosos," *El País*, March 27, 2010. According to the bank of Spain, Banks and *cajas* have given 445 billion in loans to real estate and construction companies, of this the risk is estimated at 402 (the rest is in investment in roads and trains, port infrastructures, oil, and electricity).

[22] In a speech that he gave in Valencia on March 25, 2010, MOFO demanded a reform of the financial system and called on the *cajas* to seed up the merging processes and close some branches, and demanded a new law to regulate the sector. See "El gobernador saca el mazo," *El País*, March 28, 2010; and "El Banco de España insta a acelerar el cierre de oficinas," *El País*, March 26, 2010.

cajas needed between 24 and 34 billion euros from the FROB to recapitalize.[23] Moody's, concerned about the economic prospects, announced in December 2010 the possibility to reduce the ratings of 30 Spanish financial institutions.

Spanish *cajas* started to respond to these pressures and merge initiatives accelerated after March of 2010, with the regional governments of Catalonia, Extremadura and Galicia taking the lead. By the beginning of the year, 27 *cajas* (of 45) were involved in merge processes. Many *cajas* also started to downsize their staff and reduce costs (for instance, *Caixa Girona* announced in March 31, 2010 that it would reduce its staff by 5% and would close 15 non-profitable branches). The president of one of the leading *cajas*, Rodrigo Rato (president of *Caja Madrid*) stated that he expected that by the end of the crisis only 20 *cajas* would exist in the country.[24] In Galicia, the regional government announced at the end of March the predisposition to merge *Caixanova* and *Caixa Galicia*; and *Caja de Extremadura* initiated a merge process toward a so-called virtual merge (*fusión virtual o fusion fria*) using the Institutional Protection System (SIP, under the Spanish acronyms), which did not force the merged institutions to eliminate their administrative boards or the branch network, but forced them to merge their risk models, computer systems, capital, and at least 40% of the benefits. It also forced them to sign a mutual solvency commitment, and to be part of the merged institution for a minimum of ten years; as well as to combine some activities and functions like the politics to manage risks. In Catalonia, the boards of the cajas *Catalunya*, *Tarragona* and *Manresa* approved their merge in May 17, 2010, and they received the approval to merge from Brussels. *BBK* started working to create a SIP with *Caja del Mediterraneo* (CAM) to establish the third largest *caja* in Spain, which failed; and later *Cajastur*, *CAM*, *Caja de Extremadura* and *Caja Cantabria* announced a SIP merge. *Caja Navarra*, *Cajacanarias* and *Caja Burgos* created *Banca Cívica*; and *CajaSol* and *Caja Guadalajara* also started negotiating a SIP. In December 2010, *Unicaja* (which had already absorbed *Caja Jaén*) and *Ibercaja* announced negotiations to merge; and *La Caixa* announced its

---

[23] According to a Citi's report the only institutions that did not need funds were *BBK, Cajastur, Kutxa, Unicaja y Vital*. See "El FROB dará a las cajas entre 24,000 y 34,000 millones de euros para recapitalizarse según Citi," *El País*, June 7, 2010.

[24] "Rato vaticina que al final de la crisis sólo quedaran 20 cajas," *El País*, April 14, 2010.

intention to explore the possibility to become a bank. Even the unions supported the restructuring of the sector: *Comisiones Obreras* (CC.OO.) announced publicly in April 3, 2010 that it would support the entrance of private capital in the *cajas*.

Yet, despite all these developments the delinquent rate continued increasing to 5.38% by February of 2010 (1.17 point higher than in 2009), and 5.35% in August 2010 (and for the first time during the crisis the delinquent rate was higher for the banks, 5.35%, than for the *cajas*, 5.31%) [The previous record was established in April 1996, at 5.43%]. At the same time, the problems of the financial sector were reflected in the difficulties that they experienced raising funds in international markets. The European Central Bank (ECB) played a crucial role to compensate for this situation: the debt of Spanish financial institutions with the ECB reached 81.88 billion euros in March of 2010 (it was 12.3% higher than in March of 2009, and it represented 15% of the total debt of the Eurozone with the ECB); 74.6 billion in April; and 85.6 billion in May. This unprecedented level of borrowing from the ECB was a consequence of the struggle that Spanish financial institutions had raising funding from the international capital markets, and it provided further evidence of the acute tensions in the Spanish banking system. The 85.6 billion that they borrowed in May was double the amount lent to them before the collapse of Lehman Brothers in 2008, and it was the highest amount since the launch of the Eurozone (Spanish banks make up 11% of the Eurozone banking system).[25]

The Bank of Spain continued its heavy-handed approach to overcome resistance to reform and forced *Cajasur* to continue with is merge discussion with *Unicaja*. MAFO complained publicly about the delays in mergers and takeovers and reminded Spaniards that they could intervene and take control of a lender if it had 'dire problems of viability.'[26] Despite these pressures, the board of *CajaSur* rejected in May the merge with *Unicaja*, a decision that caused enormous consternation across the financial sector, and led the Bank of Spain to intervene it to sell it to the highest bidder after re-capitalizing it with the 523 million euros that it needed to reach the legal minimum.

[25] "Spanish Banks Break ECB Loan Record," and "Turmoil in Spain Sparks Fear of Crisis Spreading," *Financial Times*, Wednesday, June 16, 2010, p. 15.

[26] "Madrid Push for Faster Savings Bank Consolidation," *Financial Times*, March 18, 2010, p. 14.

The Bank of Spain was using every opportunity to issue careful warn-ings about the financial system's 445 billion euros exposure to the domestic property and construction sectors. In May 2010, in order to accelerate the sluggish restructuring efforts, the Bank of Spain announced plans to tighten rules on how banks could value acquired properties and nonperforming loans. It decided that lenders would have to provision for 100% of a bad loan within a year of it being listed as doubtful (the previous period was between two and six years). At the same time, banks would now be able to offset loan collateral (typically land or property) against the provisions but only after a significant write-down on the value of the guarantee. Finally, banks would also be obliged to increase their write-downs on assets acquired in payment of debts, unless the owner receives a higher bid, they must be written down immediately by at least 10%, and increase it to 20 and 30% over the following two years.[27]

By March 2010, bad loans associated with the construction/real estate sectors had reached 9.6% of the total and 14% of the credits were considered 'substandard.' The leaders of the Bank of Spain leaders were increasingly concerned about the possible hiding of bad loans while most analysts agreed that the level of nonperforming assets was higher than the official numbers, especially among small- and medium-sized *cajas*, which had tended to carry out assets swaps by buying real state, taking control of property developers, and extending loan maturities to protect borrowers from bankruptcy, instead of writing off bad debts (for instance, in the previous crisis of the early 1990s the reported ratio was 9% in 1994 but the real ratio was closer to 13 or 15%). Furthermore, although home construction has slowed down since 2007, it was expected at the time that it would take another five to six years to absorb the surplus stock.[28] As a result of these doubts, Standard & Poor moved in March 2010 the Spanish banking industry rating down a level to its group 3, putting it on a par with those of the United States, the UK, and Portugal, and Moody's also warned of a possible downgrade.[29]

[27] "Banks Become the Country's Biggest Landlords," *Financial Times*, Wednesday, June 8, 2010. Special Section on Spain, p. 9.

[28] A recent survey from idealista.com found 600,000 completed, unsold new homes in the country, and the crisis is also affecting commercial property.

[29] "Crunch Time Looms for Spanish Lenders," *Financial Times*, Thursday, March 18, 2010, p. 14.

Meanwhile, the results of the *cajas* continued deteriorating during the first quarter of 2010: *Caja Madrid* announced a fall of almost 80% in revenues, and *La Caixa* 11.4%. The larger banks, however, continued growing, albeit many with lower profits. Banesto's profits reached 211.54 billion in the first quarter of 2010, a 0.3% increase compared to the same period in 2009; and Santander benefited from its geographic diversification increasing its revenue by 5.7% compared with the first quarter of 2009 (in Spain, it decreased by 8.2%, but it increased in Brazil by 38%), while it continued with its expansion abroad (in June 2010, it announced the purchase of the 24.9% that Bank of America had on its Mexican subsidiary-Santander Mexico). The delinquent rate, however also increased: from 2.94 at the end of 2009, to 3.12% in the first quarter of 2010. Yet, these figures are disputed: for Santander Spanish operations analysts put the ratio between 7.5 and 8.5% of assets against the reported 4.4%.

Against this background, Spanish authorities continued pushing for the restructuring of the sector. In March 17th Elena Salgado, Vice president and Minister of Finance, joined the calls for faster consolidation of the weaker *cajas*, admitting in parliament that a third of the lending institutions could face 'solvency problems' and stating that 'the government is going to do everything it can to accelerate [banking reforms].'[30] By May of 2010, 34 of the 45 *cajas* were involved in merging and restructuring discussions that was expected to lead to the creation of 11 *cajas*. In June, *Caja Madrid* and *Bancaja* announced their merge through the creation of a SIP, which led to development of the largest Spanish *caja*, and were joined by 5 smaller *cajas* (*Caja Insular de Canarias, Caja Avila, Caixa Laietana, Caja Segovia,* and *Caja Rioja*). It also became apparent that simply merging or combining weaker institutions would not fully address the sector's woes. The Bank of Spain also pushed for drastic restructuring processes that would deliver efficiency savings and cost cuts.

As we have seen in May 22, 2010, the Bank of Spain was forced to seize *CajaSur*, a *caja* based in Cordoba, in southern Spain, a small (it accounted only for 0.6% of Spain's banking assets) but peculiar institution controlled by the Catholic Church and chaired by a priest. By end of 2009, *CajaSur*'s bad loan ratio reached 10.2% of its assets, double the Spanish average, and its tier one capital ratio was just 1.94% below

---

[30] "Madrid Push for Faster Savings Bank Consolidation," *Financial Times*, March 18, 2010, p. 14.

the legal minimum requirement and it faced a capital shortfall of 523 million euros. With this decision, the central bank sent a clear message that it would not tolerate further delays to the mergers and restructuring processes among the unlisted *cajas*. The established deadline was June 30, 2010.

The implications of this intervention were severe: Bank stocks were marked down globally, and the euro was also pushed down. This crisis showed the global implications of a local crisis and highlighted how slow European authorities had been in cleaning up the banking sector. The failure of *CajaSur* also showcased the dire state of Spain's property market, to which the saving banks are severely exposed, as we have seen. However, it illustrated not merely the financial institutions' exposure to economic conditions in individual countries, but also the interlinked nature of their exposure, because many of these bad loans were packaged as covered bonds or mortgage-backed securities and sold them to German, French and other investors.[31] This seizure rushed other *cajas*, who were eager to unveil co-ordination agreements in an attempt to stave off a similar fate. Within days of the seizure decision four *cajas* (*Caja Mediterráneo, Cajastur, Caja Extremadura* and *Caja Cantabria*) announced plans to pool their operations in a joint holding group; and *Caixa Girona* started negotiating with *La Caixa*.

Nonetheless, deteriorating economic conditions led to downgrades of the country's debt (in April 2010 *Standard & Poor* downgraded Spanish sovereign debt because of the country's sluggish outlook for economic growth, a decision that was followed by *Fitch* in May), which forced the government to approve a 15 billion euros package in spending cuts at the end of May 2010 in an effort to bring the deficit down to 6% of GDP by 2011, from (11% in 2009), including pay cuts of about 5% for civil servants and 15% for government ministers among other measures. The Socialist government wanted to demonstrate with this decision that it could cut its bloated deficit and avoid emergency aid.

The government also enacted reforms aimed at bolstering the *cajas* and designed to dilute the influence of local governments in these institutions. On July 9, 2010 the government approved a royal decree of the *Ley de Órganos Rectores de Cajas de Ahorros* (the LORCA, the Law of the Governing Organs of Saving Banks), which represented the most

---

[31] "Leaning lenders," *Financial Times*, Friday, June 4, 2010, p. 9.

sweeping reform of the sector in 170 years (it was approved by the Spanish Congress on July 20th). The objective of this law was to make the *cajas* "more professional and democratic" and to push for the depoliization of the sector (at the time between 25 and 35% of the managers were political appointees from the regional governments, city halls and *diputaciones*). As a result of these reforms, elected and government officials were banned from sitting in the boards of these *cajas* (they were given three years to implement it), but the regional parliaments could still name their representatives in the *cajas*, so they would still maintain some political influence (although the maximum quota was reduced from 50 to 40% of the board's voting rights). Furthermore, if the *cajas* wanted to stay in banking they need to demutualize and establish shareholders structures in which the largest stakes would be capped at 40%. *Cajas* were also able to sell up to 50% of their equity to private investors (under previous rules the *cajas* were restricted to selling non-voting securities known as 'participative quotas' as well as normal shares in listed industrial holding subsidiaries).[32]

This reform established four possible models for the *cajas*: Under the first one, they could maintain their condition of *caja*, with the new quota regime and adapt their statutes to the corporate governance regulations. Under the second one, they could become part of the *Sistema Institucional de Protección* (SIP, Institutional System of Protection), popularly known, as we have seen above, as a 'cold merge.' Under the third one, they could maintain its status as *caja* but transfer all their banking businesses to a bank, while maintaining 50% of the subsidiary's stock and keeping the social work and the industrial portfolio in the *caja*. Under the fourth one, a *caja* could become a foundation transferring all its business to a bank in which it held a participation of less than 50%. Those *cajas* that opt for a 'cold merge' would have a central body, which would be a bank, which would be participated by at least a 50% by the *cajas*. Finally, if they sell more that 50% in the stock market, they would also lose their *caja* status and should be transformed into foundations.[33]

This new law also affected Spanish banks. Indeed, the government also took advantage of this reform to introduce a significant change in

---

[32] "Spain to Let *cajas* Sell 50% of Equity," *Financial Times*, July 9, 2010, p. 17.

[33] "El Gobierno abre la puerta a la privatización total de las cajas," *El País*, July 10, 2010.

banking regulations: The law gave powers to the Bank of Spain and the Ministry of Finance to impose "the obligation to have available a minimum amount of liquid assets that would make it feasible to confront potential fund exits derived from commitments and liabilities, including in the case of stress, and to maintain an adequate structure of financial sources and maturities." In addition, the Bank of Spain could impose "a maximum limit to the balance between the proprietary resources of the institution and the total value of its exposure to the risk derived from its activities" to avoid excessive risks in relation to the banks or cajas' capital that may jeopardize their future. The liquidity coefficient and the exposure limit would not be the same for all banks and *cajas*: those institutions that had more difficulties to raise capital in the markets (most likely the smaller institutions), would have higher requirements than those who have easier access to capital. The aim of this reform was to avoid a repetition of the situation in which banks were facing liquidity tensions due to the closure of international financial markets. This measure was considered an advance of the Basil III regulations, approved later in the summer.[34]

In June, the Spanish government and BBVA admitted that the country's banks were on the brink of a funding crisis because they were struggling to gain funding from the international capital markets. There were particular concerns surrounding the *cajas*, still very exposed to the ailing property sector, as shown in the rising cost to borrow for Spain in the debt markets, and the rise in the costs for the country and the banks to insure their bonds against default. In June 2010, it cost Spain twice as much to borrow from capital markets over two years as it did in the middle of April. The increase in costs fueled concerns that the country would be forced to use emergency funds from the recently created EU 440 billion euros European Financial Stability Facility (EFSF). However, any move by the country to tap the fund would further undermine the already shaky confidence of investors in the market, which led Spanish and European leaders to stress, yet again, that "financial markets shouldn't make the mistake of establishing and equivalence between Greece and

---

[34] "El Banco de España exigirá a la banca un nuevo coeficiente de liquidez," *El País*, July 14, 2010.

Spain."[35] Fortunately for the country, markets were not too alarmed and Spanish bond yields remained relatively stable in June: the benchmark 10-year bond yields rose 6 basic points to 4.73%.

Concerns over the strength and solvency of its financial sector led Spanish authorities to push for the publication of the 'stress tests' that the main European financial institutions had undergone in 2010. To counter speculative news, like the news from *Frankfurter Allgemeine Zitung* which published in June 2010 that Spain was planning to request funds from the EU stabilization fund, MAFO announced in June 2010 that the Bank of Spain would make public the results of these tests, and that, contrary to what happened in other European countries they would 'cover all financial institutions.' Partly as a response to this pressure the Spanish (and German) governments, the EU agreed to publish the results of the Spanish banks' 'stress tests.'[36]

When the results of these tests were made public on Saturday, July 23, 2010, only 5 of the 17 *cajas* had failed them: *Banca Cívica, Banca Espiga, Caja Catalunya's group, Unimm,* and *Cajasur*. It is important to note, however, that although five of the seven institutions that failed the tests were Spanish saving banks, it was partly because all Spanish lenders were subjected to the tests. Indeed, as noted before contrary to other European countries, Spain subjected 95% of its banking system to these stress tests (the average in the EU was 50%). The overall good result of the Spanish financial system confirmed the benefits of the countercyclical capital regime described above, which proved to be the best rescue plan for the Spanish banks because it allowed many of these institutions to pass the 'stress tests' (see Table 5.1). The 27 banks and *cajas* that were included in the tests still held 19,796 million euros in this kind of provisions, which would be available to them in the case of a deepening crisis. The response of the markets to these results was very positive and helped to push down the country's risk premium.[37]

---

[35] Jean-Claude Juncker, chairman of the Eurozone finance ministers group, in "Turmoil in Spain Sparks Fear of Crisis Spreading," *Financial Times*, Wednesday, June 16, 2010, p. 15.

[36] "New Concerns as Spain's Borrowing Costs Increase," *New York Times*, June 16, 2010, p. 1.

[37] "Todos los bancos españoles aprueban los exámenes de resistencia en Europa," *El País*, July 23, 2010; "Los bancos y cajas disponen aún de un 'colchón' extra de

Table 5.1 The provisions of Spanish banks (In millions of Euros)

	Assets weighted at risk	Specific provisions	Specific assets (%)	Generic provisions
BANKS				
Pastor	18.713	724	3.9	304
March	9.488	120	1.3	140
Bankinter	30.665	482	1.6	397
Guipuzcoano	7.814	200	2.6	99
Santander	585.346	14.052	2.4	6.727
BBVA	311.126	7.152	2.3	2.995
Popular	92.571	2.337	2.5	850
Banco de Sabadell	57.958	1.719	3.0	407
Total banks	1.113.681	26.786	2.4	11.919
SAVINGS BANKS				
Colonya-Pollensa	183	4	2.2	3
Ontinyent	688	16	2.3	11
Ibercaja	25.291	588	2.3	380
Caja Sol	21.237	548	2.6	312
BBK	19.202	286	1.5	272
Unicaja	21.909	911	4.2	309
CAM, Cajastur and others	83.865	3.091	3.7	1.072
Duero - España	28.852	1.106	3.8	353
La Caixa - Girona	162.979	2.582	1.6	1.874
Sabadell, Terrassa, Manlleu	18.349	559	3.0	201
Murcia, Penedes and others	44.854	1.436	3.2	430
Banca Civica	30.09	820	2.7	251
Caja Madrid, Bancaja and others	213.929	5.435	2.5	1.713
Círculo, Badajoz, CAI	14.994	472	3.1	110
Kutxa	16.100	454	2.8	94
Catalunya, Tarragona, Manresa	49.108	2.200	4.5	267
Vital Kutxa	6.652	155	2.3	36
Galicia - Caixanova	46.890	1.880	4.0	162
Cajasur	12.141	794	6.5	27
Total savings banks	817.313	23.337	2.9	7.877
GENERAL TOTAL	1.930.994	50.123	2.6	19.796

Source Bank of Spain

Yet, the crisis continued affecting the bottom line of the financial institutions: Spanish *cajas* earned 26% less in the first six months of 2010 than in the first semester of 2009. Furthermore, the consolidation process of the *cajas* intensified throughout the summer. In July, *CAM* announced its intention to move forward with the merge with *Cajastur* despite the opposition of the labor unions. In Catalunya, *Caixa Catalunya, Tarragona* and *Manresa*, created *CatalunyaCaixa*; and in September the assemblies of seven cajas (*Caja Madrid, Bancaja, Caja Insular de Canarias, Caja Segovia, Caixa Laetana, Caja de Ávila*, and *Caja Rioja*) approved their integration and the creation of a bank that would start operations in 2011. As of December of 2010, there were 16 *cajas* left versus the 45 that existed just six months before when the merging process started.

CONCLUSION

The performance of Spanish banks during the initial stages of the crisis (2007–2010) served to highlight the benefits of the 'Spanish Method,' which is now generalized all over the world. There was growing consensus that banks were under-capitalized in the run-up to the crisis, and hence most counties were now subjecting their banks to higher capital requirements. At the same time, they made regulatory rules less pro-cyclical, tightening liquidity requirements; trying to make sure that governments and regulators pay closer attention to the build-up of risk across the financial system as whole, and finally preventing off-balance sheet activities

However, it is necessary to acknowledge that no model is perfect. Spain suffered a property-linked banking crisis exacerbated by financing obstacles from the international crisis. Indeed, as we will examine in detail in Chapter Six, the countercyclical capital regime did not prevent the exposure of financial institutions to the housing bubble. Mortgages with low down payments can be uninsured, which may be particularly problematic in the case of a less than robust mortgage market with a prevalence of variable-rate and interest-only mortgages. The crisis exposed the vulnerability of the financial sector to the real estate market: Banks as *cajas* faced 96,934 million euros in questionable credits by the end of 2009. Furthermore, fund managers, hedge funds, and broking industries also suffered

20,000 millones," *El País*, Monday, July 26, 2010, p. 15; and "Premio a la transparencia española," *El País*, Tuesday, July 27, 2010, p. 18.

closures and setbacks; they were also affected by Madoff scandal; private equity firms had difficulties floating, selling and refinancing leveraged companies; and banks dependent on interbank and wholesale markets for financing had to resort to the government and the ECB.

Furthermore, as we have seen, the crisis was particularly severe for the 45 *Cajas de Ahorros*. These institutions had very close ties to regional governments and often faced political pressure to fund projects. Consequently, they served credit as the housing bubble inflated and they gained market share. By September 2009, they held $330 billion in loans to developers (up from $50 billion in 2000). As a result of the crisis, two of these institutions, *Caja Castilla la Mancha* and *Cajasur*, had to be bailed out and taken over by the Bank of Spain. Nearly half of their $1.8 trillion in assets are mortgages or other real estate loans, and they were exposed to the 325 billion euros stock of dubious loans. These fears materialized as 7% of their loans went bad in 2010 (5.1% in 2009) due to growing defaults, and they lost $3.4 billion that year, in the face of pummeling housing prices. The collapse of many of these institutions led to the country's financial bailout in June 2012. We will examine next the causes for this collapse which was rooted in the political bargains in which incentives and a lax regulatory framework favored developers, property owners, and bankers.

REFERENCES

Barrón, Iñigo. *El Hundimiento de la Banca*. Madrid: Catarata, 2012.

Blanco, Roberto. "The Securization Market in Spain: Past, Present and Future." In M. Chavoix-Mannato "Working arty on Financial Statistics: Proceedings of the Workshop on Securitisation." OECD Statistics Working Papers (2011/03), OECD Publishing, 2011.

Carballo Cruz, Francisco. "Causes and Consequences of the Spanish Economic Crisis: Why the Recovery Is Taken so Long?" *Panoeconomicus* 3 (2011): 309–28.

Deeg, Richard, and Sofía Pérez. "International Capital Mobility and Domestic Institutions: Corporate Finance and Governance in Four European Cases." *Governance* 13, no. 2 (2000): 119–53.

Fernández Ordóñez, Miguel. "The Challenges to the Spanish Banking System in the Face of the Global Crisis." Lecture on the occasion of the 50th anniversary of ESADE. Barcelona, October 30, 2008.

Lukauskas, Arvid. *Regulating Finance*. Ann Arbor: Michigan University Press, 1997.

McGuire, Patrick, and Goetz von Peter. "The US Dollar Shortage in Global Banking." *BIS Quarterly Review*, March 2009.

Pérez, Sofía A. *Banking on Privilege*. New York: Cornell University Press, 1997.

Pérez, Sofía A., and Jonathan Westrup. "Finance and the Macroeconomy: The Politics of Regulatory Reform in Europe." *Journal of European Public Policy* 17, no. 8 (2010): 1171–92.

Royo, Sebastián. "A 'Ship in Trouble' The Spanish Banking System in the Midst of The Global Financial System Crisis: The Limits of Regulation." In *Market-Based Banking, Varieties of Financial Capitalism and the Financial Crisis*, edited by Iain Hardie and David Howarth. New York: Oxford University Press, 2013.

A 'Ship in Trouble': The Spanish Banking System in the Midst of the Global Financial System Crisis (2010–2012)

INTRODUCTION

It is rare that central bankers refer to shipwrecks when talking about the institutions they oversee, however, Miguel Angel Fernández Ordoñez (MAFO), governor of the Bank of Spain (BoS), described his job dealing with the financial crisis as 'doing a double job on *a ship in trouble*, at the same time as ordering the evacuation of the passengers it was necessary to repair the lifeboats.'[1] This description summed up the situation. As we examined in the previous chapter, in the run-up to the global financial crisis, the Spanish financial system earned itself a nice reputation. A culture of invasive supervision from the regulators, coupled with a Conservative policy and regulatory frameworks that required higher provisioning, meant that most Spanish banks were prepared with plump cushions to absorb the losses caused by the outset of the global financial crisis.

From: "A 'Ship in Trouble': The Spanish Banking System in the Midst of the Global Financial System Crisis: The Limits of Regulation," in *Market-Based Banking, Varieties of Financial Capitalism and the Financial Crisis*, ed. Iain Hardie and David Howarth (New York: Oxford University Press, 2013a); and Royo 2013 (Chapter 6).

[1] As quoted in "Spain Bank Chief Faces Reform Battle," *Financial Times*, April 16, 2012.

However, this proved short lived. When the crisis intensified, it exposed the weakness in the policy and regulatory frameworks and an over-reliance on wholesale funding, specifically for mortgages and construction; the financial system was unable to escape its dramatic effects.

This chapter aims to examine changes in Spain's national financial system and the extent to which it was affected by the international financial crisis of the late 2000s. It addresses the following question: What factors explain the way in which the Spanish financial system has been affected by the crisis and has reacted to it? This chapter argues that a key feature of market-based banking (MBB), namely the degree of "financialization" of national financial systems, defined in this chapter as the trading of risk, is instrumental in understanding the impact of the financial crisis. Financialization is operationalized by looking at Spanish banks' assets (i.e., the size of the trading book and the presence of toxic assets), and liabilities (i.e., the funding base of banks, their reliance on wholesale market rather than retail deposits for funding, the securitization of lending, and the use of structured investment vehicles (SIVs).

In Spain, banking exposure to wholesale markets had increased significantly over the decade that preceded the financial crisis and, therefore, it had become "market-based banking," with significant implications for the nature of credit provision and the character of the Spanish national financial system. The lower reliance on bank lending proved to be beneficial at the beginning of the crisis. As we have seen, while most market-based systems faced further problems in the wake of the Lehman Brothers collapse, Spanish banks performed relatively well. However, the crisis eventually had a major impact, once more "traditional" problems with government debt and lending emerged. This chapter analyzes this reversal and explains its causes.

As noted on Chapter 5, Spanish regulators placed emphasis upon pre-existing regulatory and supervisory frameworks, which initially shielded the Spanish financial system from the direct effects of the global financial crisis. This contributed to its initial positive performance compared with their European counterparts. However, as in Ireland, the collapse of the real estate market eventually led to a traditional banking crisis fueled by turbocharged lending by "market-based" banking on the liability side of the Spanish bank's balance sheets.[2] In order to explain the Spanish version

---

[2] The concept of "market-based banking" (MBB) is driven by market pressures on both the asset and liability sides of the banks' balance sheets. The global financial crisis and the

of MBB, this chapter focuses on the following factors: first, the regulatory framework; second, the institutional features of the banking system, including the role of the BoS; and third, the impact of macroeconomic developments, notably the real estate bubble. All in all domestic political institutions, the rules of the game, and the role of domestic players operating within those institutions were crucial to explain the crisis, which was the result of political bargains in which incentives and a lax regulatory framework favored developers, property owners, and bankers.

MBB has often been examined as the dependent variable. In the case of Spain, it is explained as a dependent variable, but it is also examined as an independent variable that can be wielded to help explain Spain's supposedly "traditional" bank crisis as well as the country's devastating sovereign debt crisis. This chapter shows that MBB allows us to better understand what happened during the crisis. MBB explains the dramatic increase in lending, which was also crucial to explain the Spanish crisis. In addition, MBB also explains the ongoing credit crunch in Spain: When the crisis hit, international and wholesale funding dried up affecting bank lending.

However, this chapter also argues that the outcome is historical and contingent: There have been other causal independent variables of Spanish political and economic life that can be wielded to explain the crises. For instance, in Spain current account deficits (which reached close to 10% of GDP just before the crisis) explain why MBB grew (i.e., why Spanish banks became so dependent on wholesale banking and international lending). Finally, one of the interesting aspects about the Spanish crisis is that MBB was having its impact prior to much awareness regarding the crisis.[3]

Finally, as noted throughout the book, Spain is unique among European countries in the fact that the largest banks did not face problems (and indeed, as described on Chapter 4, they were buying foreign banks

---

subsequent credit crunch had a simultaneous effect: On the one hand, it reduced banks' ability to borrow and therefore to continue lending, thus leading to the property collapse; on the other, the property collapse contributed to reduce banks' ability to borrow.

[3] Lending figures from the BoS suggest a tightening due to cost and availability of funding since mid-2009, and they have mostly have remained tight since them. The more gradual but constant tightening from late 2007 to mid-2009 was clearly more significant—which shows that despite all the talk at the time about the limited problems in Spain, beneath this there was a tightening that was hitting mortgage/construction lending hard and contributed to the bursting of the bubble.

early in the crisis). The real problem was largely at the next level down, with the *cajas* (or unlisted savings and loans). In this regard, Spain is different from Germany, where both the large private banks and the *Landesbanks* were badly impacted by the global financial crisis.

THE CANARY IN THE COAL MINE: HOW MORE "TRADITIONAL" PROBLEMS LED TO A GREATER CRISIS?

As we have seen in Chapter 5, in the run-up to the financial crisis, the BoS and the regulators earned themselves a fine reputation. The regulatory framework described on Chapter 1 and a strong culture of invasive supervision meant that most Spanish banks had significant cushions to absorb the initial losses caused by the global financial crisis. MAFO stated in October 2008, after the collapse of the banking system in many other countries, that 'the sector is in a strong starting position, and the BoS has the capacity and the tradition to know how to resolve the most complex situations without trauma for the depositors, or the overall economy.' He added: 'I am convinced that our entities, which knew how to manage the expansive cycle, will also know how to adopt the management decisions and implement the adequate strategies that will allow them to confront the difficulties with success.'[4] They did not, and the Socialist government did not either. Spanish banks would come to regret such complacency (see Banco de España 2017).

Indeed, the relative success in dealing with the crisis proved short-lived. When the economic crisis deepened, it led to a traditional banking crisis caused by the collapse of the real estate market, record unemployment, increasing government debt, and difficulties accessing credit in wholesale markets. The real estate boom-bust cycle, which materialized in particular in the *cajas* sector, exposed weaknesses in the policy and regulatory frameworks, as well as the sector's over-reliance on wholesale funding. In the end, the financial system was unable to decouple itself from the economic cycle and the huge macroeconomic crisis that has besieged the country. This section seeks to account for this reversal.

The deteriorating situation is illustrated by the interest margin of the Spanish banking system—the difference between what banks earn on

---

[4] He also warned of the need to clean up the bad loans, in "Tres años, cuatro reformas y varios cadáveres por el camino," *El País*, May 11, 2012.

loans and the cost of funding—which fell about 20% since July 2007 to 0.86 percentage points at the end of 2011, the lowest since 1970 when the BoS started to compile the data. For the first time since 1985, return on equity was negative at the end of 2011, as a result of the economic crisis and increasing provisions against losses on real estate debt.

However, it is important to highlight that by 2012 Spanish banks had made significant progress in trying to address the consequences of the crisis. By the summer of 2012, banks had accumulated 112 billion euros in extra provisions for bad property loans since 2007, which increased to 147 billion by the end of 2012, or the equivalent of 14% of GDP; they had higher capital ratios; 5700 branches had been eliminated, a 12% cut (as a result 2656 little towns lost their branches and 4.8 million citizens did not have branches in their municipalities), and a 10% cut in staff (30,172 jobs were lost between 2008 and 2012 in the financial sector, and the consolidation efforts described in the previous chapter resulted in the absorption through mergers of 30 weaker institutions, all but two of them *cajas*: As of September 2012 there were only 9 *cajas* left, as a result of this consolidation process (from the original 45), and all of the them but two small ones (*Caja de Ontinyent* [Valencia] and *Pollença* [Mallorca]) had transferred to banks their financial businesses.[5]

Yet, by September of 2012 the problem with real estate 'toxic' assets had resulted in the intervention and nationalization of eight financial institutions. Between 2009 and 2010 two *cajas* were intervened (*Caja Castilla la Mancha* and *Cajasur*), in 2011 four more were nationalized (*CAM*, which was also intervened, *Unnim, Catalunyacaixa, and Novagalicia Banco*). *Banco de Valencia* (controlled by *Bancaja*) was intervened in 2011 and nationalized in 2012. Finally, *Bankia* was nationalized in 2012. Altogether by May 9, 2012, the reorganization of the sector had involved 115 billion euros from the Spanish government. Of this amount, half were guarantees, 19.3 billion were used to buy assets, 14.346 were direct assistance from the FROB, and 400 million were the losses from *Cajasur*, which were absorbed. Finally, the financial institutions themselves used 119 billion euros of their own capital to

5 "Spain bank chief faces reform battle," *Financial Times*, April 16, 2012; and "El cierre de cajas dejará sin oficina a casi cinco millones de españoles," *El País*, July 16, 2012.

clean up their accounts and comply with new provision requirements.[6] As we have seen, the main problem was with the *cajas* (see Table 6.1).

Furthermore, at that time Spanish banks continued to suffer from a range of problems:

1. *Adverse operating conditions*: consequence of the second recession in two years, the real estate crisis, and persistently very high levels of unemployment.
2. *Reduced creditworthiness of the Spanish sovereign*, which impacted banks' standalone profiles and affected the capacity of the government to support banks.
3. *Rapid asset-quality deterioration*: nonperforming loans to real estate companies increased rapidly and mortgages also deteriorated.
4. *Restricted market funding access in wholesale markets:* intensified as a result of the Euro area debt crisis, and growing investor concerns about Spanish banks.

On Wednesday, May 9, 2012, the government was forced to nationalize *Bankia* the country's largest real estate lender (and the result of the merge of several *cajas*), which again validated the concerns about insufficient regulatory oversight, and the perception that banks and the BoS had played down the risk posed by real estate loans. This was the largest bank nationalization in the country's history. *Bankia*'s nonperforming loan rate was running at 8% at the time of the nationalization. The cost of the nationalization was expected to reach 23 billion euros. This nationalization showed that some of the *caja* mergers that the BoS and the previous government had pushed for were dysfunctional and not viable. Billed as the "leader of the new banks" and as evidence of the success of the Socialist government's crusading banking reforms, *Bankia* proved to be a monumental fiasco and it showed the refusal to acknowledge the true extent of the country's real estate problem. Ultimately, concerns about Spanish banks' undercapitalization led to the announcement of an EU banking bailout for Spain in June 2012 (see Chapter 1).

---

[6] "Parte de guerra: ocho entidades intervenidas o nacionalizadas en España," *El País*, May 9, 2012.

Table 6.1 A chronology of crony capitalism among *cajas*

Caja	Collapse	Features	Problems	Controversies	Outcome
Caja Castilla-La Mancha	• March 29, 2009 • It had 26.38% of defaulted loans; deposits down by 12.51%, 1.68 billion in landholdings • Costs of collapse: 2.475 billion in loan guarantees and 1.3 billion in capital • Lost 75% of the assets accumulated for decades • It has broken the legal requirements regarding risk	Group of 5 **cajas**. Ranked 12th in the sector, and with 19 billion euros in assets	Inefficient and outdated risk assessment systems, excessive investment in the real estate sector; rapid expansion outside of its region, particularly in Levante; very dependent on wholesale funding	• Rumors about its collapse led to deposit withdrawals in Yébenes • Campaigns attributed to the PP against the solvency of the *caja* and to promote deposit withdrawals • Merge attempts with Ibercaja and Unicaja failed • PwC report found a 3 billion whole in the accounts • BoS refused to lend it 2 billion to merge with Unicaja • Confrontation between BoS and the Junta de Andalucía over the merge with Unicaja • Fondo de Garantía de Depósitos (FGD) paid for the losses (FROB did not yet exist) • The risks were known but the supervisors did little to address them	Nationalized and later absorbed by **Cajastur** • FGD paid 4.125 billion, and later when it was absorbed by Cajastur (which is now part of the **Liberbank** group), it received 1.493 billion from the FROB

(continued)

Table 6.1 (continued)

Caja	Collapse	Features	Problems	Controversies	Outcome
Cajasur	• May 21, 2010 • 596 million in losses in 2009, and 1.1 billion in losses in 2010 • In 2009 1.7 billion in the real estate sector (4000 houses and 2 million square meters in landholdings	Controlled by the Catholic Church: It had six priests in the board, and it was presided by a priest for 30 years (Miguel Castillejo, who was replaced in 2005 for another one, Santiago Gómez Sierra)	• Excessive investment in real estate • As early as 2004 the default rate was 3.5% (the average in the sector was 0.6%). In 2010 it reached 10.4%, doubling the average • Excessive concentration of credit in questionable individuals (Rafael Gómez: 400 million in 2004), and construction companies (Prasa: 209 million, and Sánchez Ramade: 143 million), exceeding the legal concentration of risk (25%)	• Massive deposit withdrawals • The board refused to heed the BoS request to merge with Unicaja • Its president requested the BoS intervention, in effect killing the caja, instead of approving the merge with Unicaja (he had a difficult relationship with Braulio Medel, president of Unicaja) • Worked with promoters, like Rafael Gómez, who were later indicted for their involvement in real estate scandals • BoS opened expedients against 38 former board members • Four CEOs between 2007 and 2009 • BoS inspections (4 between 2005 and 2008) not enough to force changes	Sold to the Basque **BBK** • 752 employees dismissed • It received 800 million in loans from the FROB that it has to repay • BBK requested 392 million from the FROB to confront the losses from bad loans

Caja	Collapse	Features	Problems	Controversies	Outcome
Caja Mediterraneo-CAM	• July 22, 2011 • From 244 million in benefits in 2009 to 2.713 billion in losses in 2010	Ranked 4th in the sector. 137 years old and the result of 20 mergers	Excessive investment in real estate sector (doubled it between 2003 and 2005, participating in 66 societies and 104 projects in that sector); lack of professionalism on top leadership; insufficient internal controls; and political instrumentalization Only *caja* that issues *cuotas participativas* (participative quotas, or tittles similar to stock but without political rights)	• Former managers indicted • Questionable investment in amusement park (the ruinous Tierra Mítica), and the Ciudad de la Luz • Soft loans to board members (161 million between 2004 and 2010 at a 0% interest rate) • Failed merger attempt in 2010 with Cajastur, Caja Cantabria, and Caja Extremadura • Multi-million salaries and payoffs to managers: the former CEO Carlos López Abad and another five top managers retired with 15.5 million in pensions. His successor, Maria Dolores Amorós, had a 600,000 euros salary and a life-long pension of 369,497 without the approval of the appropriate supervisory boards	• Intervened and Nationalized • It received 2.3 billion in capital from the FROB (it acquired 80% of the institution, and then 100%), and 3 billion in liquidity • **Banco Sabadell** absorbed it in December 2011 with an initial aid package of 5.24 billion from the FGD

(continued)

Table 6.1 (continued)

Caja	Collapse	Features	Problems	Controversies	Outcome
				• Amorós was fired on September 2011, for among other reasons, falsifying financial accounts • The last president, Modesto Crespo, did not receive an official salary, but he had arranged to receive an annual retribution of 300,000 euros through a CAM affiliate, *TI Participaciones* • 75,000 people affected by the problems with preferential stock and subordinated debt	

Caja	Collapse	Features	Problems	Controversies	Outcome
Novacaixagalicia (NCG)	• In 2011 169 million in losses • It needed a 1.162 billion loan from the FROB for the merger; and later on another one for 2.465 billion to meet the recapitalization requirements imposed by the Zapatero administration	Result of the merge between **Caixa Galicia** and **Caixanova** Ranked 5th in the sector 70 billion in assets They held **40%** of the market in Galicia	Excessive investment in the real estate sector; rapid expansion outside of Galicia; risky investment in the Mediterranean coast; associated with controversial constructors like El Pocero; politicized boards; insufficient controls; dependence on wholesale markets for funding	• After the merge, it announced a gross benefit of 2.67 billion between 2010 and 2015 • Purchase of the Sálvora Island • Compensation and rewards scandals for their top managers: José Luis Méndez (sailboats, and a 16.5 million pension package) • Julio Fernández Gayoso, leader of Caixanova, was able to manipulate the rules with the support of the Galician government to stay in the job past the legal age (he was 81) • The merged institution was valued at 1.714 billion euros. After accounting adjustment this was revised to 181 million	Nationalized. FROB owns 93% of the capital (the rest was left to the *caja* to fund the foundation activities) **Novacaixagalicia Banco** • 1200 people pre-retired and 300 branches have been closed • Still needs 4.5 billion euros

(continued)

Table 6.1 (continued)

Caja	Collapse	Features	Problems	Controversies	Outcome
CatalunyaCaixa	• September 30, 2011 • CatalunyaBank had 1.33 million in losses in 2011 • Toxic assets estimated at 12 billion • Caixa Catalunya (CC) closed 2008 with the highest loan default rate in the system growing from 1 to 5.28% in one year	Second *caja* in Catalonia. It resulted from the merge in 2010 of **Caixa Catalunya, Manresa,** and **Tarragona**	Excessive exposure to the real estate sector; dependence on wholesale markets for funding; very lax risk control systems (sometimes the collateral was simply the expected re-evaluation of the real state asset, and 32% of the mortgages were granted for an amount that was more than 80% of the value of the real estate asset); expansion outside of their territory; corporate board played very limited role in decisions	• CC planned to promote more than 4 million square meters in the Mediterranean arch and Madrid working with local companies, in which it participated with 50% of the investment (Prasa, Armilar, and Jale). It also expanded to Portugal. These alliances imploded when the real estate market collapsed and the *caja* assumed 100% of the investment • CC opened 300 new branches between 2003 and 2007 in other regions (Madrid, Andalucía, Valencia, and Murcia) • BoS has postponed its sale until they know for certain the capital needs of the institution (estimated in at least 5 billion)	Nationalized. FROB owns 89.9% (injected 1.7 billion to increase the group's capital) **CatalunyaBank** The Catalonian government is interested in having **Banco Sabadell** buy it

Caja	Collapse	Features	Problems	Controversies	Outcome
				• The former president of CC, Narcis Serra, has stated that when he took over the position in 2005 he had to impart 1 h classes to the members of the board because "they did not have the basic skills to understand the *caja*'s accounts" • Compensations: former president left CC with a 10 million package and his successor, Todó, negotiated a "post-occupation package" of 3.55 million	
Unmim	• It received 380 million loan from the FORB to merge, and 568 after nationalization • 1.75 billion in real estate assets; and 3.598 billion in loans to developers • 55 real estate societies • 469 million in losses in 2011	Resulted from the merge in 2009 of the **Sabadell, Terrassa,** and **Malleu** *cajas*	Expansion outside of their territory; excessive investment in the real estate market; excessive opening of branches; left aside their industrial participations; political meddling; lack of financial skills among top managers (many came from cultural entities and the municipalities)	• In 2004 Sabadell and Terrassa decided to invest on each other's territory (breaking an informal understanding that had lasted over a century)	Nationalized (FROB controlled 100%), and later acquired in March 2012 for 1 euro by the **BBVA** (and 953 million in aid from the FROB). 530 employees have been dismissed. BBVA has proposed an additional reduction of 1265 employees (it had 3076 in summer of 2012), and the closing of 314 branches

(continued)

Table 6.1 (continued)

Caja	Collapse	Features	Problems	Controversies	Outcome
				• In 2004 Caja Sabadell with 12.4 billion assets had 284 branches all in Catalonia; four years later it had opened 95 more, 16 of which were in Madrid, 9 in the Valencia region, and one in Andalusia; Caja Terrasa with 11.7 billion in assets grew from 232 branches in Catalonia to 268 with 11 in Madrid and 2 in Aragón; and Caja Malleu grew from 92 in Catalonia and 1 in Madrid, to 101 in Catalonia and 3 in Madrid • **Caixa Girona** was going to be part of the merged institution but the local leaders of two nationalist parties rejected the merger because it would dilute their share (despite the fact that it would be 23% with only 7.8 billion in assets). Shortly after it was absorbed by **La Caixa**	

Caja	Collapse	Features	Problems	Controversies	Outcome
				• As the crisis intensified and its capital needs increased, Unnim negotiated possible mergers with Banca Cívica, Ibercaja, and the Basque cajas, but they all failed • After it was nationalized, it still rewarded its loyal customers in 2011 with trips to Turkey in high standing hotels	
Banco de Valencia	Nationalized in 2011 after identifying a 548 million whole in its books Valued at 1.09 billion	Founded in 1900. It funded industrial, electrical, railway, and urban projects. Bought by the **Caja de Valencia** (later **Bancaja**) in 1994 It was part of the Bankia group when Bancaja merged with Caja Madrid	Politicized leadership; excessive concentration of risk in the real estate sector: it reached 65.8% (the average in the sector was 59%); political intervention from the regional government that replaced technical criteria for political one; close links with real estate developers; expansion of the branch network	• Loan default rate of 16.4%, double the average in the sector, solvency ratio of 1.7% (well below the legal limit of 8%) • Ongoing criminal accusations for illicit enrichment against the former CEO, Domingo Parra • Branch network increased 82% between 2000 and 2008, well above the sector	Nationalized. FROB injected 1 billion in 2012 to acquire 91% of the capital

(continued)

Table 6.1 (continued)

Caja	Collapse	Features	Problems	Controversies	Outcome
				• partnership with dubious characters like Eugenio Calabuig (it bought its company Costa Bellver for 107 million, when it only had assets of 10 million)	
				• It gave loans to Jaume Matas, former president of the Balearic government, who was later sentenced for corruption	
				• Dispute with Rodrigo Rato, president of **Bankia** when he requested an audit in February 2012 after the merge between Caja Madrid and Bancaja	
				• Compensation scandal: Aurelio Izquierdo former president of Banco de Valencia had negotiated a compensation package of 14 million (later he declined part of it and 'only' received 7.6 million)	

From Royo (2013, 190–95, Table 6.1). *Sources* "Agujeros negros' del sistema financiero 1–7," *El País*, June 24th–29th, 2012 and "Parte de guerra: ocho entidades intervenidas o nacionalizadas en España," *El País*, May 9, 2012

THE BANKIA NATIONALIZATION: CHRONICLE OF A COLLAPSE FORETOLD[7]

The nationalization of *Bankia* in the spring of 2012 symbolized most of what had been wrong with the financial system and Spanish *Cajas*, and it represented the ultimate symbol of *the game of bank bargains* that took place in the country in those years. As we have seen, *Bankia* was the result of the merge between *Caja Madrid*, *Bancaja*, and five smaller *cajas*. Its problems, however, started well before the creation of the new institution (see Barrón 2012).

By 2009 *Caja Madrid* was already in financial difficulties and the institution was in the midst of a political fight that symbolizes the political bargains/interferences that were so characteristic at the *cajas*. The president of the Madrid regional government, Esperanza Aguirre, was trying to oust *Caja Madrid's* president Miguel Blesa who had been in that position for 13 years, and wanted to have him replaced by her own vice president of the Madrid regional government, Ignacio González. Blesa had gotten the job because of his personal friendship with Prime Minister Aznar (they went to school together), and he had engineered the ousted of his predecessor, Jaime Terceiro, who had been named president of the *Caja* in 1988 by the Socialists. But Blesa had no previous experience in the financial sector. The Major of Madrid, Alberto Ruiz-Gallardón, opposed Aguirre's decision and tried to block it (both the regional government and the city had representatives in *Caja Madrid's* board), while supporting Rodrigo Rato's candidacy (who had been a VP and Minister of Finance in the Aznar's government, and a former Manager Director at the International Monetary Fund [IMF]).[8]

In the end, Rato won the battle and was appointed president of *Caja Madrid*. This was presented as a non-political decision, notwithstanding the fact that he had been one of the leaders of the Popular Party for years (and the Popular Party controlled both the Madrid regional government and the Madrid city hall). He continued the pattern initiated by Blesa to

---

[7] From Royo (2013, pp. 196–201). This section borrows heavily from "Así fue la caída del coloso," *El País*, May 13, 2012.

[8] The battle was so nasty that at some point Aguirre was inadvertently recorded saying that "we have given another seat [in the board] to *Izquierda Unida* [United Left, the political party], and we have taken it away from the son of a bitch [referring to Gallardón]."

increase his salary (he made one million more than Blesa, and Blesa had multiplied his salary by 18 compared with Terceiro), as well as the salaries of his top executives, while doubling the per diem of the members of the board. This was just another example of the kind of abuses that became so pervasive among the *Cajas*.

At that time, *Caja Madrid* was already experiencing severe difficulties marked by a significant decline of the institution financial margins, its enormous indebtedness with international lenders, as well as its dependence on the constructions sector and real estate mortgages, because it had given huge loans to construction and promotion companies. By 2009 its operating profits had fallen by 68%, and the *caja* was obligated by the BoS to increase provisions by 500 million euros. It was only the *caja's* investments in *Telefónica*, the selling of its stock in *Endesa*, other financial and treasury operations (which were even more profitable than its retail network), and its profitable investments in industrial companies and *Mapfre*, that gave the institution some breathing room. The rating agencies were already alarmed by the situation, and Moody's classified its bond emission as B2 (junk) out of concerns about the ability of the institution to pay the interests.

Rato inherited this problematic situation. His first priority was to recapitalize the institution, which was floundering under the weight of insufficient capital. His tenure was marred by this problem, which in the end led to the nationalization of *Bankia*. During his first year, he charged 4 billion euros against capital to avoid losses, with the approval from the BoS. At the same time, given that he was not a finance expert, it was expected that he would name one as his CEO. However, it took him a year and a half to do so, when he named Fernando Verdú from *Banca March*, and this happened just a month before *Bankia's* public offering in the stock market. This problem was compounded by the fact that Rato surrounded himself with a group of people at the top, none of which came from within the institution or were financial experts, and that he largely ignored the board. These decisions proved to be a significant handicap as the crisis intensified.

Six months after Rato became president of *Caja Madrid*, he announced the merge with *Bancaja*, another large *caja* based in the Valencia region, and five smaller *cajas*. Through this merge they created *Bankia*, the third largest financial group in Spain behind Santander, and BBVA. However, the financial situation of *Bancaja* was even worse than that of *Caja Madrid* as it was highly leveraged with construction loans, and was also short in provisions and liquidity. The reasons for the merge remain to be explained. Many in the PP government blamed the BoS for forcing it in order to shrink the number of *cajas* and build larger institutions that would be able to withstand the crisis better. At the same time, the BoS was also accused of not knowing the real magnitude of *Bancaja's* problems. The company responsible for auditing both *cajas'* accounts, Deloitte, was also in charge of estimating the value for the merger, and this may have been a problem as it may have had vested interests in the decision. The reality was that two *cajas* with significant problems merged, and one of the lessons, nowadays finally widely accepted, is that you do not create a solid bank by merging two bad ones. At the time of the merge, however, there was little opposition. The blame game intensified after the nationalization of *Bankia* and Aguirre has gone as far as to claim that the merge took place "under a barrel of a gun."

The merge decision also shows the high degree of political interference in the *cajas'* decisions, one of the main reasons for their problems. Both regional governments, Madrid and Valencia, were controlled by the Popular Party and the party leaders wanted to create a large national bank. At that time the regional governments were power centers of their own, and there was little question of their actions and decisions. In that context, some of the *cajas* had become essentially, political instruments to achieve political goals. Only the crisis brought these issues to the fore.

In this regard, it is worth highlighting that Rato had been a candidate to replace PM Aznar when he decided to step down and not run for a third mandate in 2004. The internal contest within the PP pitted him against Mariano Rajoy, and Aznar eventually handpicked Rajoy as his successor. Rato was the loser but according to many people he never quite accepted the outcome. Henceforth Rato never stopped his political activities during his tenure in *Caja Madrid/Bankia* and he continued participating in some of the party's public events. Many observers interpreted his actions as dominated by political considerations, and as part of a political strategy to try to emerge as a potential alternative to PM Rajoy. According to this perspective, the merger would give him the platform do so.

However, the merger did not address the subjacent problems, the banks' exposure to toxic real estate assets, which continued growing as the crisis intensified and the construction and real estate sectors continued deteriorating. In March of 2011 *Bankia* requested 4465 billion euros from the FROB and intensified the process of slashing costs by reducing branches and staff, which reduced costs by 550 million euros and was considered a significant achievement. However, it proved to be insufficient to make up for the bank's huge problems.

At the same time, Rato pushed for the decision to move ahead with a public stock offering, which took place in the midst of the economic crisis and was also controversial. It announced its arrival to potential investors with the advertising slogan "our future together," a message that just a year later proved prescient when the bank was rescued. This decision proved to be ruinous for the institution.

Again, the rationale for this decision is mired in controversy. Some claim that it was required in order to fulfill the conditions of the financial reform approved by the Socialist government, which requested less capital from entities listed in the stock market. However, others interpreted the decision in political terms, yet again: Rato was convinced that PM Zapatero wanted to hurt him because he was from the PP. Hence, he announced to the board that he would not wait for the government demands and that he had decided to become independent and not rely on government's funds, by issuing public stock. This decision took place while the merging process was still ongoing and it was perceived as a huge gamble, as there were serious concerns that the public offering could be a failure. To avoid that, the government went as far as to pressure institutional investors to buy stock from the new institution in order to avoid a public relations fiasco for Spain.

Immediately following the public offering *Bankia* had to confront the collapse of Banco de Valencia, a subsidiary (see Table 6.1). The problems of this entity had not been recognized until a few weeks before its collapse. In an unprecedented decision, the BoS decided to allow the collapse of a subsidiary: *Bankia*, the matrix company would not rescue it, and it would be absorbed by the FROB. This fiasco led to the final confrontation between Rato and the president of *Bancaja*, Olivas, who was forced to resign.

The beginning of 2012 did not bring positive news for the institution either. Rato tried to negotiate a possible merger between *Caja Madrid* and its main historical competitor, the Catalonian *La Caixa*, which had the support from the Rajoy government. News of the possible merger were filtered to the press, and it was perceived as a positive solution given the healthier status of *La Caixa*, which would have made it easier for the new institution to absorb the losses from the construction sector and would have alleviated *Bankia*'s shortage of capital. Yet, in the end Rato decided not to move forward out of concerns for his own role in the new institution (La Caixa was the larger institution and it was the stronger *caja*) and over the control the new entity. Nevertheless, Rato still looked for other alternatives and pushed for new mergers. He proposed one with the Asturias' *Liberbank* and later another one with the Andalusian *Unicaja*. Both were rejected by the BoS.

On May 11, 2012, Finance Minister Luis de Guindos ordered banks to set aside provisions equivalent to 45% on the nation's 307 billion euro ($387 billion) book of loans linked to real estate developers. This new additional demand proved to be devastating for *Bankia*, as it did not have the capital to meet it. On May 18, 2012, it requested government assistance, and the Spanish government was forced to nationalize it in an effort to restore investor's confidence in the Spanish banking sector. Even though the government had made a sweeping declaration that no more public money would be put into Spain's banks as late as February of 2012, it changed tack in light of *Bankia*'s problems and recognized that it would have collapsed had it not been rescued. The government approved and injection of 23.5 billion euros of state aid in exchange for a 90% control of *Bankia*, in what has become Spain's biggest bank nationalization. The bulk of the money would be used to boost provisions against real estate losses to get a coverage ratio of nearly 49%, in line with other Spanish banks. Following the nationalization, *Bankia*'s parent group BFA restated its 2011 results to reflect an astonishing 3.3 billion euros loss, rather than the reported 41 million profit.

Initially, the government announced that it would inject its own government debt into *Bankia*, issuing government guaranteed debt in return for equity, thus allowing *Bankia* to deposit those bonds with the European Central Bank (ECB) for cash. However, the ECB rejected this plan because it would potentially breach the European Union (EU) ban on "monetary financing," or central bank funding of governments, and informed the Spanish government that it would have to proceed with a proper capital injection for *Bankia*. This was, yet another public relations disaster for the Spanish government, and it reinforced the pattern of improvisation and contradictions that had characterized its decisions from the beginning of its tenure in December 2011. In the end, *Bankia*'s nationalization was the tipping point that led to an EU bailout package to rescue the ailing Spanish banking system. When Rato forced Olivas out, he stated that "all the financial leaders that have taken the *cajas* to their current predicament should assume their responsibilities." This sentence proved prescient in the end, as it was applied to him as well. It would serve as his own epitaph. Concerns about Spanish banks' undercapitalization led to the announcement of an EU banking bailout for Spain in June 2012 (see Chapter 1).

However, it is important to emphasize, yet again, that as late as May 2012, analysts and markets were still drawing a distinction between strong and weak banks, and recognizing the management and lending standards of the so-called "Big Three"—Banco Santander, BBVA, and La Caixa— as opposed to those of many *cajas*, which were often run by politicians with little (if any) financial and banking experience. While the *cajas* were struggling under the weight of bad mortgages and loans to construction companies, the "Big Three's" exposure to toxic assets associated with the Spanish real estate sector was a relatively small problem for them; BBVA and Santander diversified internationally giving them access to greater resources, funding and capital, which allowed them to liquidate the toxic property assets at a lower price than their rivals, and with limited damage to their earnings. Net profits for the first quarter of 2012 at Santander and BBVA (1.6 billion euros and 1 billion euros, respectively) were more than 10 times greater than their nearest rival, La Caixa, which lacked the international diversification of the other two and depended largely on a powerful branch network that was located in a market mired in recession.[9] *How do we account for the deteriorating performance of Spanish banks?*

The Impact of Economic Conditions

As described on Chapter 4, between 1996 and 2008 Spain witnessed one of its most spectacular periods of economic growth. During that period, Spain grew at a fast pace; 3.8% annually (whereas the EU15 average was 2.5%) and it created eight million jobs in 10 years, almost 30% of the jobs created in the EU. Unemployment decreased from 22% in 1996 to 7.4% in 2007. The flip side was the low-productivity growth (it grew a meager 0.5% between 1997 and 2006 versus the 1.3% average in the EU); the loss of competitiveness marked by record-high trade deficits (over 10% of GDP); and the dependency in the construction sector (see Royo 2013 and Royo 2013a).

---

[9] "Two Tiers, One Crisis for Spanish Banks," *Financial Times*, Friday, May 18, 2012.

However, the global financial crisis that started in 2007 exposed these weaknesses and brought an abrupt end to the unprecedented era of economic growth. By the summer of 2008, the effects of the global crisis were evident in Spain, and afterward, the country suffered one of the worst recessions in history. In 2010, Spain's GDP fell 3.6%, the worst performance since data have been compiled; the public deficit reached over 11%; public debt increased from 36 to 50%; and housing construction fell 20%. In 2012, the deficit reached over 5%; unemployment stood at over 24% with more than 5.6 million people unemployed; and the debt reached over 79% of GDP.

The deteriorating economic conditions had a severe impact on the banks' balance sheets. Indeed, the second recession in three years and the record-high unemployment, close to 25% in mid-2012 (and an astonishing 50% for those 25 years old and younger), triggered new loan losses in the Spanish mortgage market and increased the percentage of non-performance loans (see below).

This collapse was not fully unexpected. Yet, it is important to emphasize, again, that in Spain the crisis did not originate with wildly mismanaged public finances; Spain had a budget surplus between 2005 and 2007, and public debt stood at 36% of GDP before the crisis. On the contrary, as explained on Chapter 4, it was largely a problem of ever-growing private sector debt, compounded by reckless bank investments and loans, particularly from the *cajas* (to developers and construction companies rather than to home buyers), as well as aggravated by competitiveness and current account imbalances.

Indeed, while Spain had a relatively low ratio of public debt (70% in 2011, compared with 165% for Greece or 120% for Italy), the nonfinancial private sector debt was 134% of GDP, higher than any major economy in the world with the exception of Ireland (and in the case of Ireland, figures were skewed because of the outsized presence of foreign multinationals). Yet, it is worth emphasizing that the debt of financial institutions was lower than the Euro zone average (102% of GDP versus 128%). Their main problem was not the level of debt, but where it was invested: principally in real estate assets.

Finally, the austerity policies implemented since May 2010 aggravated the fiscal position of the country. The ratio of Spain's debt to its economy was 36% before the crisis and reached 79% in 2012, and 84% in 2013. The shrinking economy continued raising the debt, as well as reducing government revenue. In addition, the premium Spain paid to borrow over

Germany reached a Euro-era record of 485 basis points in June 2012, and yields on government bonds also increased to over 7%.

In sum, Spain seemed to have fallen into the 'doom loop' that had already afflicted Greece or Portugal and led to their bailouts. The sustainability of the Spanish government debt came back to haunt Spanish banks (including BBVA and Santander) because they had been some of the biggest buyers of government debt in the wake of the ECB long-term refinancing operation liquidity infusions (the percentage of government bond owned by domestic banks reached 30% in mid-2012, while the percentage of bonds held by foreign investors declined from a high of over 56% in 2011 to less than 39% in mid-2012).

The Collapse of the Real Estate Sector

An additional factor that helped explain the financial crisis was the collapse of the real estate sector. Land prices increased 500% in Spain between 1997 and 2007. However, the country's credit boom peaked in 2008, when the supply of cheap, and largely external, finance began to dry up. The real estate sector fitted like a glove with a financial sector that was seeking rapid growth. At the end of 2010, the IMF reported that Spain had the largest real estate bubble in the developed world. But the global financial crisis burst it. Indeed, by the end of 2011, land prices, adjusted for inflation, had fallen about 30% from the 2007 peak, and home prices were off about 22%. House prices fell by 11.2% in 2011 alone, while in Madrid, prices were down by 29.5%. By the end of the crisis, real estate prices fell to the levels of the mid-1990s.

The implosion of the real estate market exposed the vulnerability of the banking sector to that market, which constituted 60% of the banking loans (i.e., loans to families, enterprises in the real estate sector, or direct real estate assets). Consequently, Spain suffered a property-linked banking crisis exacerbated by obstacles to financing created by the international crisis.

Four years after the crisis started, the quality of Spanish banks' assets continued to plummet. The losses for banks were in the range of 130 billion–300 billion euros. The banking sector ended up needing over 200 billion euros in recapitalization in 2012 to maintain the 9% minimum required by the European Banking Authority. But Spanish banks still held 308 billion euros in total real estate assets. As property prices continued to fall and bad loan ratios to increase, the BoS classified 180 billion euros

Table 6.2 The financial sector and the exposure to real estate risk

	Consolidated assets (billion euros)	Exposed to credit to developers (million euros)	% of credit	Total at risk (million euros)	% of credit	Total covered by provisions (billion euros)	% covered
Total Cajas	1255	151,648	19.3	128,601	16	42,373	33
Total banks	2187	69,774	11.4	59,375	9.7	18,420	31
Total	3442	221,423	15.8	187,976	13.4	60,793	31

Source *El País*, Wednesday, May 9, 2012, p. 24

as troubled assets at the end of 2011, and in 2011 alone, they added 45 billion euros to their problematic assets. In addition, banks were sitting on 656 billion of mortgages and 2.8% were classified as nonperforming. The *cajas* were the ones experiencing the most difficulties because they financed much of the real estate boom.

Spanish financial institutions at the end of 2011 accumulated 405 billion euros in loans associated with the real estate sector given to developers and companies, and, of those, 188 billion were considered toxic and at risk of default (see Table 6.2). The *cajas* had to write down 50 billion euros in their property portfolios in 2011. As of December 2011, banks had already made provisions for 30% of the 170 billion euros of property assets concerned, but some landholdings could not be priced or sold. The BoS announced that bad loans on the books of the country's commercial banks, mostly in the real state sector, had reached 7.4% of total lending. In May 2012, the BoS disclosed that the value of bad loans held by Spanish banks increased by a third over the previous year to 148 billion euros, and loans in arrears accounted for 8.4% of the sector's loan portfolio in March 2012. Barclays estimated that nonperforming mortgages would reach up to 5% with 16 billion in losses. In other words, the provisioning was not able to keep pace with the crisis.[10]

[10] "España, duda permanente," *El País*, Sunday, May 20, 2012.

The problem was compounded by the increase in bad loans, which reached almost 8% in 2012, the highest level in 18 years as a percentage of total lending, forcing them to offload real estate assets from their balance sheets, and hindering their ability to lend to the real economy.

As of July 2012, despite raising billions of Euros since the collapse of the real estate market, the banking sector was still perceived as woefully undercapitalized and there were still serious questions about the full extent of the sector's real estate problem. The continuing recession led to further decline in property prices and put in question the value of previously performing loans. This concern extended even to the largest banks, with the potential impact on earnings and capital reserves.[11]

Wholesale Funding

According to the ECB, Spanish banks, on average, funded 46% of total assets with deposits at year-end 2011, a relatively high level compared to other countries' banks examined in this chapter. Yet, they still relied, to varying degrees, on market funds, which made them vulnerable to growing market tensions over the Greek crisis, and the sovereign and financial crisis in Spain. In addition, Spanish banks also funded themselves in the US dollar interbank market to finance their domestic lending (McGuire and von Peter 2009, p. 51). This was a market that was particularly fragile during the crisis.

Significant levels of market-based liabilities were taken on by both the big Spanish banks and the *cajas*, but in particular by the *cajas* in the context of their efforts to expand and strengthen their national presence, as illustrated most visibly by a rapid growth in the number of employees and branches. Their market share measured in terms of total assets, increased from around 20% in the 1980s to 40% in 2010 (see Fig. 6.1). This aggressive expansion went hand in hand with increased lending to construction companies, real estate developers, and to households for mortgages, which was increasingly financed by the wholesale market. As a result, the *cajas'* share of total assets funded by domestic deposits (public and private sector, excluding credit institutions) trended downward from over 80% in the early 1980s to 64% in 2010 (IMF 2012, p. 9).

---

[11] Nonperforming rates were running at 4% at Santander and BBVA.

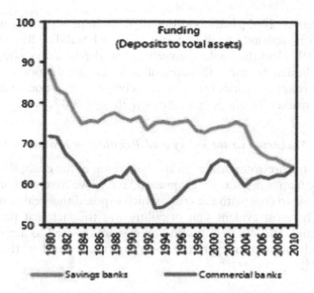

Fig. 6.1 Spain: savings vs. commercial banks, 1980–2010 (*Source* IMF 2012)

Indeed, another significant weakness was the dependence of Spanish banks on wholesale funding for liquidity since the crisis started, and, in particular, their dependence on international wholesale financing, as 40% of their balance depended on funding from international markets in 2012. Spanish banks increased their ECB borrowings by more than six times since June 2011, to the highest level in absolute terms among Euro area banking systems by April 2012. Funding from the ECB mitigated short-term funding needs; in March 2012, Spanish banks borrowed a record 316 billion euros from the ECB, 28% of the Eurozone total. However, their reliance on the ECB raised concerns over their ability to reduce their dependence on ECB funding over time, which made them very vulnerable and open to attacks from the markets.

Finally, on the liabilities side, funding was complicated by the difficulties of accessing international wholesale markets, because the securitization markets, as well as others such as the interbank or senior debt markets, were still not completely open to small- and medium-sized institutions. The BoS acknowledged as much: 'Spanish deposit institutions have been affected by the international financial crisis because of the evaporation of the wholesale funding channel, which, in the years prior to the summer of 2007, had contributed to funding a significant portion of growth in activity' (Bank of Spain Financial Stability Review, March 2010, p. 19). And the banks themselves: 'on the liabilities side, funding was complicated by the difficulties of accessing international wholesale markets through securities issues, asset securitization or borrowing on the interbank market' (*Caja Madrid Annual Report* 2007, p. 22).

Weaknesses in the Policy and Regulatory Framework

As seen in the previous chapter, at the beginning of the crisis, the financial regulatory framework was widely praised. Yet, as we have seen, in the end it was unable to cope with the crisis, which exposed the weaknesses of the system. The most evident sign of failure was the fact that the country adopted four financial reforms in three years, as well as implemented two rounds of bank mergers (and was starting a third in the summer of 2012)[12]:

1. *June 2009*: The government approved the FROB and the first round of 'cold' mergers. The FROB was funded with 9 billion euros to clean up the balance accounts of the entities most affected by the real estate crisis, in exchange for preferential stock that could be returned in five years. As a result of this process, there were seven groups of newly merged *cajas* that asked for support, and more than 30 *cajas* were involved in merging discussions that led to the reduction in the number of *cajas* from 45 to 22.

2. *February 2011*: In anticipation to the requirements that would be established by Basil III, the government approved a decree that increased to 10% the capital provisions for all entities that fund more than 20% of their needs in the markets (for the others the minimum is established at 8%). If they did not meet the new requirements, they had two weeks to present a capitalization plan to the BoS. They either found new partners or they had to issue stock. In addition, there was another round of mergers and the number of *cajas* was reduced to nine.

---

[12] From "Tres años, cuatro reformas y varios cadáveres por el camino," *El País*, May 11, 2012.

3. *February 2012*: The new Conservative government increased the capital provisions to 50 billion euros to clean up the real estate assets. This amount was the result of increasing to 80% the percentage of land covered against a possible non-payment, to 65% of unfinished houses, and 35% finished ones. The banks had until the end of 2012 to comply. The government also increased the FROB funding to 15 billion euros and capped the salaries of executives from institutions that had received public aid.

4. *May 2012*: In response to the intensifying crisis and pressures from the IMF, ECB, Brussels, and the markets, the government approved a new reform, the fourth since the crisis started. As part of this reform, the government has ordered banks to set aside an additional 30 billion euros in provisions against bad loans to ensure that banks had provisions of 52% of the value of loans made for land purchases. Those banks unable to meet the new provisioning rules would be able to borrow the additional money in the form of state-backed convertible bonds carrying a 10% interest rate. Furthermore, banks could transfer their riskiest assets to state-guaranteed asset management companies to help speed the sale of real estate assets the bank holds. Each bank would be forced to create a bad bank into which it will put physical property assets at marked down valuations, in preparation for potential sales to outside investors. Finally, the government agreed to ask two independent firms—Roland Berger y Oliver Wyman—to audit the banks real estate portfolios. Following this reform, four Spanish banks (*Banco Mare Nostrum, Liberbank, Unicaja*, and *Ibercaja*) announced that they were working on a merger to create the country's fifth largest lender, with assets of 270 billion euros.[13]

5. *August 2012*: A new financial reform was approved in response to the EU financial rescue package. It created a *bad bank* that could absorb the toxic assets from the real estate sector and had the authority to buy and sell all kind of assets and to issue bonds.

---

[13] The reaction from the markets to this reform was quite mixed. While there was a sense of relief that Spain was finally coming to grips with its banking difficulties, there was the sentiment that it failed to sway investors: Banks shares fell sharply and borrowing costs rose again over levels before seen as unsustainable (6.3% for 10-year bonds). Investors seemed to think that this reform was not the definite cleanup that the market expected. *Standard & Poor* downgraded 11 of the country's largest banks on April 30, 2012, and *Moody's* downgraded 16 of the banks on May 18, 2012.

The reform reinforced the role of the BoS in the creation of the *bad bank*. The BoS was also assigned a central role in the decisions regarding the assets that would be transferred from each individual financial institution to the *bad bank*, and more importantly in the pricing of these assets before their transfer. It also established a new process to restructure and liquidate financial institutions and it gave a central role in that process to the FROB and the BoS. Finally, the reform reduced the role of the regional governments in the restructuring and liquidation of *cajas* and saving cooperatives.

The results of the first four reforms were questionable as best, and these efforts all proved inadequate in addressing the magnitude of the crisis (see Table 6.3). They were largely perceived as "too little and too late," and they failed to rebuild investors' confidence in the Spanish financial sector. Indeed, the implemented "low-cost" reforms did not solve the problems. Both Socialist and Conservative governments were reluctant to admit the depth of the liquidity and solvency problems of many institutions (particularly the *cajas*, where, as we have seen, politics had

Table 6.3 Restructuring of the financial sector with FROB assistance (as of May 8, 2012)

Bank	Process	Guaranteed assistance (billion euros)	Type of assistance
Catalunya Caixa	Integration	1250	Preferential stock
	Recapitalization	1718	Ordinary stock
NovacaixaGalicia	Integration	1162	Preferential stock
	Recapitalization	2465	Ordinary stock
Caja España-Duero	Integration	525	Preferential stock
Bankia	Integration	4465	Preferential stock
Mare Nostrum	Integration	915	Preferential stock
Cívica	Integration	977	Preferential stock
CajaSur	Restructuring	392	Protection of risky assets
B.CAM	Restructuring	5249	Ordinary stock
			Protection of assets
			Liquidity support
Unnim	Restructuring	953	Ordinary stock
			Liquidity support
B. Valencia	Restructuring		

Source Fondo de Restructuración Ordenada Bancaria (FROB)

continued to play a role throughout the crisis) and the reforms were insufficient to address them. At the same time, the attempt to force restructuring through mergers also backfired, as proven by the *Bankia* fiasco. The governments also failed to recognize that merging weak banks does not create a strong one. The 100 billion euros assigned at that time to deal with the crisis were woefully insufficient as investors waited for additional funding (and reforms).

The loss of confidence from foreign investors was confirmed by the fact that they withdrew 31 billion euros in just one month (March 2012). There was growing consensus that the banking and securities regulators' independence needed to be strengthened and also that the BoS needed further authority to address pre-emptively the buildup of risk in the system and that the sanctioning regime in banking and securities supervision needed to be strengthened (IMF 2012).

The Role of the Bank of Spain[14]

At the beginning of the crisis, the BoS's policies, its experienced and respected professional staff, and its thorough supervisory processes were all praised and modeled by other countries. Time, however, tempered that praise and the BoS ended up being criticized for its actions and decisions (or lack thereof) during the crisis (see Ekaizer 2018).

The BoS anticipated the potential pitfall from the real estate bubble as early as 2003, when it alerted in one of its Economic Bulletins that the magnitude of the bubble could be as high as 20%. This caused a conflict with the Conservative government and Rodrigo Rato, the minister of finance at the time, questioned the report. This was the first instance of a battle that marked the course of action in subsequent years and this, despite the Autonomy Law, which protected the BoS against political interference. Indeed, the political bargains between the government and the BoS, as well as between the political and the financial powers, are crucial to understanding what has happened as well as why the BoS failed to act to burst the real estate bubble and to address the recapitalization of the *cajas* until it was too late. Naturally, any action to bust the bubble was opposed by the governments (both the *Partido Popular*, PP; and *Partido*

---

[14] This section borrows from "El día que el Banco de España se doblegó," *El País*, March 13, 2011 and "Un gobernador entre dos fuegos," *El País*, May 13, 2012.

Socialist Obrero Español, PSOE) because of the negative impact that it would have had on economic growth and consumption.

When Jaime Caruana became BoS Governor in 2000, he knew that the economy had been overheating since 1995; he was also fully aware that the banks and *cajas* were assuming large risk in the real estate sector. In his first speech to the *cajas* on April 18, 2001, he asked for four things: not to abuse industrial investments because they could be 'foreign to their and traditional nature and objectives'; to reduce the emission of preferential stock, which had become a major source of unlimited liquidity; to control the expansion of credit; and to professionalize their administrative counsels and reduce the political weight in the sector. These recommendations proved to be prescient. Some of these recommendations had to do with the MBB framework discussed throughout the chapter (as in the need to control the expansion of credit), while others dealt with other problems, but had they been heeded, the *cajas* would not have faced the crisis that later materialized. Unfortunately, ten years after Caruana's remarks, the *cajas* continued facing most of these problems, which, in many cases, led to their collapse. Indeed, most of the leaders of the *cajas* disagreed with these suggestions, which in effect led to a confrontation that only ended in April 2010 in with the selection of Isidro Fainé as president of the CECA (the Spanish Confederation of Cajas de Ahorros). Having worked in banks since 1964, he had the appropriate and extensive financial background to push for the necessary reforms.

Despite Caruana's warnings, mortgages continued growing at a fast pace: a 20% annual rate. On June 21, 2003, he warned, yet again, that 'the increase in the real cost of housing may have surpassed coherent levels,' and asked 'for a progressive reconciliation and moderation of the credit to families and companies involved in the real state sector.' He concluded that unless action was taken, there would be 'more abrupt adjustments.'[15] Unfortunately, banks and *cajas* continued to ignore his recommendations. Pushed by intensifying competition for growth and market share, *BBVA* increased its mortgages by 17%; *La Caixa* by 23.4%; *Caja Madrid* by 24.9%; and *Banco Santander* by 14.9%. The herd logic, once again, triumphed.

Compounding the problem was the lack of political will to change a model that had resulted in unprecedented levels of growth. The Socialist

---

[15] "El día que el Banco de España se doblegó," *El País*, March 13, 2011 and "Un gobernador entre dos fuegos," *El País*, May 13, 2012.

Party, at the time in opposition, often criticized the Conservative government. Rodriguez Zapatero, leader of the Socialist Party, blamed the Conservative government for the bubble several times in parliament and promised a shift in the growth model when his party won the elections. Yet, when he won the election in 2004, he did little to change the growth model and reduce the dependency on the construction sector (and this, despite a commitment, included in the Socialist Party's election manifesto).

Indeed, the government was too slow in reacting and acknowledging the depth of the crisis, something that has been established as a pattern throughout the crisis (Royo 2013). As late as May 2007, when the crisis was already brewing in the United States and the stock market was penalizing construction companies' stock in anticipation of the crisis, the Socialist minister of finance, Pedro Solbes, provided a quote for the history books: 'I do not see the construction sector affected at all, it is just a small deceleration which will allow it to adjust to a new reality that will logically result in lower demand,' and added 'of course this is an issue that has to be monitored, but I would not give it too much importance.'[16] Clearly the forthcoming election (which took place in March of the following year) was on his mind. At the same time, the regional governments and municipalities, which were very dependent on construction for tax revenues, continued resisting reining in the *cajas* (see Rodríguez Zapatero 2013; Solbes 2013).

Governor Miguel Angel Fernández Ordoñez (MAFO) has also been severely criticized. A Socialist Party militant who had served in Socialists governments in different capacities since the 1980s, he had scant financial experience when he became Governor of the BoS (this shortcoming has been used by many observers to explain the deficiencies in the design and implementation of the financial reforms). Prior to assuming the Governor's position, Governor Ordoñez played a very prominent and public role in the Spanish media, criticizing the country's growth model and its dependence in the construction sector. However, he failed to heed his own advice when he became Governor (see Fernández Ordoñez 2016).

The BoS inspection services had already alerted his predecessor, Jaime Caruana, about the dangerous emergence of a real estate bubble. While MAFO sought to address it, he found obstacles that he was unable (or

---

[16]This section borrows from "El día que el Banco de España se doblegó," *El País*, March 13, 2011 and "Un gobernador entre dos fuegos," *El País*, May 13, 2012.

unwilling) to overcome, principally the opposition from politicians and regional governments who controlled the *cajas* and were very reluctant to relinquish their power. Governor Ordoñez was unable (or did not dare) to confront them. This was a key factor in explaining the delays in the BoS' intervention, which often waited too long and/or acted when it was too late, resulting in the problems getting out of hand (as had happened with the *cajas* from the Valencia region, the Galician ones, or even with Caja Madrid).

The BoS also approved mergers that proved to be economically absurd (i.e., *Bankia*). Moreover, the BoS, under his leadership, focused too much on economic reforms (the Popular Party accused him of acting like a minister of finance rather than as a governor of the BoS) and did not push enough for the necessary financial reforms, which were often delayed and largely insufficient. For instance, the decision to intervene in Caja Castilla la Mancha (CCM) was interpreted more as a signal and as an example (it was controlled by the Socialist Party, then in power at both the regional and national level) rather than as a decisive step to address the problems (if that was the case there were other *cajas* that should have been intervened even earlier: *Cajasur*, *CAM*, *Bancaja*, or others that were absorbed through mergers). The official line from the BoS was that it 'could not intervene until there were solvency problems,' claiming 'banks do fulfill the coefficient requirements, and interventions are the most traumatic and expensive options, therefore the bank sought other alternatives.'[17] This approach helps explain what happened with *CCM* and *Cajasur*. The positive result of the stress test may have also influenced the actions of the BoS and contributed to the decision to delay the interventions.

The arguments often mentioned by the BoS—that it lacked control of monetary policy (in the hands of the ECB), that it would have been counterproductive to burst the bubble because it was already declining, and that it did not have the tools to burst the bubble—seem disingenuous at best.[18] It is true that the establishment of the countercyclical regime

---

[17] "Un gobernador entre dos fuegos," *El País*, May 13, 2012.

[18] The BoS defended itself with seven arguments: It lacked the appropriate instruments to address the crisis until 2009; the economic deterioration was far longer and intense than anticipated; the government had decided to reject the creation of a 'bad bank'; international banks were unable and unwilling to participate in the merge and acquisition process that was taking place in Spain because of the crisis; they could not use the traditional tool to liquidate banks, namely forcing bondholders to take loses because of the

was costly for the BoS and led to significant confrontations with the banks that complained bitterly arguing that it placed them at a disadvantage with their international competitors because it reduced their profit margins. There was also a sense of complacency that the regime would provide a sufficient cushion if the real estate sector fell down. Yet, it was the BoS itself that approved the high concentration of risk in the real estate sector. Moreover, there were other actions that the BoS could have taken to reduce the risk from and dependence on the construction sector, such as increasing the required provisions together with the ECB, or strengthening the regulatory framework. Political considerations (both Caruana and MAFO wanted to avoid a confrontation with the PP and the PSOE) and concerns over potential litigation drove the agenda and precluded more decisive action to address the problem. The government also had options; it could have eliminated housing tax breaks and/or established higher stamp duty on property sales, or raised capital gains tax on second properties.

Even MAFO's successor as Governor of the BoS, Luís María Linde, admitted in Congress (in the summer of 2012) that the BoS had responded "insufficiently or inadequately" to the crisis, and recognized that it "was not successful in what we call *macro-prudential* supervision." He added that initially, "we did not made the decisions that now we finally realize would have been necessary to address the large increases in indebtedness, and later on, we failed to contain and correct the strong deterioration of the banks accounts, which followed the collapse of the real estate bubble," while recognizing that the fact that the BoS was "not the only European supervisory entity that made these mistakes, that was not of much consolation." Under questioning from members of parliament (MPs) he acknowledged that "it would be absurd not to recognize mistakes in the oversight of banks," and that the euphoria associated with the economic boom made it more difficult to "see the risks that we were accumulating. It was as if no one wanted to foresee the possibility of a recession, of interest rates increases, or the collapse of funding." He still gave credit to his predecessors for introducing the anti-cyclical provision system in which the BoS

potential contagion effect to other banks and *cajas*; that the corporate governance system of *cajas* was very deficient and politicized; and finally, the autonomous governments refused to approve the mergers during the restructuring process. See "El Banco de España se defiende de su lentitud con siete argumentos," *El País*, June 11, 2012.

"was a pioneer," but recognized that the main flaw of that system was "its timidity and its insufficiency in containing the excessive growth of credit," and admitted that the BoS had not gone far enough and had to "curtail its aspirations." He was also very critical of the *Sistemas Institucionales de Protección* (the so-called cold mergers) introduced by his predecessor, because they were not "the solution" and did not produce the "desired results." Adding that "they tried to overcome the political difficulties that emerged from the regional governments, as well as other difficulties raised by the cajas themselves to the mergers and integration" of these institutions. Yet, the resulting mergers were not very positive and contributed, contrary to the original goal, "to delay decisions and adjustments." Finally, he admitted that the banking stress tests were not sufficiently strong either, but argued that "almost no one was able to anticipate the depth of the crisis, something that affected the quality of the stress tests." He even admitted the possibility (which had already been mentioned before by Joaquín Almunia, the EU Commissioner, and rejected by the PP government) to proceed with an "orderly resolution" of institutions that do not have a "strong pulse." But he was immediately corrected by the minister of finance, Luís de Guindos, who reaffirmed the government's position against the "liquidation of any Spanish financial institution."[19]

In the end, the BoS chose the path of least resistance: Alerting as to the risks but failing to act decisively, even though some people within the bank had been very critical of such inaction and lack of leadership. The Inspectors Association of the BoS published in May 2006 a letter to the minister of finance, Pedro Solbes, criticizing Governor Caruana's passivity in the face of the accumulated and growing risk associated with the real state sector. They attributed the extraordinary growth of real estate prices to the 'excessive growth of bank lending,' fueled by MBB liabilities, and accused the governor of lacking determination in demanding 'rigor' from financial institutions 'when assuming risks.'[20]

This experience shows that, in the absence of appropriate leadership supported with the necessary regulatory and financial controls, the actions of the BoS were influenced by political considerations that ended up

[19] "El Nuevo gobernador critica la gestión de sus predecesores en el Banco de España," *El País*, July 17, 2012.

[20] "El día que el Banco de España se doblegó," *El País*, March 13, 2011 and "Un gobernador entre dos fuegos," *El País*, May 13, 2012.

undermining its role. If anything, it shows the increasing concentration of power away from the BoS, in the hands of the government, and the banks.

By the end of 2012 the calamitous consequences were clear. As we described on Chapter 1, on November 25, 2012, Minister Guindos requested 37 billion euros for the nationalized *cajas* (*Bankia*-almost 18 billion; *Novagalicia*, *Catalunya Caixa*, and *Banco de Valencia*), and four additional institutions (*Ceiss*, *BMM*, *Caja 3* and *Liberbank*) ended up requesting an addition 5 billion euros, bringing the total to approximately 45 billion euros. In exchange for this support the European Commission requested draconian measures: A 60% cuts in the assets from nationalized *cajas*, very painful restructuring plans that ended up requiring thousands of dismissals, the selling off of affiliates and their participation in industrial companies, the prohibition of loans to constructions developers, and a request to retreat to their original region of operations. By the end of 2012 the nationalized *cajas* has already transferred 36 billion euros of their toxic assets to the newly created "bad bank" (called *Sociedad de Gestión de Activos Procedentes de la Restructuración Bancaria*-SAREB), and it was estimated that they would have to sell the adjudicated assets for an average discount of 33% of their gross value.[21]

CONCLUSION

Contrary to Greece or Italy, the financial crisis in Spain originated from private sector over-indebtedness (as in Ireland). Yet, Spanish financial institutions (unlike those of the Anglo-Saxon countries) were originally sound and well capitalized. The relatively low degree of financialization sheltered the Spanish economy from the initial stage of the crisis, which was particularly severe in countries (like the United States or the UK) that were heavily exposed to "toxic" financial products.

Indeed, as we examined in Chapter 5, in the first stages of the global financial crisis the Spanish financial system was one of the strongest and better regulated financial systems in the world. Initially, Spanish financial institutions were less "financialized," had a stronger retail focus and relatively low leverage, and were soundly capitalized. However, the implosion of the real estate bubble, their dependence on MBB liabilities, and

---

[21] "La banca nacionalizada traspasa al banco malo activos por valor de 37.110 millones," *El País*, December 26, 2012.

the depth of the economic crisis weakened them and made them far more vulnerable.

There is consensus that the stern regulations of the BoS played a key role in the initial performance of Spanish banks—and these clearly contained the development of Spanish MBB, explaining notably the absence of off-balance sheet securitization. However, regulation was not enough, particularly for the *cajas*. As the international crisis dragged on, the banking sector could not escape its dramatic effects. Deteriorating economic conditions, the implosion of the real estate market, the dependence on wholesale funding, weaknesses in the regulatory framework, and the role of the BoS all help to explain this reversal.

In the end, Spain could not escape the ensuing effects of the crisis: the liquidity and credit crunch as well as the collapse of the real estate sector. Initially, the strong fiscal position of the country (particularly in comparison with Italy or Greece), combined with the some of the features that characterize the Spanish "model of capitalism" (namely a relatively strong public sector, the informal economy, and the role of the family) (see Royo 2008) contributed to shelter the country from the effects of the crisis.

As we have seen, the funding gap was high and MBB was largely concentrated on the liabilities side of the balance sheet, with a significant element of bank lending prior to the crisis funded on wholesale markets. While the financial crisis did not have a devastating direct impact on most Spanish banks during the first phase of the crisis, the difficulties accessing wholesale funding eventually had a knock-out effect on Spanish bank lending, thus leading to the bursting of the asset price bubble and a seemingly rather more "traditional" banking crisis.

In the end, as we develop further in the Conclusions, the crisis was the outcome of a *political bargain* (see Calomiris and Haber 2014, pp. 280–81): In Spain, *cajas* were allowed to grow and expand. This afforded them economies of scale and scope, as well as increasing market share and market power, and thus a larger too-big to fail protection. In exchange, however, they had to share some of their rents with other groups, in this case new homeowners eager for cheap credit to buy their homes, and politically influential real estate and construction companies. These groups seized a portion of the banks benefits for themselves and their members, and leveraged their positions to influence the rules of the game. For instance, they used their influence with local, regional, and national politicians to appoint political (and often uneducated and unskilled) managers to the *cajas* to do their bidding, and they pushed to loosen underwriting standards and to maintain low capital requirements. For their part policy-makers and regulators, although they knew what was happening, chose to acquiesce and to go along. They could have intensified prudential regulation and/or demand higher capital requirements, but they chose not to do so because they also accrued benefits (not only electoral but often personal and/or financial ones) from this game. On the contrary, politicians from all political parties of the right and the left (and even trade unions) were actively involved in the *cajas*, and they actively supported this game and allowed it to unfold. Finally, although there is now a tendency to blame politicians and bankers, and this crisis originally emerged as a result of the dominance of a particular political coalition, it is important to stress that many others soon joined in. Indeed, by the time that the bubble burst the majority of mortgage borrowers were playing and benefiting from the game. Most of them, however, would end up paying a high price when the crisis hit the country.

Furthermore, this chapter shows the importance of focusing on the concept of MBB and the need to examine the activities undertaken by financial institutions on the assets and liabilities sides of their balance sheets. On the assets side, the deterioration in the quality of loans caused by the collapse of the real estate market led to the repeated downgrade of Spanish financial institutions and a financial bailout, turning a banking crisis into a sovereign debt crisis. At the same time, the financial crisis intensified the credit crunch and worsened the recession. On the liabilities side, initially Spanish financial institutions were sheltered by their reliance on domestic deposits, and their limited exposure to "toxic" assets. Yet, they were very vulnerable to wholesale funding and ABSs, which they issued for funding purposes to close the gap between the net flow of loans and the flow of deposits. This dependence, contributing to the real estate bubble, led to the downgrading (and collapse) of several *cajas*, as well as to the financial bailout.

Additionally, the degree of financialization, key to MBB, is also an important element to account for the way in which the crisis played out in Spain, specifically the response of the Spanish authorities. The collapse of the real estate markets eventually led to a traditional banking crisis fueled by turbocharged lending by MBB on the liability side of the Spanish bank's balance sheets. MBB explains the dramatic increase in lending, and the ongoing credit crunch in Spain; when the crisis hit, international and wholesale funding dried up and it affected bank lending. In this regard, the main "toxic" assets held by Spanish banks were domestic bad loans and mortgages.

Finally, one of the interesting aspects about the Spanish crisis is that MBB was having its impact prior to keen awareness regarding the crisis and it was largely overlooked as a source of the bursting bubble. In this sense, although the Spanish banking crisis has been labeled as a traditional bank crisis, it was more than that.

REFERENCES

Banco de España. *Report on the Financial and Banking Crisis in Spain, 2008–2014*. Madrid, 2017.

Bank of Spain. *Financial Stability Review*. Madrid, March 2010.

Barrón, Iñigo. *El Hundimiento de la Banca*. Madrid: Catarata, 2012.

Caja Madrid. *Annual Report*. Madrid, 2007.

Calomiris, Charles W., and Stephen H. Haber. *Fragile by Design: The Political Origins of Banking Crises & Scarce Credit*. Princeton: Princeton University Press, 2014.

Ekaizer, Ernesto. *El Libro Negro. Como Fallo el Banco de España a los Ciudadanos*. Madrid: Espasa, 2018.

Fernández Ordóñez, Miguel. *Economistas, Políticos y Otros Animales*. Madrid: Ediciones Península, 2016.

International Monetary Fund (IMF). *Spain: The Reform of Spanish Savings Banks Technical Notes*. IMF Country Report No. 12/141, 2012.

McGuire, Patrick, and Goetz von Peter. "The US Dollar Shortage in Global Banking." *BIS Quarterly Review*, March, 2009.

Rodríguez Zapatero, José Luís. *El Dilema: 600 Días de Vértigo*. Barcelona: Planeta, 2013.

Royo, Sebastián. *Varieties of Capitalism in Spain*. New York: Palgrave, 2008.

Royo, Sebastián. *Lessons from the Economic Crisis in Spain*. New York: Palgrave, 2013.

Royo, Sebastián. "A 'Ship in Trouble': The Spanish Banking System in the Midst of the Global Financial System Crisis: The Limits of Regulation." In *Market-Based Banking, Varieties of Financial Capitalism and the Financial Crisis*, edited by Iain Hardie and David Howarth. New York: Oxford University Press, 2013a.

Solbes, Pedro. *Recuerdos*. Bilbao: Deusto, 2013.

CHAPTER 7

Bank Bargains and Institutional Degeneration

INTRODUCTION

This book examines the historical circumstances that have shaped the formation of political institutions and coalitions that control banking outcomes in Spain. The central argument of the book is that banking systems arise from a process of political bargaining in which actors with differentiated interests come together to form coalitions that determine how banking systems are created and how they operate (Calomiris and Haber 2014). However, it is crucial to emphasize that these bargains are structured by a society's fundamental political institutions because they are mediated by these institutions, which in the case of Spain were not able to insulate the Spanish banking system from populist policies. This chapter shows that the 2012 banking crisis was just another symptom of the institutional degeneration process that took place in the country in the years prior to the crises and argues that the banking crisis cannot be explained without examining that process of institutional degeneration.

As we have seen throughout the book, the 2008 economic crisis had an earth-shattering effect in Spain at all levels: economic, political,

From: Royo, Sebastián. "Institutional Degeneration and the Economic Crisis in Spain." Special issue of *American Behavioral Scientist. The Economic Crisis from Within: Evidence from Southern Europe.* Anna Zamora-Kapoor, and Xavier Coller (eds.) 58, no. 12 (2014): 1568–91.

institutional, and social. While there have been different interpretations about the causes of the crisis, with most of the analyses concerned with phenomena like mismanaged banks; excessive debts; the bubble in the real estate sector; the loss of competitiveness; or the European Monetary Union (EMU) structural shortcomings, this chapter moves beyond those explanations and argues that a central reason for the crisis was rooted in the process of institutional degeneration that preceded the crisis (see Ferguson 2013) and the failure of the Spanish elites. It argues that, as in many previous financial crises (such as the Savings and Loans crisis in the United States or the Asian crisis, in the 1980s, or the Irish financial crisis of 2008), governance played a critical role in the development of the Spanish crisis.

This chapter seeks to explore the consequences that the institutional degradation that took place in the country in the years prior to the crisis brought to the Spanish economy. It argues that institutional degeneration led to a Spanish version of crony capitalism characterized by the misgovernment of the public; an outdated and inadequate policy-making process; an inefficient state; and an often corrupt and inefficient political class. Indeed, the excesses that led to the crisis marked by mismanaged *cajas*, excessive private debt, bubbles in the real estate sector, or competitiveness losses were all symptoms of an institutional malaise that intensified in the years prior to the crisis. We cannot understand any of those outcomes without referencing the institutional divergence in the rule of law between Spain and the EU core, which led to a real estate bubble, a competitiveness divergence, and a financial crisis. Moreover, it was this institutional divergence that made it more difficult to implement the reforms that EMU membership demanded. Political and economic institutions resisted reforms because they would jeopardize the existence of the extractive rent-seeking mechanisms that became the main source of rent for the economic and political elites that controlled them.

In the end, the economic success of the 1980s, 1990s, and early 2000s, which was spurred by the country's modernization and European Union membership, was not sustained because local, regional, and the national governments became less accountable and responsive to citizens. In terms of causal mechanisms between institutional degradation and economic crisis, this chapter shows (following Acemoglu and Robinson's terminology, 2012) that political and economic institutions across the country became more 'extractive' and concentrated power and opportunity in the hands of only a few, and they came short in empowering

Spanish citizens to innovate, develop, and invest, thus failing to foster the degree of 'creative destruction' that is so vital for innovation and sustainable growth. On the contrary, those institutions promoted an unsustainable growth model based on a real estate bubble that inflated prices and fostered a sense of wealth that propelled private debt and consumption to unsustainable levels.

The chapter is organized as follows. The first section analyzes some of the causes of the crisis. Sections two and three discuss the process of institutional degeneration and institutional divergence that led to the crisis. It finishes with a conclusion.

UNDERSTANDING THE CRISIS

Chapter 4 examined the economic crisis and explored its main causes and consequences, but *who was responsible for the crisis?*[1] As we have seen, fingers have been pointed in all directions, and people in Spain were quick to blame bankers, politicians, the elites, or the *caste*, but in reality it was a collective failure both at the national and European levels. While it is undeniable that to a significant degree Spain squandered the privileges of EMU membership, and a lot of the resentment during the crisis years from its wealthier European neighbors came from that, it was not just the Spaniards' fault. It is true that a series of Spanish governments and bank leaders bear a large part of the responsibility for mismanaging the economy and finances of the country (particularly at the local and regional levels), but there were also other culprits. European banks, for instance, also fueled the real estate bubble because they continued to lend money to Spaniards, and the subprime crisis in the United States ignited an unprecedented global financial crisis with severe effects on the European and Spanish economies for which Spain shared little responsibility.

In addition, the fundamental institutional design problems of the EMU, which did not include a fiscal union, a European joint bank regulator, or a system to deal with financial institutions in stress, also went a long way in explaining the crisis of the Eurozone. Spain, like Ireland, suffered the consequences of these deficiencies, and the country's financial crisis precipitated a national debt crisis because it could afford to bail

---

[1] This part of the chapter borrows from Royo (2013).

out its own financial institutions. The EMU also lacked effective mechanisms to control the member states' finances and to penalize those who violated its budget rules. And it is worth remembering that these violations of the rules did not start with the 2007 crisis. On the contrary, when the Euro was first introduced, the Germans and the French both violated the deficit rules, and were the first ones to argue vehemently for the rules to be watered down to avoid spending cuts and being penalized (as Portugal was). Eurozone countries minimized these risks and had to deal with the consequences of their hubris, while they continue trying to make the necessary institutional changes (toward a fiscal union, a banking union and a European-level bank regulator), but in a much more difficult political and economic environment.[2]

Furthermore, investors also share some of the blame; they underpriced the risk of Spanish debt before the crisis, but later they overreacted to even minor events causing havoc with the borrowing costs for the country and thus on its economy. European governments, including the German and French, also lent money to Spain that then could be used to buy German and French exports. As we have seen in Chapter 6, the financial regulators also overlooked the problems of the banks and *cajas*, and failed to act decisively to address them; and successive national governments in Spain did little to address the real estate bubble or to shift the existing economic growth model based largely on construction. What we ended up witnessing at the national and European levels was a blame game that underscored the erosion of the principles of cooperation and solidarity that had underpinned the process of European integration since its inception.

Nevertheless, as much as there has been a fixation with the economy, with economic policies and with the economic responses to the crisis, it is impossible to understand what happened without focusing on the politics. In the case of Spain (and the Eurozone as well), it is only appropriate to update the famous campaign slogan, "it's the economy, stupid," that James Carville coined during the first Clinton presidential campaign with a new version, "it's the politics, stupid." The politics of the crisis, both at the domestic and European levels, were simply abysmal. Indeed, years after the crisis, it was still perplexing that the political leaders of the country did not have a clear and coherent diagnosis of the crisis, that

---

[2] See "In Euro Crisis, Fingers Point in All Directions," *New York Times*, August 25, 2012.

they keep blaming others for what was happening to the country and that they largely failed to assume any responsibilities,[3] that they lacked a credible long-term plan on how to get the country out of the crisis, and that they insisted to the bitter end on austerity and undermined the decisions (such as investment on Research-Development-Innovation R-D-I and education) that would have set the path for a more equitable and sustainable future growth. In this regard, the crisis exposed the deficiencies of a political class, supported by its political parties, which had developed its own particular set of interests and instruments to sustain it through a system of rent-seeking based on crony capitalism.

Politicians in Spain have been blamed for the real estate, infrastructure, and the renewable energy bubbles, and for the collapse of the *cajas*, which ultimately led to the EU bailout. According to an influential newspaper article titled *"Theory of Spain's Political Class,"* published by Cesar Molinas in 2012 in the Spanish daily *El País*[4] (which became the basis of a book published in 2013), the roots of the problems originated in the transition to democracy when politicians adopted a proportional representation voting system with closed, blocked lists that sought to consolidate the party system by strengthening the internal power of the party leaders, and also adopted the decentralization of the Spanish state (see also Molinas 2013). The consequences of these decisions have been enduring. The choice of voting system and blocked lists resulted in a professional political class that owes its allegiance to the leaders of their political parties (which are the ones that place candidates in the voting lists and/or give them public jobs in exchange for their compliance and submission). This has led to a structure, still largely in place, in which there is very little contestation within parties and in which loyalty, rather than merit or competence, rules. At the same time, the decentralization process—originally designed as a top-down process, became a bottom-up one led by local and regional elites—has led to the creation of 17 regional

---

[3] See Interview with Soraya Rodríguez, Socialist speaker "Quien da motivos para que salgamos a la calle es el PP," *El País*, March 1, 2012. After almost eight years in power, Ms. Rodríguez in a response to the question "many say that the PSOE demonstrates against [the PP] measures to confront a situation that has been in fact created by the PSOE government," answered that "when there was a Socialist government, the clear economic difficulties that we confronted were the consequence of a global crisis, not of poor management," and added later in the interview "surely we could have done things better, we do not say that we were perfect."

[4] From Cesar Molinas, "Theory of Spain's Political Class," *El País*, September 12, 2012.

governments, as well as thousands of public agencies and companies that became instruments of political patronage. The subsequent decentralization of political parties that followed led to the emergence of regional, local elites that took over the local and regional institutions, including the *cajas*, whose boards were quickly filled with political appointees who used their position for their own personal gain and/or as a clientelist instrument to finance their projects. These elites have been central to the *political bargains* that led to the banking crisis.

These developments, according to Molinas (2013), can be largely attributed to the collusion between the political and economic elites who developed a system to "extract" resources from taxpayers for their own benefit. They developed the rent-seeking mechanisms that allowed them to extract these resources, and they colonized the institutions that made these decisions, as well as the ones (i.e., the *cajas*) that provided the funding to implement them. In other words, their selfish interests took prevalence over the general ones. This explains why they failed to articulate a clear diagnosis of the crisis (except blaming each other and/or other external forces—the global crisis, the EU, Brussels, or Germany), why they failed to assume any responsibility for the crisis, not even an apology, and why they failed to develop a clear strategy to overcome the crisis beyond waiting for others (the European Union/European Central Bank, EU/ECB) to come up with solutions, and/or wait until it is over. This collusion also explains the resistance from the political and economic elites to reforms because they would jeopardize the existence of the extractive rent-seeking mechanisms that became the main source of rent for most of them. As a result, reforms were often equated during the crisis with fiscal consolidation (budget cuts and tax increases) while making decisions (like the cuts in education, research, development, and innovation) that would be detrimental to the future competitiveness of the country (Molinas 2013, pp. 165–84).[5] The capital flight that took place during the crisis, also symbolized the selfishness of the country's elites: in one month (July 2012), they withdrew 74.2 billion euros from Spanish financial institutions, the largest amount since 1997, and between July 2010 and 2011, the net fall in deposits reached 55 billion.[6] If the elites did

[5] See Cesar Molinas, "Theory of Spain's Political Class," *El País*, September 12, 2012.

[6] "La banca española sufre en julio una fuga record de depósitos," *El País*, August 28, 2012 and "El Banco de España cifra la caída real de los depósitos en 55.000 millones," *El País*, September 19, 2012.

not trust the future of the country and pulled out their capital, can the regular citizens believe in its future?

But the problem, as convenient as this "theory of Spain's political class" is, is not only with the political class and the extractive political and economic elites, but also with a civil society that tolerated such abuses and repeatedly voted for politicians accused of corruption. As widespread and popular as it became across the country to blame politicians (who became favorite scapegoats for everything that was wrong with the country), this fixation also offered a convenient excuse to overlook Spanish citizens' collective responsibility for the crisis. In many ways, it would be far easier if political parties and the political class could be blamed for everything. However, this explanation fails to account for citizen and voter behavior. Were these parties and these politicians imposed? Who voted for them? Who failed to hold them accountable? Who ignored the rampant corruption and continuing scandals? Who questioned their decisions? Who wondered where the funds for those scandalous projects came from? These are all important questions that also need to be answered. The crisis exposed a passive society that failed to hold its political class accountable, that was not vigilant, and that was largely more interested in perpetuating and living the "fiesta" than in asking the tough questions challenging the status quo. As long as society benefitted, it did not question the situation.

Many see in this behavior one of the enduring legacies of Francoism. Political participation and political dissatisfaction in Spain have been a subject of extensive research (Pérez Díaz 1993; McDonough et al. 1998). These studies emphasize the exceptionally low rates of civic engagement and political participation in Spain, the very low understanding of important political issues among Spanish citizens, and the systematic distortion of the public sphere. Fishman (2004) brilliantly explored the paradox of a country in which collective protests end up "disengaging" rather than "engaging" and in which the public sphere often elicits disappointment, despite the high number of expressed grievances.

The crisis, if anything, proved the conclusion that Spain had a largely apolitical society, that the "emperor had no clothes." Indeed, a European poll from *Eurobarometer* (June 2013) showed that Spain was, together with Portugal, the European country in which there was the least interest in politics: Only about 30% of the population showed some interest in politics. According to Vallespín in studies about political commitment, Spain scored high only in the "demonstrations" variable.[7] In other words, Spanish citizens were "reactive"; they mobilized ad hoc demonstrations when specific interest that affects them are touched but later disconnect very rapidly at the same speed with which they were compelled to demonstrate on the streets. The expectations created by the May 15 (M-15) movement (in many ways similar to the "Occupy Wall-Street" and "We are the 99" movements in the United States), with its indictment of the political class, did not prove lasting. Despite the emergence of new political parties (notably the leftist *Podemos*-We Can-and the centrist *Ciudadanos*, as discussed below), voters continued voting for the traditional parties, and the percentage of absenteeism or null votes increased very marginally in immediate elections that followed the crisis. This lack of citizens' responsibility, not just the politicians', also needs to be accounted for. If there is an important lesson from the crisis, it is that citizens need to assume their responsibilities, that it is time for Spanish citizens to stop merely protesting and to start engaging, that they have to channel the high level of popular energy toward becoming real citizens who hold their governments and politicians accountable for their acts and decisions. As of fall 2019, it is still not very clear whether the crisis has had any such strong lasting effect.

Spanish society confronted one of its worst crises in its recent history, yet it did so from a position of pessimism. Polls showed that the crisis led to a profound sense of demoralization and a crisis of self-esteem, with Spaniards being increasingly pessimistic about their future. Indeed, according to polls from the Center for Sociological Research (CIS), the assessment of the future evolution of economic conditions deteriorated markedly in less than one year: 32.8% of the respondents felt that it would worsen by November 2011 and 40.5%, by July 2012, and 48.6% felt in July of 2012 that their personal economic situation was worse than before.[8] In some ways, the crisis led observers to look at what happened in 1898, when Spain lost its last colonies (Cuba and the Philippines) in the war against the United States.[9] This "disaster" was the spark of a profound reflection about the country's political structures, and it ignited

---

[7] See Fernando Vallespin, "¿Súbditos o ciudadanos?" *El País*, September 14, 2012.

[8] See CIS' *Barómetro Julio 2013*, and "La evolución de la percepción económica y los principales problemas," *El País*, September 11, 2011.

[9] From Andrés Ortega, "La desmoralización de España," *El País*, September 11, 2012.

a debate about the need to reform them to allow Spanish people to fulfill their political aspirations. It would take almost a century of political turmoil, and a bloody civil war, to fulfill that goal. With the transition to democracy in the 1970s, it seemed that the country had finally found the political system that would allow it to move forward and complete the processes of democratization and modernization, while resolving some of the country's historical challenges (such as dealing with the aspirations of its three historical regions: the Basque Country, Galicia, and Catalonia). The crisis placed this achievement into question. The enormous progress of the previous three decades stalled and was now being questioned (particularly by *Podemos* and the pro-independence parties in Catalonia). The crisis has been viewed as a deep humiliation by the Spanish people and has had a profound demoralizing and debilitating effect, both at the individual and collective levels. One of the manifestations of this phenomenon was the generalized resignation and sense of apathy against the crisis. Another was the yearning for the past, particularly for the transition years in which politicians were willing to work together and overcome their differences in pursuit of the common good, something lacking during the crisis.

Institutional Degeneration

As we have seen throughout the book Spain experienced between 2008 and 2013 not only a political and economic crisis but also an institutional one. Indeed, a salient feature of the crisis was the extent to which the country's institutions, many established during the democratic transition of the 1970s, were battered. One of the worst consequences of the crisis, beyond the dramatic social and economic costs, was the delegitimization of institutions at all levels (see Table 7.1). The discredit of institutions was running wide and deep, partly as discussed above, not only because they had been colonized by the political elites regardless of qualifications, but also because Spanish citizens tolerated it: They were willing to live with dysfunctional institutions as long as they benefited from them, rather than risking institutional changes that could work against them.[10]

---

[10] Ignacio Torreblanca stated it very succinctly: "All sides prefer to live with malfunctioning institution rather than one that may work against them," in "Spain Post-Franco Institutions Battered by Contact with Politics," *Financial Times*, July 23, 2012.

Table 7.1 Consideration of different institutions

	2007		2011	
	Mean	Std. deviation	Mean	Std. deviation
The King	7.20	2.83	5.79	3.32
The Constitutional Court	6.08	2.58	5.00	2.91
The Congress	5.54	2.47	4.87	2.76
The Senate	5.45	2.49	4.07	2.78
The Spanish Government	5.46	2.87	3.93	2.97
The Judiciary	4.79	2.58	4.28	2.81
The Armed Forces	6.72	2.66	6.83	2.67
The European Union	6.67	2.15	5.79	2.59
The NATO	5.20	2.71	5.17	2.79
The Autonomous Parliament	5.67	2.52	5.08	2.81
The Autonomous Government	5.66	2.71	4.83	2.86
The European Parliament	6.06	2.24	5.16	2.56
Your City Hall	5.45	2.82	5.29	2.86
The Church	4.19	3.25	3.76	3.30
The National Policy and Civil Guard	7.03	2.50	7.01	2.56
NGOs	6.17	2.53	6.37	2.64
Political parties	4.22	2.39	3.38	2.59
Trade Unions	4.58	2.61	3.26	2.80
Business Associations	4.92	2.12	4.32	2.70
The Media	–	–	5.05	2.58

Source Encuesta de Cultura y Representación Política en España (CSO2009-14381C03-01). CIS, Junio-Julio 2011

However, although much of the focus from Spanish citizens and Spanish media was on the impact of the crises on institutions, which had been battered by the crisis, in this chapter I make the case that the process of institutional degeneration preceded the crisis, and was in fact a contributing factor to the crisis. Indeed, in the years prior to the crisis Spain seemed to have fallen into the category of countries in which institutions had become extractive and concentrate power and authority in the hands of a few. In a recent seminal work, Acemoglu and Robinson (2012) show that nations thrive when they develop "inclusive" political economic institutions and fail when they have extractive ones that concentrate power and opportunity in the hands of the elites (pp. 73–79). They show that inclusive economic institutions "that enforce property rights, create a level playing field, and encourage investments in new technologies and skills are more conductive to economic growth than extractive economic institutions that are structured to extract resources from the many by the few ... [inclusive ones] are in turn supported by, and support, inclusive political institutions, [which] distribute political power widely in a pluralistic manner and are able to achieve some amount of political centralization." They conclude that "it is politics and political institutions that determine what economic institutions a country has" (p. 43). According to their view, the role of inclusive political institutions is to promote sustainable economic growth, which requires innovation. Therefore, they need to protect and empower citizens to innovate and invest in order to foster Schumpeter's process of "creative destruction," which is conducive to innovation.

Unfortunately, the crisis in Spain exposed an institutional model that in many cases unleashed an extractive model of crony capitalism. A central problem, as noted by Molinas (2013), was the colonization by the new political elites that emerged since the transition of critical institutions, such as the Constitutional Court, the Bank of Spain, the General Council of the Judiciary (CGPJ), or the National Stock Market Commission (CNMV). These institutions became politicized and largely lost legitimacy (see Table 7.1). As a result, institutions that should have played a central role holding politicians accountable (the judiciary, parliament, regulatory bodies...) largely failed to fulfill that responsibility and instead have become transmission belts of the politicians (and worse, in many cases part of a patronage system), validating their (often questionable) decisions, and giving them a free pass.

The problem extended to political institutions such as the Spanish Congress, which failed to fulfill its accountability role in the years prior to the crisis. Not surprisingly, Spaniards largely believe that the political system does not represent them (see Table 7.2).

The Senate is predominantly considered useless (see Table 7.1) (for instance, despite the fact that it is considered the regional chamber, in the 2008 legislature it did not introduce a single amendment to the half dozen regional statutes that were approved by Congress), and Congress (the lower chamber) is dominated and instrumentalized by the party that holds a majority. Since both the Socialist and the Conservative governments often resorted to decrees for most of their measures, and the prime minister (PM) rarely spoke (particularly Rajoy) in parliament, both chambers largely appeared as spectators of the crisis. Parliamentary rules allow the government with an absolute majority (or with a sufficient

Table 7.2 Spaniards opinion about the degree of representability of their political system

In your opinion: In Spain the PP and PSOE...		Does the congress of deputies represent the majority of Spaniards?	
	June 2011		February 2013
Represent the interest of Spaniards	18.9	Yes	23.5
Represent only the interests of part of the citizens	24.9	No	73.9
Represent their own interests as political parties	51	Do not know/do not answer	2.6
Do not know/do not answer	5.2		

Source Metroscopia

parliamentary majority in case of coalitions) to avoid debates or even full disclosure, and they can veto calls from the minority parties requesting the presence of cabinet ministers for questioning. During its first eight months in power since the 2011 election (in which it won an absolute majority), the PP submitted to Congress 27 royal decrees that only required the validation of Congress without any opportunity for amendments. The PP government allowed only eight of them to be processed as law proposals. Since 1978, there had never been a year in which a government approved as many decrees as the PP in 2012. During that period, Prime Minister Rajoy went to Congress only when obligated (i.e., during the control sessions and after the two European summits). All the other parliamentary requests to question him were rejected. When the 65-billion euros bailout package was rammed through Congress in July 2012, he was absent from parliament (and during the debate, one of the PP deputies caused a political storm when she greeted the cuts for unemployed by saying '*que se jodan*'—'let them screw themselves').[11] And in 2012, there was not even a State-of-the-Nation debate. In one hour, Congress approved 10 billion worth cuts in health and education. In the summer of 2012, while the German Bundestag was debating the conditions for the financial rescue to Spain, the Spanish Congress was on recess, and it did not even vote on that memorandum (the only way it was

[11] David Gardner, "The Silent Rajoy Is Deaf to the Spanish Emergency," *Financial Times*, August 6, 2012.

vetted was in a commission attended by Minister Guindos, but without a vote). The Spanish daily, *El País*, submitted a request for information on official trips from members of Congress; initially, it received a positive response, but three months later, it was informed that the secretary-general had decided that such information could not be released, and that it would have to wait for the implementation of the new Law of Transparency. And there were no firm plans to change any of this: The Socialists, now that they were out of power, presented some proposals to change the electoral law and the internal functioning of Congress, but the PP did not make it a priority to address them. Five legislatures before (!) a group finished a proposal with the support from all parties to reform the Congress' bylaws, but the initiative was stopped by the Spanish Socialist Workers' Party (PSOE) and the PP, and it has not been reopened. In sum, the walls that surround the Spanish Congress are the metaphors of separation between the legislative powers and the citizens.[12]

In addition, time after time investigative commissions worked as forums to express grievances, but rarely as places in which people were held accountable, or that led to any significant political and/or criminal responsibilities. The commission that looked at the collapse of *Bankia* was just an example of this diluted (if not nearly useless) role: Former leaders of the bank, former political leaders (including the former Minister of Finance Salgado), and the former Governor of the Bank of Spain were all questioned about the disaster. They all gave their own reasons, largely blaming others for the outcome, yet nothing substantive came out of the process (which was also under judicial review). For instance, *Bankia*'s former chairman, Rodrigo Rato, questioned by the parliamentary panel, underlined the extent to which his decisions were regularly approved by auditors, financial consultants, regulators, and the Bank of Spain's inspectors, and directly contradicted the testimony of Miguel Ángel Fernández Ordoñez (MAFO), the Socialist-appointed Bank of Spain governor, who had stepped down a few weeks earlier, and who had claimed during his testimony that he had not pressed Mr. Rato to merge *Caja Madrid* with the Valencian *Bancaja* to form *Bankia*. Mr. Rato declared that in June 2010 he was called to the Bank of Spain and in effect was forced to negotiate with *Bancaja*.[13] Reports showed that *Caja Madrid* gave loans to

---

[12] From "Parapetados tras las vallas," *El País*, September 15, 2012.

[13] Bankia announced losses of 4.4 billion euros in the first half of 2012, which forced the government to inject cash. "Ousted Bankia Chief Blames Central Bank and Politicians,"

customers who lacked the resources to pay back.[14] The fourth largest financial institution collapses and is no one responsible?[15] How can we be surprised that a government that seems incapable of engaging its own citizens, or institutions such as the Congress, can inspire any confidence on markets and investors? That summed it all.

This colonization of institutions from politicians extended to public companies and agencies, including the *cajas*, which were staffed (and led) by acolytes of the politicians with no educational background or professional experience in the field, who were appointed for their loyalty and allegiance to their political patrons. Under their leadership, they became piggy banks and instruments for regional and local leaders of all parties to distribute patronage. As noted in Chapter 1, a study from Vicente Cuñat and Luis Garicano (2010) shows that the main difference between banks and *cajas* was not so much the latter's political nature but the lower level of professionalization of their managers: Only 31% of their presidents had postgraduate degrees—half of them had banking experience and half of them had occupied political positions before becoming presidents. According to them this development means that *cajas* could have saved 12,000 million euros have they had better prepared and qualified managers without a political past. *Cajas* with a political president had on an average 0.93 points more of delinquent loans than those that did not, 0.98 points more if the president did not have postgraduate degrees, 0.93 points more when they had no financial experience, and 2.84 points more if they met all the three conditions.[16] This was not an accident; it was an outcome of the *political bargains* that we have examined throughout the book.

It should not be surprising, therefore, that many of their decisions led to the financial bailout.[17] The real estate boom at the heart of the crisis

---

Financial Times, July 27, 2012 and "Spanish Bank's Ex-Leader Defends Record There," *New York Times*, July 27, 2012.

[14] "Caja Madrid concedía prestamos a clientes sin capacidad de pago," *El País*, September 23, 2012.

[15] Andrés Ortega, "La desmoralización de España," *El País*, September 11, 2012. As of January 2020 we are still waiting for the sentencing of the *Bankia* leaders who were put in trial.

[16] See "La Politización eleva la morosidad de las cajas en 12.000 millones," *El País*, October 31, 2010.

[17] Andrés Ortega, "La desmoralización de España," *El País*, September 11, 2012.

was fueled by the *cajas*, and the property burst blew holes in their balance sheets that are the heart of the current financial crisis. The government's response to the crisis of the *cajas* also leaves many questions unanswered: Why did it give 16 billion to *CAM* or 23 billion to *Bankia* instead of letting them fail? Why did it give 5 billion to *Bankia* instead of waiting for the EU funds?

Other institutions that should have played a vital role to build up confidence, both domestically and internationally, such as the Bank of Spain (which, as we discussed on Chapter 5, was initially widely praised), had already been dragged to the mud of partisan warfare prior to the crisis, and its governor MAFO was compelled to resign before his term expired amid the controversy of *Bankia*'s collapse and his failures in managing the financial crisis. This was another instance of damaging politization of the institutions and dereliction of responsibility from the two leading parties: by appointing as governor someone with a strong reputation but who did not have the appropriate background and was politically closely linked to the Socialist government that appointed him (he had long been associated with the Socialist Party and had served in several government positions throughout his career), and also by failing to provide the necessary regulatory oversight over the institution.

The role of the BoS in the financial crisis cannot be underplayed. As we discussed on Chapter 6, the banking crisis could also be blamed on the *actions (and inactions) of the Bank of Spain.*[18] As we have seen before (Chapter 5), at the beginning of the crisis, the Bank of Spain's policies were all praised and were taken as model by other countries. Time, however, has tempered that praise and the BoS are now widely criticized for its actions and decisions (or lack thereof) during the crisis.

The BoS anticipated the potential pitfall from the real estate bubble as early as 2003, when it alerted in one of its Economic Bulletins that the magnitude of the bubble could be "as high as 20 percent." This caused a conflict with the Conservative government, and Rodrigo Rato, the minister of finance at the time, questioned the report. This was the first instance of a battle that marked the course of action in subsequent years and this, despite the Autonomy Law, protected the BoS against political interference. Indeed, the conflicts between the government and the BoS, between the political and the financial powers, are crucial to

---

[18] This section borrows from "El día que el Banco de España se doblegó," *El País*, March 13, 2011 and "Un gobernador entre dos fuegos," *El País*, May 13, 2012.

understanding what has happened as well as why the BoS failed to act to burst the real estate bubble, and to address the recapitalization of the cajas until it was too late (see Ekaizer 2018). Naturally, any action to bust the bubble was opposed by the governments (both the PP and PSOE ones) because of the negative impact it would have had on economic growth and consumption.

Despite the BoS' warnings, mortgages continued growing at a fast pace: a 20% annual rate, while banks and *cajas* continued to ignore its recommendations. Pushed by intensifying competition for growth and market share *BBVA* increased its mortgages by 17%; *La Caixa* by 23.4%; *Caja Madrid* by 24.9%; and *Banco Santander* by 14.9%. The herd logic, once again, triumphed.

Governor MAFO was also severely criticized. As mentioned before, he was a Socialist Party militant who had served in government in different capacities since the 1980s, he had scant financial experience when he was appointed Governor of the Bank of Spain (this shortcoming has been used by many observers to explain the deficiencies in the design and implementation of the financial reforms). Prior to assuming the governor's position, Governor Ordoñez had played a very prominent and public role in Spanish media criticizing the country's growth model and its dependence in the construction sector. However, he failed to heed his own advice when he became Governor. Facing strong opposition from the politicians and regional governments that controlled the *cajas* and were very reluctant to relinquish their power, he was not able (or did not dare) to confront them. This was a key factor in explaining the delays in the BoS' intervention, which often waited too long and/or acted when it was too late, resulting in the problems getting out of hand (as it happened with the *cajas* from the Valencia region, the Galician ones, or even with *Caja Madrid*).

The arguments often mentioned by the BoS—that it lacked control of monetary policy (which was in the hands of the ECB), that it would have been counterproductive to burst the bubble because it was already declining, and that it did not have the tools to burst the bubble—seem disingenuous at best.[19] While it is true, as noted on Chapter 6, that the

---

[19] As noted on Chapter 6, The BoS defended itself with seven arguments: that it lacked the appropriate instruments to address the crisis until 2009; that the economic deterioration was far longer and intense than anticipated; that the government had decided to reject the creation of a "bad bank"; that international banks were unable and unwilling

establishment of the countercyclical regime was costly for the BoS and led to significant confrontations with the banks that complained bitterly that it placed them at a disadvantage with their international competitors because it reduced their profit margins. There was also a sense of complacency that the regime would provide a sufficient cushion if the real estate sector fell down. Yet, it was the BoS itself that approved the high concentration of risk in the real estate sector. Moreover, as noted on Chapter 6, there were other actions that the BoS could have taken, and should have taken, to reduce the risk and dependence from the construction sector, such as increasing the required provisions together with the ECB, or strengthening the regulatory framework. Political considerations (both governors, Caruana, and MAFO, wanted to avoid a confrontation with the PP and the PSOE) and concerns over potential litigation drove the agenda and precluded more decisive action to address the problem.

In the end, as we discussed on Chapter 6, the BoS chose the path of least resistance: alerting about the risks but failing to act decisively, despite the fact that there were people within the bank who had been very critical of such inaction and lack of leadership. Even MAFO's successor as Governor of the BoS, Luís María Linde, his successor as Governor of the BoS, Luís María Linde, recognized in Congress in 2012 that the Bank of Spain "was not successful in what we call macro-prudential supervision," and admitted that the BoS had acted "insufficiently or inadequately" to the crisis, and recognized that it "was not successful in what we call macro-prudential supervision."[20] Furthermore, the Inspectors Association of the Bank of Spain published in May 2006 a letter to the minister of finance, Pedro Solbes, criticizing Governor Caruana's passivity in the face of the accumulated and growing risk associated with the real state sector. They attributed the extraordinary growth of the real estate prices to the "excessive growth of banks," and accused the governor of "lack of determination" in demanding "rigor" from financial institutions "when

---

to participate in the merge and acquisition process that was taking place in Spain because of the crisis; that they could not use the traditional tool to liquidate banks, namely forcing bondholders to take loses because of the potential contagion effect to other banks and *cajas*; that the corporate governance system of *cajas* was very deficient and politicized; and that the autonomous governments refused to approve the merges during the restructuring process. See "El Banco de España se defiende de su lentitud con siete argumentos," *El País*, June 11, 2012.

[20] "El Nuevo gobernador critica la gestión de sus predecesores en el Banco de España," *El País*, July 17, 2012.

assuming risks." This experience showed the increasing concentration of power away from the BoS, in favor of the government, the cajas and the banks.

And the battering also reached previously untouchable institutions like the monarchy, which no longer seemed like the unifying force that it had been in the past (see Table 7.1). King Juan Carlos was revered for his role in bringing democracy to the country. During the crisis, however, Spaniards started questioning their king, and scrutinizing his lifestyle (and that of his family) and the lack of transparency. The controversy over the king's elephant-hunting trip to Botswana in the spring of 2012, in which he broke his hip, generated a public outcry because it exposed a murky world of business contacts and details about the king's lifestyle (including speculation about his relationship with a German princess who accompanied him in the trip) at a time of national crisis. It led to an unprecedented public apology, but the damage was already done. The royal family's estimated fortune at up to 1.79 billion euros had been the subject of growing scrutiny. The safari was organized by a Syrian magnate who had worked together with the king on a 9.9 billion bullet train contract that the king had helped broker in 2011 for a Spanish consortium in Saudi Arabia. This controversy was compounded by an influence-peddling case aimed at the king's son-in-law, Iñaki Urdangarín, who was sentenced to jail for using a non-profit foundation to embezzle public money for sporting events and to use his position to bypass standard bidding procedures. A growing number of Spaniards, including some smaller parties, used the crisis as a further reason to challenge the monarchy.[21] The compounded scandals ultimately led to King Juan Carlos' abdication in June 2014. His son Phillip was crowned and has been able to restore some support for the institution. According to polls, most Spaniards still support the monarchy and value the role of the king as a representative of the country and a unifying force, yet they yearn for more transparency.

Another crucial institution, the family, which historically has been a stabilizing force and has played a crucial role in the provision of welfare,

---

[21] "Chastened King Seeks Redemption, for Spain and His Monarchy," *New York Times*, September 29, 2012. Polls from *Metroscopia* (April 2013) showed that the king's popularity rating had fallen from a positive 21 points to a negative 11 (among those aged 18–34 the figure was minus 41). He polled below 27 other social institutions, including tax inspectors. The *Centre of Sociological Research* in 2011 showed for the first time that the monarchy's popularity fell below 5, achieving 4.89 out of 10.

also seemed to be fraying as unemployment increased and any members of the family unit lacked any source of income. More and more families had to lean on their elderly relatives (pensions were among the very few benefits that were not slashed). The situation was so dire that many families were removing their relatives from nursing homes, so they could collect their pensions. A survey from *Simple Lógica* found a sharp increase in the number of older people supporting family members: In February 2010, 15% of adults, 65 years or older, said they were supporting at least one relative, and in the survey conducted two years later, the number had increased to 40%; the association of private nursing homes has reported that 76% of its members had vacancies in 2009, while that number had increased to 98% in 2011. Retired people willing to share their pensions to support their families were the silent heroes of this economic crisis. This may have been one of the reasons why conflict in the streets was not even more intense.[22]

Institutional Divergence

A central theme of this chapter is that the crisis in Spain cannot be explained without reference to the process of institutional degeneration that took place in the country in the years that preceded the crisis. This institutional degeneration has had economic consequences. Membership in a monetary union would have required a series of structural reforms to ensure that the economic structure of the country was prepared to compete in the context of a monetary union. However, successive governments failed to address the imbalances of a growth model based largely on construction (it was an economic 'miracle' based on bricks and mortar) and tackle the core problems of the Spanish economy: inflation differentials, the loss of competitiveness, and the relatively low productivity of labor.

While factors such as inertia driven by economic success, the process of European integration, social bargaining, and ideological/programmatic consensus, all help explain continuities in economic policies in the years prior to the crisis and the failure of successive governments to address the shortcomings of the Spanish economy (see Royo 2013, pp. 105–18),

---

[22] From "Spain's Jobless Rely on Family, a Frail Crutch," *New York Times*, July 29, 2012.

the process of institutional degeneration was also a crucial factor. Political and economic institutions resisted those reforms because they would have jeopardized the existence of the extractive rent-seeking mechanisms that became the basis of the *political bargains* examined in this book, as well as the main source of rent for the economic and political elites who controlled them. Indeed, the economic divergence in productivity and competitiveness between Spain and the EMU core that preceded the crisis were consequences of the institutional divergence in the rule of law between the core and Spain. In other words, competitiveness divergence was rooted on rule of law divergence. Finally, this institutional divergence made it harder to implement the structural reforms that EMU membership demanded.

Competitiveness, broadly defined, includes the ability of the government to approve and implement effective laws; the protection of intellectual property; the efficiency of the legal framework; lack of corruption; the ease to set up new business; and predictable and effective regulations.[23] In the case of Spain, there is ample evidence that the country had been suffering an institutional loss of competitiveness in the years that preceded the crisis. Here is some evidence from the *World Economic Forum*'s annual competitiveness' report (see Table 7.3).

According to the *World Economic Forum*'s annual 2012 competitiveness ranking of 144 countries, Spain was placed thirty-sixth. The report highlighted the vast differences between the best and worst performing nations in the Eurozone, stressing that the process of convergence among Eurozone economies had reversed and is one of the causes of the current difficulties of the Eurozone. According to the report, "one of the shared features of the current situation in all these [Southern European] economies is their persistent lack of competitiveness and their fore their inability to maintain high levels of prosperity ... Over all, low levels of productivity and competitiveness do not warrant the salaries that workers in Southern Europe enjoy and have led to unsustainable imbalances, followed by high and rising unemployment."[24]

While it is true that these surveys are based largely on survey data and therefore are subjective, other sources seem to confirm that premise. The

[23] See *Global Competitiveness Report* from The World Economic Forum.

[24] See http://reports.weforum.org/global-competitiveness-report-2012-2013/ (p. 24). See also "Competition Gap Grows in Europe, Study Says," *New York Times*, September 6, 2012.

Table 7.3 WEF competitiveness ranking. Spain 2006–2010

	Ranking			
	2006–2007	2007–2008	2008–2009	2009–2010
Public institutions	42	45	44	50
Public trust in politicians	35	46	39	50
Judicial independence	60	68	56	60
Favoritism in decisions of government	65	66	43	46
Transparency of government policy-making	70	67	89	80
Burden of government regulation	45	60	94	105
Quality of the educational system	53	52	52	78
Quality of math and science education	74	69	78	99
Hiring and firing practices	110	115	116	122
Exports as a percentage of GDP	99	104	102	104
Total tax rate	172	97	110	109
Pay and productivity	67	74	84	91
Ease access to loans	38	43	47	71
Capacity for innovation	29	35	30	34
OVERALL	29	29	29	33

Source *Global Competitiveness Report.* Various Years

International Finance Corporation's *Doing Business,* which reports data on the ease of doing business, ranked Spain number 30 in 2006, 39 in 2007, and 44 in 2013. In 2004 in variables like "enforcing contracts" the number of procedures was 20, the number of days 147, and the cost (as % of income per capita) 10.7. In 2006, it was 23 procedures, 169 days, and 14.1% (see Fig. 7.1).

In the World Justice Project's *Rule of Law 2013 Index* (which provides data on nine dimensions of the rule of law) Spain scored relatively low in providing mechanisms for public participation, in effectively enforcing government regulations (ranked 22), in judicial delays, in effective enforcement of civil justice, and in police discrimination. The World Bank indicators of *World Governance* suggested that between 1996 and 2005 Spain had suffered a decline in the quality of governance in dimensions such as: control of corruption (83.9–82.5%), government effectiveness

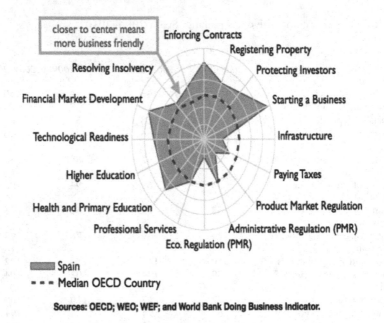

Fig. 7.1 Business environment. Selected ranking within OECD, 2013 (*Source* IMF, *2013 Article IV Consultation*, p. 22)

(from 90.2 to 89.3%), regulatory quality (from 84.8 to 88.2%), the rule of law (from 90.9 to 84.2%), and voice and accountability (from 89.9 to 84.1%). Transparency International's *Corruption Perception Index* ranked Spain 22 in 2001, 30 in 2012, and 40 in 2013. All this data seemed to confirm a decline in institutional quality. And this deterioration was even more meaningful in comparative terms with the core EMU members, thus leading to a process of institutional divergence.

The performance of the education sector, another key institution, also left much to be desired in the years prior to the crisis: Almost one in every three people between the ages 18 and 24 were early school dropouts (this was double the EU average), and according to the Organization of Economic Cooperation and Development (OECD), results of Pisa's test in reading, math, and scientific knowledge were poor; there was no university in the top 150 in the main rankings; up to 35% of university students dropped out before graduation and only a third completed it on time.[25] In 2012, 47% of Spaniards had only completed primary education (24% in the EU and 26% in the OECD): 22% high school (48% in the EU and 44% in the OECD) and 31% university degrees (28% in the EU and 30% in the OECD); the graduation rate in high school was 48% and in professional training, 28%; and 26.5% of youths between 18 and 24 years old did not study beyond mandatory education.

It is important to highlight that the problem was not just insufficient resources: Average annual public spending per student in public education in Spain was comparatively higher—$10,094 in Spain, $8307 in the EU, and $8329 in the OECD.[26] Research, development, and innovation spending, at 1.38%, were significantly lower than the EU average. The 2013 OECD's Pisa report showed that in math and reading outcomes there had been virtually no improvements during the prior decade (i.e., Spain was ranked 33 of 65 countries in math), and this despite the fact that the budget on education had increased 35% during that decade.

Finally, the situation of the judiciary, a central institution to apply the law and hold people accountable, has also been deteriorating. One structural problem has been that Spain's courts are unprepared and do not have the means to deal with the quantity and the complexity of cases that have emerged during the last decade. Courts were particularly drowned with corruption cases since the beginning of the 2008 financial crisis (more than 800 as of summer of 2013), some of them affecting previously untouchable institutions like the royal family (Villoria and Jiménez 2012).[27] According to Joaquim Bosch, a magistrate who was also the national spokesman for Judges for Democracy, their case load grew twofold to threefold between 2008 and 2013, leaving some

[25] William Chislett, "Expect the Pain in Spain to Continue," *Financial Times*, April 12, 2011.

[26] Data from the OECD. See "Estado de la educación 2012," *El País*, September 14, 2012.

[27] The 2007 CIS's 2671 survey asked participants "Please tell us to what extent you feel that corruption is widespread in politics: very, quite, not very, not at all." The response showed that 51.9% of those surveyed responded with 'quite' or that nearly all politicians were involved in corruption. This is moderate compared to the data from the CIS's June 2011 barometer (Survey 2905) where the "very or quite widespread" figure reaches 86.6%; and the 2826 survey which showed that 79% thought it was very or quite widespread and a mere 6.5% thought that it was not very or not at all widespread. See more at: http://phys.org/news/2012-06-perception-corruption-spain-eur opean-average.html#jCp.

Table 7.4 The justice system treats wealthy and poor people the same (%)

	2009	2016
Agree or strongly agree	12.1	10.7
Disagree or strongly disagree	82.1	88.8

Source CIS and Metroscopia

specialized courts which deal with special crimes and corruption (like the *Audiencia Nacional*) on the brink of collapse. By summer 2013 the judiciary had accumulated a backlog of more than three million unhandled complaints, a problem compounded by the government's decision to make budget cuts. As a result, courts were overwhelmed with caseloads and were prone to unreasonable delays (the majority of cases languish in courts for years). In addition, they were perceived as easily politicized and public trust was undermined by persistent leaks (according to a poll from *El Pais*, 92% of people surveyed in January 2013 agreed that the slowness of the Spanish courts made it harder for them to fight corruption). And citizens did not perceive that the judiciary treated them fairly (see Table 7.4).

Courts' decisions have been largely determined by the preferences of the particular judges in charge of any given cases (in Spain judges take the lead in investigations, rather than prosecutors, and they have the power to stall or speed up cases and to decide how far to pursue any investigation), and some of them became media stars, upstaging politicians as decision-makers. Finally, the constant judicial reforms seem to run in one direction; increasing political control over the judiciary. As a consequence, the country was also suffering a judicialization of politics, which has compounded the traditional perception of politicialization of justice.[28]

Conclusions: The Political Consequences of the Crisis

In the years prior to the crisis, the country seemed to be in 'a place of suspended effects,' an Indian summer characterized by a series of

---

[28] From "Political inquire Makes Judge a Star, as It Reveals Flaws in Spain's Courts," *The New York Times*, Wednesday, July 31, 2013.

bubbles; the housing bubble, the stock market bubble, the alternative energy bubble … One by one they all burst, and their bursting showed that they had been merely temporary solutions to long-term problems (Packer 2013, p. 382). The bubbles disguised the reality that things were fundamentally not working, they were the excuse not to address the fundamental structural challenges that the Spanish economy faced since it joined the EMU, and they provided distractions and evasions from those long-term challenges. But during that time key institutions continued to erode.

Sordid practices had penetrated Spanish society at all levels as shown by the still ongoing *Villarejo scandal* that involves allegations of illegal corporate espionage (as well as rivalries within the country's power elite). This case is hanging over some of the country's best known banks, companies, and business leaders, and it is likely to leave a mark on the country's business and political landscape for years to come. BBVA had already been placed under formal investigation by a court in July 2019 in a case involving phone tapping and bribery, and its former chairman, Francisco González, who led the bank for 18 years has also been placed under formal investigation. The BBVA hired Villarejo, a seedy former police commissioner with close links to the Franco regime who also run a corporate security company, to help frustrate a planned takeover that was perceived to be backed by Zapatero's Socialist government, and continued using his services until 2017 when he was imprisoned on allegations of bribery, money laundering, and operating a criminal organization. His trial is still pending (as of January 2020), but so far it has exposed hundreds of encrypted recordings that he has been making for over twenty years, some of which have been leaked to the media. The scandal has exposed the lax practices of Spain's old order, and much more may be exposed before the case is closed. But its implosion hopefully highlights a new era for Spanish companies.[29]

Indeed, political factors shaped banking outcomes in Spain. The erosion of institutions had profound consequences for the Spanish economy and its banking system, because they set the rules (Calomiris and Haber 2014, pp. 13–14): They determined which groups were included in the government-banker partnership, and which ones were left out. These institutions produced a set of rules that resulted in abundant credit

---

[29] See Daniel Dombey, "Legal Cloud Hangs over Some of Spain's Best Known Companies," *Financial Times*, January 2, 2020.

for some sectors of the economy (e.g., real estate), but scarce credit for others (see Fishman 2012). Furthermore, the coalitions that emerged as a result of the process of institutional degeneration described in this chapter determined the rules governing banking as well as the flow of credit and its terms, and they established how the banking game was played, prioritizing some actors (e.g., the leaders of *cajas*, developers, and their political allies) at the expense of others (e.g., other productive sectors of the economy). Not surprisingly, the groups in control ended up receiving the larger share of the benefits with disastrous consequences for the country once the crisis hit. And the crisis persists, *Transparency International,* still ranked Spain number 58 in 2018 on its *Corruption Perception Index* (it had been ranked 58 since 2025), and as late as July 2017 corruption was still listed as second main concern by Spaniards (see Table 7.5).

Moreover, the impact of the institutional crisis has also led to important political developments in the country. Spanish society has experienced profound changes in the years that followed the economic and political crises that have crystallized in a crisis of representation characterized by a deep breach between citizens, and politicians/political parties. This breach has been deepened by the perceived lack of responsiveness to citizens' needs, as well as by the absence of clear diagnoses and plans to address those needs from the main political parties (Urquizu 2016). This crisis of representation has led to collapse of the traditional two-party system and the further fragmentation of the political system, which is making not only the formation of governments more difficult (the country has had four general elections between 2015 and 2019), but also

Table 7.5 What is the main problem that currently exists in Spain? (% of the population that selects each option as the main problem)		
	Unemployment	71.2
	Corruption and fraud	49.1
	Economic problems	21.7
	Politicians. political parties. politics	20.9
	Health system	10.3
	Education system	9.2
	Social problems	9
	Problems related to the quality of employment	7.6
	International terrorism	5.8
	Civic insecurity	3.5

Source CIS

more challenging to take decisive action to address the problems that the country faces.

Indeed, initially Spain followed the pattern of other countries such as Greece, Ireland, France, or the United States, which punished severely the parties that were in government when the great recession started (see Royo 2020). The Socialist Party (PSOE) suffered a humbling defeat in the November 2011 election and lost 59 seats in parliament (from 169 in 2008 to 110 in 2011), while the PP, under its leader Mariano Rajoy, obtained an absolute majority with 10.8 million votes (or 44.5%) and 186 seats in parliament (up from 154). That election marked a transformation in the country's party system. Indeed, from a political standpoint one of the most important consequences of the crisis has been the rupture of the two-party system that had characterized the country's political system since the 1980s in which the Socialist Party (PSOE) and the Conservative Popular Party (PP) had been alternating in power. As noted before, the pain caused by the crisis, the endemic corruption in which largely all traditional parties (both national and regional) were implicated to different degrees, and internal divisions within the traditional parties (notably the PSOE) all fueled a wave of discontent that led to the erosion of support for the traditional national parties (PSOE and PP) and the emergence of new political parties, notably the leftist-populist *Podemos* and the centrist *Ciudadanos*, which have led to the fragmentation of the political system (Royo 2020).

This fragmentation started in 2011: In 2008 the PP and the PSOE controlled 92% of the seats in parliament (the rest were controlled by smaller regional parties), and after the 2011 election 'only' 84.5%; and it crystallized in the December 2015 election in which both *Podemos* (with 69) and *Ciudadanos* (with 40) entered the Spanish Parliament. Voters looking for alternatives to the traditional parties flocked to the new parties, *Podemos* and *Ciudadanos*. These electoral results introduced a new political landscape characterized by increasing fragmentation, volatility, unpredictability, and the need for parliamentary agreements to form a government and get any initiative approved in parliament. The 2016 election produced a similar result: the PP got 137 seats, the PSOE (ravaged by internal divisions) 85, *Podemos* 71, and *Ciudadanos* 32. However, this time the PP was able to negotiate an investiture agreement with *Ciudadanos* and other smaller parties, and Rajoy was re-elected as PM with the abstention of the PSOE, which led a rupture within the party and a painful leadership election process that brought back its leader, Pedro Sánchez. Following a period of stability, a judicial sentence that confirmed the PP's involvement in a corruption scheme and sentenced some of its leaders to long-term prison precipitated a motion of no-confidence in June 2018 that led to the defeat of the Rajoy government and the unexpected election of the Socialist leader Pedro Sánchez as PM, with the support of *Podemos* and other smaller regional parties. While Sanchez's tried to plow through despite his exiguous majority, the failure to approve a new budget precipitated a new general election in April 2019. That election, which was won by the Socialists with a relative majority, resulted again in a fragmented parliament that made very difficult the formation of government and led to a new general election in November 2019 which was won, again, by the Socialists, and led to the formation of a new minority coalition government (the first since the transition to democracy) between the Socialist and *Podemos* in January 2020. In the end, the economic crisis has ushered a new era in Spanish politics that will likely be characterized by the continuing erosion of the two-party system, increasing polarization, enduring fragmentation and hence, more instability.

Furthermore, as we have seen, the crisis also led to the creation of new political parties. The Left-wing populist *Podemos* emerged in January 2015 following the surfacing of the 'indignados movement' and the impasse over the so-called Regime of '78. Indeed, popular and social discontent over the dual crisis came to the fore on May 15, 2011, which became a key day in the political transformation of the country. That day thousands of citizens took over their cities and towns' squares, and over *la Puerta del Sol*-the Sun's Gate (one of Madrid's most representative squares, located at the geographical center of the Iberian peninsula, and came to symbolize this mobilization). This movement was called the 'indignados movement,' or the M-15 movement. Although the movement did not last long (in early August, taking advantage of the summer lull, the police removed them from la Puerta del Sol), it conveyed a powerful message, not just in Spain but all over the world: The Spanish people were fed up, they wanted their voices heard, and they were ready to fight to have their problems addressed. The movement continued operating in neighborhoods, communal assemblies, and in many social movements. And a year later, thousands of citizens returned to la Puerta del Sol to commemorate the movement's first anniversary.

The M-15 movement was very relevant. On the one hand, the movement seemed to show that something had finally changed in Spain. As noted before, the crisis exposed a passive society that had failed to hold its political class accountable, that was not vigilant, and that was more interested in perpetuating and living the "fiesta" than in asking the tough questions challenging the status quo. As long as the society benefitted, it did not question the situation. However, the expectations raised by the May 15 movement, with its indictment of the political class, changed all that. The crisis had a profound effect on many Spanish citizens, as they seem to be behaving differently (Urquizu 2016; Pereira-Zazo and Torres 2019). They became more engaged: Political participation has increased (there was a 71.8 turnout in the April 2019 general election, up 5.3% from the previous one); they seem to have higher expectations of their leaders, as well as lower tolerance for corruption (according to polls, the Spanish population considers corruption their second biggest problem, only behind unemployment, and 1378 officials were prosecuted for corruption between July 2015 and September 2016); and they are willing to hold their leaders more accountable (as evidenced, as we have seen, by the fall of the Rajoy government in 2018 following an accumulation of corruption scandals, and the subsequent electoral disaster for the PP in the April 2019 election in which it lost 15.9% of its votes).

At the same time, the M-15 movement was the precursor of a political movement that transformed the electoral map of the country in 2014 and 2015. At the local and regional level, it was closely connected with the formation of new political parties and coalitions like *Ganemos* o *en Común*, and at the national level it had a crucial impact in the development of *Podemos/Ganemos*, as recognized by the party's founding leaders like Elvira Villa, one of the speakers of *Ganemos Madrid* who acknowledged in an interview with *20 Minutos* that "the M-15 is in our genes, although we are more that the M-15" (see Weiner and López 2018; Pereira-Zazo and Torres 2019).

Podemos is both a political party and a movement (Iglesias 2015). As we have seen, it has its roots in the politization of the masses into a national popular movement (the M-15-'March for Change'), but it has the aspiration to become a majoritarian party. Podemos follows a populist strategy, and it views populism as a way to create a collective will that unifies and identities. But it also follows a new approach, as political conflict moved from the traditional Left-Right cleavage, to a new people-elite one, and it seeks 'transversality' to frame the party's message to the widest range of voters as a new way of doing politics, which moves away from the traditional Left-Right divide, or from parties a strong ideological entity. As a result, it has a somewhat ambiguous political program. Podemos is also an anti-establishment party that reclaims the role of people as principal political subject and has developed binary oppositions: El *pueblo* vs. la *casta*, 'us' vs. 'them.' Moreover, Podemos' narratives of historical transition are tied to the prefix "post": 'Post-regime 1978,' 'post-hegemony,' 'post-ideology,' or 'post-politics.' But it also supports more traditional leftists policies such as the modernization of social democracy; the revitalization of the welfare state; democratizing institutions; and challenging austerity. It seeks the expansion of the struggle in multiple fronts: economic, social, cultural, and political (García Agustín and Briziarelli 2018; Pereira-Zazo and Torres 2019).

Another consequence of the crisis has been the emergence on the far-right side of the political spectrum of Vox, a right-wing nationalist party founded in 2013 by former members of the PP who were disenchanted with the centrist stance of the Rajoy government, and unhappy with its unwillingness to follow more socially Conservative and economically liberal policies. For a long while it was remarkable that despite the increases in unemployment and the rise in inequality and immigration, there was no right-wing anti-EU party in Spain with the success of Golden Dawn in Greece, the National Front in France, or the Lega in Italy. However, the electoral success of the far-right Vox in the 2018 regional elections in Andalusia (and the subsequent general, European, and local/regional elections in 2019) represented a turn in Spanish politics that further aligned it with its European neighbors. Vox is a 'national-populist' party that has focused on issues of sovereignty, security, and the economy. Vox supports the anti-immigrant and xenophobic fare typical of European far-right parties. They champion nativist fears of lost national identity and rally against the establishment for selling people out. Their proposals include the erection of walls (in Ceuta and Melilla in northern Africa), and the banning of mosque-building and of teaching Islam in state schools. Vox advocates for Spain's return to fully centralized government, the revocation of the 2007 Law of Historical Memory, which extended reparations to the victims of the Spanish Civil War and the Franco dictatorship; as well as state protection for hunting and bullfighting, and lower taxes. In the November 2019 elections, Vox became the third largest block in parliament with 15.09% of the votes (and 52 seats).

Indeed, in the last 4 elections the main parties (the Socialist and the PP) have been hobbled by the rise of smaller, more radical groups, and the trajectory of the country's politics shows that centrist parties have lost ground to more extreme factions, "which often pursue their particular interests at the expense of democratic norms and institutions." This has led Freedom House to include Spain in a list of 25 of 41 established democracies that have suffered overall declines over the past 14 years in the functioning of government, freedom of expression and belief, and rule of law (Freedom House 2020, 10) (Fig. 7.2).

Finally, the crisis also led to the emergence at the national level of a centrist reformist party, *Ciudadanos*, which also contributed to the fragmentation of the party system, but is in the antipodes of populism. Indeed, *Ciudadanos* is even more pro-European than the traditional parties, but also more right-wing/Spanish nationalist when it comes to the Catalan issue, which has been since 2017 the most divisive issue in Spanish politics. In economic policy, *Ciudadanos* has repeatedly argued that Spain should accelerate the path of economic reforms, accepting

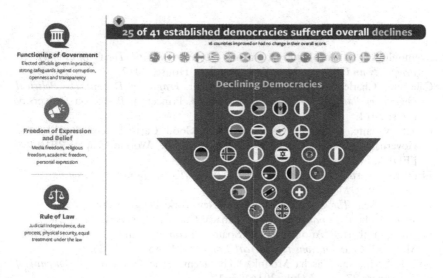

Democracies in Decline

More than half of the world's established democracies deteriorated over the past 14 years. Functioning of government, freedom of expression and belief, and rule of law are the most common areas of decline.

Fig. 7.2 Spanish democracy in decline (*Source* Freedom House. *Freedom in the World* 2020, p. 10)

virtually all recommendations coming from the European Commission, form the establishment of a single contract in the labor market to the increase in VAT and the deregulation of the services sector (Royo and Steinberg 2019). In the end, Spain seems to be part of a broader pattern in which citizens' dissatisfaction with their governments is not merely a dissatisfaction with particular governments but rather with the political system as a whole. Indeed, a serious democratic disconnect seems to be emerging in which citizens of mature democracies have become markedly less satisfied with their form of government and they seem to be open to nondemocratic alternatives (Foa and Monk 2016).

Ultimately the Spanish experience shows that causes and effects are inextricably linked. In their state of ecstasy's and oblivion, Spaniards seemed to forget that simple reality. The lives of millions of Spaniards have been transformed by the dissolution of many of the things that used to hold them together, and the crisis resulted in the breakdown of the pre-existing social compact. None of this can be explained without examining the process of institutional degeneration that preceded the crisis and the failure of the Spanish elites. The effect of that betrayal will take decades to fully unfurl.

References

Acemoglu, Daron, and James Robinson. *Why Nations Fail: The Origins of Power, Prosperity and Poverty*. New York: Random House, 2012.

Calomiris, Charles W., and Stephen H. Haber. *Fragile by Design: The Political Origins of Banking Crises & Scarce Credit*. Princeton: Princeton University Press, 2014.

Cuñat, Vicente, and Luís Garicano. "Did Good Cajas Extend Bad Loans? Governance, Human Capital and Loan Portfolio." Working Papers 2010-08, FEDEA, 2010.

Ekaizer, Ernesto. *El Libro Negro. Como Fallo el Banco de España a los Ciudadanos*. Madrid: Espasa, 2018.

Ferguson, Neil. *The Great Degeneration*. New York: Penguin Press, 2013.

Fishman, Robert. *Democracy's Voices*. Ithaca: Cornell University Press, 2004.

Fishman, Robert. "Anomalies of Spain's Economy and Economic Policy Making." *Contributions to Political Economy* 31, no. 1 (2012): 67–76.

Foa, Roberto, and Yascha Mounk. "The Democratic Disconnect." *Journal of Democracy* 27, no. 3 (July 2016): 5–17.

Freedom House. *Freedom in the World 2020*. 2020. https://freedomhouse.org/sites/default/files/2020-02/FIW_2020_REPORT_BOOKLET_Final.pdf.

García Agustín, O., and Marco Briziarelli. *Podemos and the New Political Cycle*. New York: Palgrave, 2018.

Iglesias, Pablo. *Politics in a Time of Crisis: Podemos and the Future of Democracy in Europe*. New York: Verso, 2015.

McDonough, Peter, Samuel Barnes, and Antonio López Pina. *The Cultural Dynamics of Democratization in Spain*. Cambridge: Harvard University Press, 1998.

Molinas, César. *Qué Hacer con España*. Madrid: Imago Mundi, 2013.

Packer, George. *The Unwinding*. New York: Farrar, Straus and Giroux, 2013.

Pereira-Zazo, Oscar, and Steven Torres. *Spain after the Indignados/15M Movement*. New York: Palgrave, 2019.

Pérez Díaz, Victor. *The Return of Civil Society*. Cambridge: Harvard University Press, 1993.

Royo, Sebastián. *Lessons from the Economic Crisis in Spain*. New York: Palgrave, 2013.

Royo, Sebastián. "Institutional Degeneration and the Economic Crisis in Spain." Special issue of *American Behavioral Scientist. The Economic Crisis from Within: Evidence from Southern Europe*. Anna Zamora-Kapoor, and Xavier Coller (eds.). 58, no. 12 (2014): 1568–91.

Royo, Sebastián. "The Causes and Legacy of the Great Recession in Spain." In *Oxford Handbook of Spanish Politics*, edited by Diego Munro and Ignacio Lago. New York, NY: Oxford University Press, 2020.

Royo, Sebastián, and Federico Steinberg. "Using a Sectoral Bailout to Make Wide Reforms: The Case of Spain." In *The Political Economy of Adjustment Throughout and Beyond the Eurozone Crisis: What Have We Learnt?*, edited by Michelle Chang and Federico Steinberg. New York: Routledge, 2019.

Urquizu, Ignacio. *La crisis de representación en España*. Madrid: Los Libros de la Catarata, 2016.

Villoria, Manuel, and Fernando Jiménez. "La Corrupción en España (2004–2010): Datos, percepción y efectos." *REIS* 138 (abril-junio 2012): 109–34.

Weiner, Richard, and Iván López. *Los Indignados: Tides of Social Insertion in Spain*. Winchester, NY: Zero Books, 2018.

CHAPTER 8

Conclusions: The Implications of 'Bank Bargains' for Democratic Politics

BANK BARGAINS

This book has attempted to answer the central question posed in Chapter 1: *Why couldn't Spain build a banking system capable of providing stable and abundant credit?* Following Calomiris and Haber's analysis of other countries (2014), it has identified the role of political factors and shown how they have determined banking outcomes and caused banking crises in Spain. Throughout its chapters, the book has examined the effects of political conditions on banking system outcomes, and it has done so from a historical perspective understanding that such outcomes are contingent and also depend on changes to the historical context. For instance, in Spain (like in many other countries), fiscal affairs have been influenced by the need to finance wars, close widening fiscal deficits, or finance large infrastructure or industrialization projects, which have historically driven the funding of banks (and central banks). The book has taken a panoramic historical perspective to try to identify systematic patterns in the run-ups of crises.

Indeed, in Spain, shifts in banks' outcomes have reflected dramatic political (and economic) changes, and banking crises have arisen primarily from political and/or economic crises that have produced substantive changes in the structure of the Spanish banking system. These changes, in turn, have impacted dominant political coalitions in charge of banking policy and often have led to their replacement by a new coalition. The examination of the Spanish case shows that domestic social, political, and

© The Author(s) 2020
S. Royo, *Why Banks Fail*,
https://doi.org/10.1057/978-1-137-53228-2_8

economic factors are crucial to understand coalition formations and policy choices. These coalitions are not neutral, and they influence the stability and resilience of the banking system and its ability to provide credit. It is therefore crucial to examine their objectives and strategies.

The central argument of the book has been that political conditions underpin what Calomiris and Haber call the *'Game of Bank Bargains'*: a process of political bargains in which parties with different interests come together to form coalitions that determine what banks will create and how they will function. These outcomes in turn determine the level of access to credit and the stability of the banking system (Calomiris and Haber 2014, pp. 477, 479). The examination of Spanish banking crises confirms that thesis. Spanish banks have been the outcome of political partnerships that included coalitions between the government and citizens that gained control over the banking system. And these coalitions have set the rules of the banking game: How they are chartered, how they are regulated, and how they interact with the state.

Prior to the late 1970s when democracy was finally established in Spain (with the notable exception of the 1930s, in which the country had its first brief experience with democracy but it ended in a tragic civil war and a 40-year dictatorship), the country's autocratic regimes produced a mixed banking system that included both public and private banks (and *cajas*), whose main function was to finance the government and the interests of the sectors that controlled those institutions. This banking system was the result of a partnership between the governments and a group of financers (mostly domestic) in which the *'Game of Bank Bargains'* was played by a small group of government officials and businessmen who were often closely linked by a dense network of personal, economic, and political ties. As we have seen, throughout Spanish history these coalitions have been conditioned by the state chronic financial needs. The country needed banks and has sought partners that provided those funds. This *'Game of Bank Bargains'* led to rent-seeking coalitions that ultimately undermined access to credit and the stability of the system.

The advent of democracy coincided in time with a major banking crisis in the late 1970s–early 1980s. While the new democratic regime made the supply of credit more readily available, it did not bring full stability to the banking system as proved by the most recent crisis of the second half of the 2010s, which was also the result of the historical circumstances that shaped the formation of new political coalitions and institutions that determined that outcome. Two crucial elements (see Calomiris and

Haber 2014, p. 459) that have separated Spain from other successful countries (like Australia, Canada, or New Zealand) have been the country's comparatively new democratic institutions that emerged after the transition to democracy in the late 1970s, and even more importantly the weakness of the institutions that limit rent-seeking. Indeed, both crises were underpinned by rent-seeking coalitions between bankers and populists that ended up undermining the stability of the banking system.

As we have explained, within the complex bargains among politicians, bankers, shareholders, depositors, debtors, and taxpayers, it is crucial to describe the identities and motivations of the players in explaining their policy choices and decisions. Indeed, banking crises in Spain were not pre-ordained, nor were they accidents due to unforeseen circumstances. On the contrary, they have been a function of choices made by bankers and regulators regarding how much cash to hold, how much equity to raise, how much risk to assume, and how to diversify that risk across different kinds of loans and assets. While Spanish governments have been (nominally) committed to prudential regulation, establishing safety nets to protect banks and consumers, the rent-seeking coalitions that we have examined throughout the book have undermined that system and allowed bankers to be less cautious in the management of risk. Ultimately, Spanish banking crises show that the extent of the safety nets and prudential regulation were also political choices made by individuals who were part of those coalitions, and who were largely motivated by maximizing their own short-term interests (which not always coincided with society's ones). In other words, the banking structure failed to insulate the banking system from populist politics and the rent-seeking interests of the members of those coalitions of politicians, bankers, and governments, making it feasible for them to fashion a banking system that suited their interests. And these coalitions have been quite durable as their members have gained wealth and political power, thus reinforcing and entrenching their bargaining power. In the end, Spain got the banking system that the country's dominant coalitions and political institutions permitted.

The Banking Crisis of the 1970s

The transition from dictatorship to democracy in Spain that began in the late 1970s showed how banking-system outcomes can have significant and unanticipated political consequences. Spaniards were denied the right to effective suffrage until the mid-1970s. During the previous four decades, the country was governed by an authoritarian regime. As examined in Chapter 3, that period provided an opportunity to examine how authoritarian political leaders formed coalitions with other groups to create a banking system. Calomiris and Haber (2014) show that autocracies can generate stable banking systems when governments are strong enough to centralize decision making, but not so strong that it can weaken property rights of bankers and shareholders with impunity. As in Mexico (Calomiris and Haber 2014, p. 21), the examination of the Spanish case showed that the Spanish government regulated bank entry tightly to increase rates of return sufficiently to compensate bank insiders and shareholders for the risk of expropriation. Political bargains gave big banks rents in the form of market power and lax prudential regulation in exchange for their commitment to share those rents with favored constituents. The transition to democracy in the 1970s led to the establishment of a new democratic regime, which led to new political bargains regarding the banking system.

Indeed, the transition to democracy helps illustrate the impact of regime changes on these coalitions and their bank bargaining. Spain's Francoist authoritarian regime fitted into Calomiris and Haber's "centralized autocratic network" taxonomy (2014, pp. 42–44). Franco lacked absolute power but was strong enough to control the levers of power while building a network of alliances, including bankers, to hold onto power. In Spain, the course of financial and banking policies was largely influenced by a contest within the country's policy-making elite that preceded the democratic transition, and with long roots in Spanish history. Bankers allowed the regime to finance expenditures in excess of tax revenues. This coalition did not lead to the development of a competitively structured banking system. On the contrary, high expropriation risk constrained entry into the system, which was characterized by entry restrictions and privileges to favored bank insiders who in turn provided a portion of their rents to the regime, hence giving it a vested interest in the favored banks and thus reducing the risk of expropriation.[1] Franco and the country's financers crafted a set of institutions designed to attract capital into the banking system by limiting competition.

The big joint-stock banks that operated during the Franco regime dealt largely in commercial banking, attracting deposits and making loans, but

---

[1] According to Calomiris and Haber (2014, p. 46), under such regime returns to equity holders are high, loans to insider firms are subsidized, governments and banks insiders extract significant rents, and periodic fiscal firms result in expropriations.

many of them also actively promoted industrial firms and public utilities, and controlled industrial groups. This banking schema was accompanied by active banking repression, as well as "tight government controls of interest rates, restriction of competition, government intervention in banks' policies such as distribution of dividends, fields of investment, [and] creation of branches" (Tortella and García Ruiz 2013, p. 3). This compact was accepted by bankers because it granted them almost riskless profits.

As in other autocratic regimes like Mexico and Brazil (see Calomiris and Haber 2014, pp. 44–45, 332), the rents generated by that oligopolistic system were split among the bankers (who received dividends, directors' fees, and used the banks to fund their nonfinancial enterprises), bank minority shareholders (who earned healthy benefits in the form of dividends and benefitted from above-normal stock returns), the government (who got access to cheap capital in the form of low-interest loans), the individuals who controlled the government (who obtained board seats for themselves and their acquaintances, as well as cheaper loans), the dictator (who received a source of public finance that also helped to cement alliances with other groups), and the bank insiders (who earned high rents in a non-competitive market), at the expense of depositors (who did not have access to a low-risk, liquid means of savings because deposits often earned negative real returns and were subject to risk of loss, but they offered taxpayer-financed deposit insurance). That coalition established a framework of banking laws and regulations that enshrined the terms of the bargain and ensured benefits to the members of the coalition. Everyone else (notably the majority of the population and those potential investors and entrepreneurs who did not have links to the banking sector) was left out, with limited access to credit (as noted in Chapter 1) and scant opportunities for economic mobility. The consequence of this system, characterized by a small number of banks and a high level of insider lending, was scarce credit and a high concentration of finance dependent, downstream industries. However, this alliance between the Franco dictatorship and Spain's bankers was always fragile. One of the main reasons that held it together was the dictatorship's need to reward the corporatist labor union for its political support. Organized workers were employed by industrial conglomerates who also owned (or were owned by) banks, which acted as their funding arm, thus providing bankers some protection because it tied up them with manufacturers.

As we have seen in Chapter 3, by the 1970s as a result of the oil shocks the country got into a recession and government expenditures started to outstrip revenues. The government could have closed the gap by increasing taxes, but the dictator was on his last legs, and he did not want to increase its political problems and pay the political price of raising taxes by alienating the bankers and industrializers that supported the regime. It was easier to expand the money supply, with the consequent impact on inflation.

Political and/or economic crises often galvanize pressures for reform and lead to new coalitions that push for changes in the structure of financial systems (Hoffman et al. 2007). However, the Spanish transition to democracy, which coincided with a global economic crisis, shows the endurance and stickiness of these coalitions and bargains, as it did not result in an instantaneous reorganization of the banking system. On the contrary, in Spain, domestic elites promoted agendas to advance their domestic interests, which were not just driven by market pressures. Following the democratic transition, Spain was still characterized by an interventionist state, and the state elites who played a crucial role in the banking sector (including elected officials, reformers, and technocrats) had a major goal: monetary control. In interventionist states, state elites sought to subsume monetary policy instruments to the government's policy objectives to slow down credit growth while boosting investment.

Indeed, as we have seen in Chapter 3, in Spain, the banking oligopoly that had emerged at the beginning of the twentieth century led to a structure dominated by the so-called Big Seven banks that controlled 72% of total bank deposits by 1957. According to Pérez (1997), an influential group of economists based at the research department of the Bank of Spain gained prominence in the major parties and pursued macroeconomic policies strongly oriented to market mechanisms and private initiatives favored by the existing players—including the large banks. These banks formed a coalition with that group of economic reformers trained in the Research Service of the Bank of Spain under the Franco regime. These reformers, neoliberal technocrats closely connected with the Catholic lay organization Opus Dei, had emerged in the late 1950s and were able to secure a significant degree of control over the regime's economic policies, which led to the launching of the stabilization plan (a series of four-year plans) and the dismantling of the corporatist autarkist policies that had dominated the country until 1957 (see Chapter 3). The neoliberal technocratic reformers, although ideologically opposed to state intervention and cheap credit, were interventionists at heart and used the state apparatus to reach an accommodation with the big banks to preserve their oligopoly and kept interest rates low to stimulate investment in certain areas and to preserve social peace. Following the transition to democracy, these reformers retained control over economic policy (Pérez 1997, p. 43). They took advantage of the unstable social and economic conditions of the 1970s, which shifted the balance of power in favor of that small network of reformers that had emerged around the Spanish Central Bank's research department and propelled them into leadership positions. These reformers opposed the traditional approach that subordinated monetary rigor and market discipline to the principles of state discretion, and sought to reverse the dominance of the dominant bureaucracy that supported planning as an instrument for development. In order to achieve their objectives, they pursued accommodation with the banking sector.

In other words, in the years that followed the democratic transition, the *Game of Bank Bargains* in Spain was marked by this interplay between state elites and the domestic banking sector, which helps explain monetary and banking outcomes. The interest of this coalition was served by a particular set of policies that had income distribution consequences and enriched the banking oligopoly. In the end, the new democratic government's attempts to dislodge the existing oligopolistic banking structure largely failed and ultimately led to the resignation of the prime minister, Adolfo Suarez, who blamed the banks and their opposition to his market-opening initiatives for his fall from power. Cheap credit provided a means to diminish social tensions and stabilize the new democratic regime at a time of worldwide recession. According to Pérez, the Spanish financial reforms that shifted away from interventionism were driven by state elites rather than economic actors. The way banking liberalization was carried out was the result of strategic choices on the part of anti-interventionist reformers (who sought to defeat the historical scourge of government interventionism in the country) for whom their first priority was to alter the institutional structure of Spanish policy-making to give greater influence and leverage to the central bank, and it was the result of a pattern of accommodation between state elites and the private banking sector that started at the beginning of the twentieth century. That development was instrumental in the shift away from interventionism and the neoliberal reforms of the 1970s and 1980s, as well as in prolonging the privileges of the banking sector in Spain and in the primacy of monetary policy considerations over other economic objectives (Pérez 1997, p. 190).

Yet, even more surprising was that the election of a new Socialist government in 1982 (which had called for the nationalization of the banking sector) did not lead to a new coalition or new banking policies. On the contrary, the new Socialist government quickly broke its electoral commitment to nationalize the banks, and it appointed the same central banker reformers to key economic policy-making positions, who sustained the accommodative partnerships with the banks, allowing the Big Seven banks oligopoly to persist well into the mid-1980s while protecting them from strong competition (as part of Spain European Community [EC] membership, they negotiated a transition period of seven years for the banks, which sheltered the banking sector from significant external competition, at precisely the same time that other sectors of the Spanish economy were exposed to brutal EC competition). While Lukauskas (1997) attributed the Socialists' banking policies to an electorally based desire to secure economic outcomes, such as growth, and please the median voter, Pérez (1997) offers a largely political explanation for the ability of Spain's large banks to maintain that state of affairs. According to her analysis, the main political actors in post-Franco's Spain choose to accept the power of the large banks or the economic consequences of that banking system (Pérez 1997, p. 149), as Spanish economic policy-makers quickly turned their emphasis to the effort to fight inflation (Pérez 1997; Royo 2000). Furthermore, the Socialist government under Felipe González maintained generally friendly relations with the large private banks for the thirteen years in which he was in power, and avoided any sustained public-ownership role of banks, going as far as even re-privatizing the nationalized *Rumasa* banks, as we noted in Chapter 3. His first two finance ministers, Miguel Boyer and Carlos Solchaga, were intimately linked to the network of academic reformers of the Bank of Spain and they supported a long-term strategy of boosting competitiveness, profitability, and growth through macroeconomic rigor, fiscal consolidation, and wage moderation, seeking to raise the rate of public savings and to shift resources away from social transfers into capital investment in infrastructure (Boix 1995, p. 2).

It is important to note, however, that this accommodation took place in the context of a democratic transition defined by efforts to avoid confrontations that would bring back the memories of the Second Republic and the Civil War, as well as by the need to avoid the perceived danger of socioeconomic over-reaching and to achieve left-right accommodation (Fishman 2010; Linz et al. 1981; Linz and Stepan 1996). In the end, the transition to democracy in Spain minimized the political relevance of social protests and the relative openness of policy-makers to such pressures (Fishman 2010). At the same time, the relatively hierarchical and centralized nature of the PSOE and its growing distance from the union movement (in the 1980s, they came to see union power as an obstacle to the liberalizing policies they favored), combined with the favored approach of the Socialist economic team, which relied largely on market pressures and competition for growth and the revival of employment, led the Socialist government toward neoliberal labor market and macroeconomic policies (Royo 2000). In this context, it is not surprising that they abandoned their commitment to bank nationalizations or that they failed to assign any significant role to state-owned banks into credit creation for small businesses, relying instead on market competition and labor contract flexibilization.

Subsequently, as we discussed in Chapter 4, throughout the 1990s and 2000s, this pattern continued and became the predominant tendency, as Spanish policy-makers from both major parties avoided any significant challenge to the large banks and continued relying heavily on market-based mechanisms and liberalization as strategies to generate growth and employment. And this approach even persisted under the more progressive\e Socialist government of Rodriguez Zapatero, widely considered the most leftist prime minister since the Second Republic (Fishman 2010).[2] In this regard, Fishman has argued that the dispositions of relevant policy-makers in Spain, and their approach to forging state policies, were conditioned by the country's path to democracy, and that the political handling of banking and financing for SMEs has been reflective of that broader pattern. According to him, Spanish policy-makers' economic approach has been shaped by their limited sensitivity—or even closure—to social pressures from below and their lack of receptivity to policy advice promoting, or discouraging, state involvement in the economy. These tendencies were put in place during the regime democratic transition and they continue to make their mark on policy-making.

---

[2] Fishman highlights the fact that in 2009 when a government minister offered critical words against the banks because of their reluctance to make credit available for firms, a more powerful member of the PSOE quickly issued a public statement friendly to the banks. *El País*, February 5, 2009, p. 17.

Those choices, however, have had consequences. The structure of the Spanish banking system produced significant (negative) economic consequences. Indeed, there is growing consensus that in order to promote economic growth and employment creation, systems of finance need to meet the credit needs of small and medium enterprises (SMEs), and that a strong state role in banking has positive consequences for SME financing (Stallings 2006). Spain came short in that regard. Emphasizing the oligopolistic nature of the sector, Pérez has argued that as a result of that characteristic of the country's financial system, "Spanish banks had far higher cost, interest and earning margins than other West European countries," and she contends that this feature of the system generated significant costs for Spanish firms outside the finance sector (Pérez 1997, p. 20). The elimination of the credit controls that had kept the real interest rates low in the 1960s, and the establishment of tight-credit policies to address the oil-shocks of the 1970s, allowed banks to increase real credit rates significantly and therefore led to higher profit margins for them at the time in which the public and industrial sectors were undertaking a brutal restructuring (which led to a then record unemployment of 21%). Yet, the combination of credit deregulation and oligopolistic control of financial and capital markets led to high interest rates that intensified the recession and adjustment process in the 1980s, and limited access to credit (Pérez 1997). It was only EC membership that led to a policy shift and dislodged the oligopoly of the Big Seven banks.[3]

The 2008 Banking Crises

The crisis of 2008 reaffirmed the instability and crises proneness of the Spanish banking system. Indeed, the latest banking crisis in Spain confirmed a long-standing tenant: Banks or banking systems collapse when they meet two conditions: They take on too much risk in their loans and investments, and they do not have sufficient capital on reserve to absorb the losses associated with their risky investments and loans (Calomiris and

---

[3] Foreign bank ownership had several advantages: They would contribute to the recapitalization of the banking system; since they did not own nonfinancial companies they would not be tempted into finance their losses; they would not be expected to be bailed out by Spanish taxpayers in the case of insolvency; and finally, they would be subject not just to Spanish regulators, but also to shareholders and regulators from their own countries. See Calomiris and Haber (2014, pp. 385–86).

Haber 2014, p. 207). Indeed, the cause of the 2008 crisis in Spain was rooted in policies that eroded underwriting standards and weak prudential regulation. Populists formed coalitions with technocrats and bankers, and then enacted banking policies to their liking.

As we have seen in Chapter 6, with a few relatively small exceptions, the Spanish financial crisis was a crisis of the *cajas*. The three most problematic Spanish *cajas*: *Bankia*, *CatalunyaCaixa*, and *Novagalicia*, had capital deficits (that have been covered partly or fully by the taxpayer) of €54 billion—the equivalent of over 5% of Spanish GDP. These institutions borrowed short term from depositors and then lend long term on fixed-rate mortgages. However, by the mid-2000s, the context in which these institutions operated had changed markedly and they were particularly vulnerable on wholesale funding. Yet, as part of the banks bargains that we have examined throughout the book, governments' protection of *cajas* had insulated them from the consequences of their own risk-taking and facilitated the reckless decisions that led to their downfall. When the real estate market collapsed after 2007, wholesale funding dried up and their funding costs skyrocketed, which caused significant problems because they had to pay more for capital, and they held mortgages (many of which went into default as a result of the crisis) that still earned only low fixed interest rates of return. This brought several of them to the point of insolvency, and many of them tried to cover up their losses reclassifying, refinancing, and extending loans during 2008, 2009, and 2010 (Cuñat and Garicano 2010). Other research has also shown that there was a clear cyclical pattern in nonperforming loans (NPL) recognition: They increased sharply in the first two months of each quarter and then became systematically negative in the third one, when the numbers had to be reported (Coterill 2010). And this happened while the stock of real estate developer loans, which represented 32% of Spain's GDP, was still growing through that period in spite of large bankruptcies in the sector, which suggests that many loans were being informally restructured or refinanced (Garicano 2012).

If the *cajas'* vulnerability, driven by their high reliance on the real estate market and wholesale funding, and their (often fraudulent) attempts to cover up their losses had been recognized and addressed in the years prior to the crises through *cajas'* closures, shrinkage, or consolidation, the crisis for the *cajas* sector would have been significant but not as devastating as it ended up being. As losses started to pile up the Spanish government, supervisory agencies should have shut down insolvent ones or forced them to raise additional capital. Yet, as we have seen, they ignored or minimized the signs and looked the other way, postponing the day of reckoning. But in doing so they ended up ensuring that the final outcome would be much worse. In many ways, the *cajas'* crises were a failure of risk management, which led to an increase in risky lending and to inadequate levels of capital cushions. But, it was not just merely a management problem.

What were the true reasons for the extent of the crisis? How was it that so many cajas ended up making so many risky loans while maintaining insufficient capital to protect themselves against insolvency? What were the processes by which cajas' portfolios became increasingly risky? And, why increased risk in their assets was not adequately matched by increasing amounts of capital in reserve?

In response to those questions, this book has argued, following Calomiris and Haber (2014), that institutional and regulatory frameworks favored both the government and other privileged actors' access to finance at the expense of an environment conducive to a stable banking system. Indeed, political institutions have structured the incentives of bankers, as well as political and economic actors to form coalitions that shaped regulations and policies in their favor. This institutional framework was the result of political choices that made it vulnerable, because prudent lending practices continued being influenced by the desires of the groups that were in control of the government, who often channeled credit to groups that were considered politically crucial. Therefore, it is not surprising that Spanish banks/*cajas* have been fragile and crises prone.

Furthermore, the existing regulatory framework for the *cajas* encouraged excessive risk-taking at the taxpayer's expense: In the absence of appropriate regulatory oversight and with the government's implicit guarantee of their debts, it was not surprising that their managers, who had little at stake, had strong incentives to lend with borrowed money and little equity capital. This was compounded by weak underwriting standards that anyone could take advantage of, which opened the doors to riskier mortgages, and increase the leverage of Spanish families. The crisis and the subsequent loss of employment (unemployment reached over 26% at the peak of the crisis) pushed thousands of them into default.

In the end, the collapse of housing prices and a recession that led to reduced employment for overindebted homeowners were enough to start a major financial crisis. However, although risky housing lending was the fuel for the financial crisis, it was weak prudential regulation (i.e., the role that regulators play to ensure the safe operation of the financial system) that made that fuel explosive. While these two factors may be perceived as independent from each other, both of them were symptoms of a larger problem: an underlying political culture that allowed for higher risk tolerance. Regulators could have imposed higher capital ratios on banks and *cajas*. Yet this decision would have met resistance from bankers and voters, because they would have increased the costs of mortgages and made it less likely for banks and *cajas* to supply risky mortgages.

In other words, stronger prudential regulation would have subverted the goals of providing cheap credit and expanding home ownership, which was a central tenant of the *banks bargains* among bankers (particularly the managers of the *cajas* who were political appointees and had strong links with the political parties), voters, regulators, and politicians. Managers were able to borrow massively in wholesale markets because the creditors took for granted that their debts would be assumed by Spanish taxpayers. This assumption proved to be correct when they were bailed out. Since this money did not belong to their 'shareholders,' it made it even easier for them to lend recklessly and invest in questionable real estate projects. This perpetuated an environment in the years that preceded the crisis in which risk-taking with borrowed money became the norm, and in which many bankers "threw any caution to the wind" (see Calomiris and Haber 2014, p. 257).

In the end, the examination of the Spanish banking crisis confirms that the crisis was the outcome of a *political bargain* (see Calomiris and Haber 2014, pp. 280–81). In the *cajas* sector, populists formed coalitions with bankers, and they enacted banking policies that allowed them to grow and assume unsustainable levels of risks in the years prior to the crisis, and during the crisis to avoid the actions of regulators and scape their troubles (for instance, covering their losses with disastrous mergers). As we have seen in Chapter 7, Spanish *cajas* were allowed to grow and expand, which afforded them economies of scale and scope, as well as increasing market share and market power, and thus a larger too-big-to-fail protection. In exchange, however, they had to share some of their rents with other groups, in this case new homeowners eager for cheap credit to buy their homes; politically influential developers, real estate and construction companies; as well as their local and regional governments. These groups used their influence with local, regional, and national politicians to appoint political (and often uneducated and unskilled) managers to the *cajas* to do their bidding, and they pushed to loosen underwriting standards and to maintain low capital requirements. For their part, policy-makers and regulators, although they knew what was happening, chose to acquiesce and to go along because they also accrued electoral and often personal and/or financial benefits from this game, or they simply did not want to pay the political and/or personal price to try to stop them.

Moreover, the failure of prudential regulation was manifested by the unintended (and perverse) role of safety nets, in the form of deposit insurance, or implicit too-big-to-fail guarantees of bank debts not covered by deposit insurance. These safety nets can be considered subsidies and their value changes according to two criteria: the size of the bank/*caja* (the larger it is, the larger the value of the safety net, and vice versa) and the amount of debt (the more the debt vis-à-vis capital, the larger the value of the safety net). As a result, the higher the leverage (or the larger the institution), the stronger the protection and the higher the default risk assumed by the government that protects the debt. Hence, the government's decision to push for the consolidation of the *cajas* paradoxically intensified moral hazard because it expanded the size of the safe net/subsidy. In other words, the reckless managers of these institutions had an incentive to build their strategies trying to maximize the value of their safety-net provisions by growing and expanding as much as they could with borrowed money, and/or by maintaining as little equity capital as possible to fund their operations. Still, as in other countries, they "were rewarded by the government safety net for having as little skin in the game as possible" (Calomiris and Haber 2014, p. 259). They, of course, also embraced the push for mergers and consolidation from the regulators and regional governments, a process as we have seen that was largely politicized and led to the creation of larger institutions, like *Bankia*, that created a larger systemic risk and intensified moral hazard.

In addition, regulators were "asleep in the wheel" when they allowed these *cajas* to keep inadequate capital cushions. Once again, the reason for the inadequate capital ratios was also the outcome of political bargaining: *Cajas* expanded mortgage lending in exchange for lower capital requirements and weak regulatory oversight (or larger safety-net subsidies). But the problem was not so much lack of regulation but ineffectual regulation. For instance, capital requirements for mortgages had not been determined for a new world in which borrowers would not put any money down in their new houses, or in which there was much lesser oversight of risk when approving loans. The question remains, *did they see these changes and refused to act for their own interests? Or did they lack the power to do anything about it?* The fact is that, despite ample warnings about the scope of the problem and the increasing building up of risk in many of the *cajas*, and despite the ample power to intervene by rejecting banks' decisions regarding their own risk or by imposing higher capital ratio requirements, the Spanish regulators (as it happened in many other countries, see Calomiris and Haber 2014, pp. 265–66) failed to control and minimize risk and did not increase capital requirements to limit banks/*cajas'* investment in risky assets.[4]

In other words in their failure to take action to limit the potential systemic consequences of the risky mortgage investment by *cajas* and banks, they chose to "hear no evil, and see no evil." Concerned that a decision to increase capital requirements may have led to a credit crunch with significant impact on the supply of mortgage credit, and aware that such decision would have been heavily resisted by banks and *cajas* (who would have fought to the bitter end to keep low capital requirements because they allowed them to generate higher rates of return), they sat idle. In the end, it was easier to stay put than to fight a bitter battle because the regulators did not want to confront the political coalition that underpinned these policies (which included not just bankers, but also governments who benefitted from the support of happy new homeowners, and happy new homeowners who finally had access to cheap credit to buy a house), and instead chose the path of least resistance and failed to act. Indeed, the main explanation for the supervisory failure of the Bank of Spain has to do with the political control of the *cajas*: Confronted with powerful and well-connected actors, it decided that it was easier to look the other way (Royo 2013a).

Indeed, the inaction from the Bank of Spain has been widely criticized, as we examined in Chapters 6 and 7 (for contrasting views, see Ekaizer 2018; Fernández Ordóñez 2016). While no evidence of corruption has been unveiled (so far), its reluctance to recognize and uncover its own previous mistakes (although some inspectors tried to do it as we have seen); the absence of an appropriate resolution framework at that

---

[4] Calomiris and Haber (2014, pp. 266–68) explain how regulators compounded their failure by subcontracting regulation to private firms, like the rating agencies, and undermining the incentives that those agencies had to provide accurate ratings. In the United States, the loose monetary policies of the Fed between 2002 and 2005 further intensified the lower pricing of risk.

time (it was created in the summer of 2012); plus the Bank's overly confidence on the dynamics provisions that forced banks to increase provisions without reference to any specific loan (which, as we have seen in Chapter 5, worked as intended in the initial stages of the crisis) help partially to explain the Bank's inaction. But the consequence of this inaction, as noted before, was that it "allowed the reality to be hidden in plain sight for longer than it would have been otherwise possible" (Garicano 2012) or desirable. In the absence of those provisions, the *cajas'* losses would have become visible much earlier and would have forced action from the supervisor.

But as noted by Garicano (2012), the main explanation for the supervisory failure had to do with the political control of the *cajas*. As we examined in Chapter 6, governance played a critical role in the development of the crisis. The Bank of Spain confronted with powerful and well-connected ex-politicians decided to look the other way in the face of obvious problems. As we have seen on Chapter 6, Cuñat and Garicano (2010) have shown how the political connection of the managers of the entities was a good predictor of poor management and subsequent problems of the *cajas*. They show that *cajas* with chief executives who had no previous banking experience, no graduate education, and were politically connected did substantially worse in the run-up to the crisis, granting more real estate developer loans, and during the crisis with higher nonperformance loans. Hence, it is not surprising that *Bankia*, which turned out to be the worst in terms of the losses imposed on taxpayers (it needed a rescue of 22,242 million euros), was the most politicized of the *cajas*. For instance, the appointment of its CEO, Rodrigo Rato (a former minister of finance in Spain under the Aznar government and a Manager Director of the International Monetary Fund (IMF) between 2004 and 2007) was the result of a formal but secret pact between the Popular Party (PP) (of which he had been a main leader for years) and the major leftist trade union, *Comisiones Obreras* (CCOO), in 1986. In February 2017, Rato was found guilty of embezzlement and sentenced to four and a half years' imprisonment. At the time of writing (January 2020) after ten months of trial, Rato and 31 other *Bankia* managers (plus *Bankia*, BFA, and Deloitte) are still waiting for the sentence on the *Bankia* case. They have been accused of falsifying the bank accounts and scam investors when the bank went public and entered the stock market.

Furthermore, the politization of the *cajas* was also crucial in diluting the role of the supervisor after the crisis started, because the *cajas'* insiders used their political connections to try to escape their troubles and avoid the wrath of the regulator. For instance, as noted above, these political connections were instrumental in the mergers that we discussed in Chapters 5 and 6, which were mostly decided based on political and geographic criteria, rather than economic rationale. The *cajas* controlled by PP politicians merged together, which meant that two of the most problematic *cajas (Bancaja and Caja Madrid) from regions in which the PP* was in power (Valencia and Madrid) were merged and led to the creation of Bankia with the results that we know. The same happened in other regions like Galicia or Catalonia with similar disastrous results. Yet again, while the Bank of Spain had reservations and was cognizant of the risks, it failed to confront the politicians (Garicano 2012).

In addition, while many analysts have blamed the European Central Bank's (ECB) policies prior to the crisis, which led to a sharp decline in the cost of capital for peripheral countries (real interest rates were negative during part of the boom, which intensified the bubble) (see Jiménez et al. 2014), it is important to emphasize again that monetary policy alone did not cause the crisis. First, because the initial growth of mortgage risk and the decline in prudential regulation preceded the loose monetary policies of the ECB. In addition, the Spanish Central bank could have countered the effect of the ECB loose monetary policies by increasing capital requirements from banks and *cajas*. Finally, although monetary policy can contribute to the overpricing of assets, notably real estate ones, and may lead to the development of bubbles, it is important to stress, as noted by Calomiris and Haber (2014, p. 271), that banking crises require that banks invest in those overpriced and risky assets and that they back those investments with insufficient capital. Therefore, it is not enough just to account for weaker lending standards and prudential regulatory failures, but also to explain differences among banks/*cajas*.

Indeed, the fact that banks and *cajas* had the opportunity to gamble and assume huge risks does not mean that they had to, and not all did (a notable example was the Basque's *Gipuzkoa Donostia Kutxa*—the Savings Bank of Gipuzkoa and San Sebastián). They all faced a choice: They could maintain high-quality lending standards, limit their risk exposure, and maintain appropriate capital cushions; or alternatively, they could exploit the lower regulatory environment assume higher risks and lower their capital cushions. In Spain, there is evidence that this choice was also conditioned by the education level of the managers of the *cajas*, as well as by how they were appointed to these positions, both an integral part of the *bank bargains* that took place in the country. As noted before, Cuñat and Garicano (2010) have shown that the main difference between banks and *cajas* was not so much the *cajas'* political nature but the lower level of professionalization of their managers: Only 31% of their presidents had postgraduate degree, and half of them had occupied political positions before becoming presidents. This evidence shows that bankers' abilities and the nature of their appointments may have encouraged them to assume exceptionally high risks. Hence, given the strong connections between the managers of the *cajas* and political leaders, it cannot be a surprise that regional and local governments encouraged the *cajas* to lend to local firms in order to encourage local/regional employment growth and also used their power over the *cajas* to generate employment in order to maximize their chances of winning elections.

As in other countries (see Calomiris and Haber 2014, pp. 276–78), another factor that helps explain which banks and *cajas* were worst affected by the crisis was *variations in the franchise value of banks* (i.e., the ability of banks management teams to develop strong client lists to provide potentially profitable investment opportunities). There is some evidence that differences in risk-taking prior to the crises reflected differences in franchise value: Banks with higher franchise value and with stronger risk-management systems tended to be more Conservative prior to the crisis and took on lower risk relative to their capital. They had a stronger institutional commitment to risk management, which conditioned their risk management decisions and prevented their senior management from taking unnecessary risks (see Ellul and Yerramilli 2010). Unsurprisingly, these banks, like Santander and BBVA, did not require bailouts during the crisis. On the contrary, those banks and *cajas* with weaker risk-management systems and weaker franchises were more tolerant of risk and assumed larger risks, which in some cases led to their default. Again, this may be another instance in which bankers' institutional capabilities may have encouraged them to assume (or may have prevented them from taking) exceptionally high risks.

The structure of the Spanish banking system and differences in the regulatory frameworks for banks and *cajas* also help account for the differential impact of the crisis. The Spanish banking system is composed of a small number of very large banks with nationwide branches. Although it is true that before the late 1980s, as noted in the previous section, when EC integration forced Spanish banks to compete with international ones and the banking system was dominated by the Big Seven, such a concentrated system undermined competition among banks and resulted in less credit at higher prices for Spanish consumers and companies, since then this structure has allowed banks to diversify, as well as to capture economies of scale and scope, while heavily competing among themselves for market share.

Furthermore, although banking regulations centralized authority in the national government and national regulators, this was not the case for the *cajas*, which were also overseen by the regional governments. This distinction had far-reaching implications for banks and *cajas*. For the *cajas*, prior to the crisis, decision making and regulation were fragmented at the regional level allowing local and regional coalitions to shape policies. These coalitions pushed for strategic decisions to expand and grow, even outside of their traditional regional markets, and for reckless lending. Spanish banks, on the contrary, were subject to centralized authority from the beginning (similar to Canada, see Calomiris and Haber 2014, pp. 297–98), which in effect meant that any political bargaining among coalitions had to take place at the national level, and at that level, all interests are aggregated, which facilitates the taking into account of rules for nationally organized constituencies. Centralized control over the banks and banking policy made it harder for populists to form coalitions with the bankers to enact policies to their liking, as it happened with the *cajas*. And the political system proved much more resilient at the national level (than the regional or local one) against pressures from *cajas* in the interest of preserving a strong regulatory supervision and competition.

While *cajas'* lax risk management reflected the inadequate oversight of regulators, the far more Conservative risk management from some of the banks reflects the appropriate oversight of regulators. Banks understood that failure to abide by the regulations would cause the loss of valuable privileges and they want to avoid at all costs getting on the wrong side of the government and the Bank of Spain. Their problems emerged as a result of their decision to fund much of their lending with money that they borrowed from wholesale markets, rather than holding onto their traditional depositors. This fateful decision made them more vulnerable because they relied on volatile wholesale markets.

Despite the perception that Spanish banks operate in a cozy oligopoly, the reality is that for the last couple of decades Spanish banks have operated in a more competitive environment than the *cajas*, despite a restrictive regulatory environment (that allows them to operate national branch networks: The country has a very large density of per capita bank branches). At the same time, they had been able to make more money than the *cajas* because they have been able to capture economies of scale in back-office operations and deploy capital more efficiently. Moreover, their national network and international expansion allowed them to distribute credit efficiently across the country and beyond (i.e., rural areas are well supplied with bank facilities and with costs of credit similar to the urban ones); lower information costs have facilitated transactions and economized the costs of credit; and international expansion allowed them to diversify their portfolio of risk and achieve large economies of scale. Finally, Spanish banks have been more efficient in managing risk than the *cajas* and have been able to achieve lower risk of default on their debts.

The stickiness and persistence of the coalitions that have underpinned the banks bargains examined in this book underscore the difficulties to change the system. While there have been changes to the elite power networks that have ruled the country and dominated its economy, there have also been substantial continuities. According to some studies, there is a relative small elite with very specific characteristics that still rules the country (Villena Oliver 2019): Their positions of influence and power are largely inherited as many of the members proceed from families that have been in these positions for decades; they are deeply interconnected through persisting links that have different origins/sources (family, schools, professional, social, business...); there is a division of labor that places members of the elites in different positions of power (the traditional division whereby a son was a priest, a second a soldier, and a third was in charge of the family's patrimony has now been replaced by a more modern distribution in which some are in the public administration, others in business, and others in politics or public institutions with the additional novelty that these positions can be interchanged throughout their careers); these elites are not only involved in politics but also in the economy, media, and all other decision-making institutions; and their intervention in the country's economic, social, and economic life is not based merely anymore on ideological considerations (on the contrary, many of these networks are represented across the political spectrum), but rather on their interests. These networks work hard and skillfully to retain their positions of privilege in order to advance their interests, guarantee their immunity, and to ensure that they may be displaced. Villena shows that rather than a traditional perceived revolving door between the public and private worlds, what we have are dense networks that have representatives on both worlds, but also that whenever necessary they co-opt external allies in public institutions (like the *cajas* or the Bank of Spain) to carry out their interest in exchange for benefits. These networks have proved to be not only compatible but also substitutive (for instance, between the PP and the PSOE). This would help explain, for instance, the significant consistency in economic policy-making (as noted in Chapter 4) that has characterized the country for the last three decades regardless of whether the PP or the PSOE was in power. According to Villena, elections have become merely a formal medium that has helped perpetuate the power of these networks that extend their influence across all sectors by virtue of their access to the budget and control of regulatory power and the judiciary. Political leaders are so indebted to these networks that they have very limited room for maneuver in policy-making. Regardless of the merit of this interpretation, Villena's perception is shared by millions of Spaniards and helps account for the crisis of representation that was referenced in Chapter 7, as well as the coalition formation that has been the focus of the book.

While everyone understood the risk that a potential collapse of the banking system would present to the Spanish economy, there was a strong sense of complacency because the country has weathered relatively unscathed the first couple of years of the global financial crisis. However, the problem preceded the crisis. Public officials (from the central government, congressmen and senators, regional congressmen, local authorities, bank supervisors, and regulators) understood the risk, but they had little incentives to change the existing rules of the games. As in other countries, the "success" of the political coalitions was marked by their ability to get a public official to go along with "something that he knows is not in the long-run public interest because it is in his own short-term interest" (see Calomiris and Haber 2014, p. 212).

In other words, the costs of their decisions would be in the future but the benefits were immediate. Public officials should have known the outcome of those decisions but they chose to sit on the sidelines and do little to prevent it. This is an important consideration because a typical reaction to the crisis has been the attempt to try to blame particular actors (bankers, central bankers, politicians, regulators...) for it. This misses the larger point, which is that people typically pursue their self-interests, and in democratic countries one of the ways in which they do that is by exercising their voting rights, and building coalitions with politicians to pursue their mutual interests and benefits. Hence, as Calomiris and Haber (2014, p. 213) suggest, no individual should be "blamed for what happens as a result of that coalition." That distracts us from examining the real problem: *How do our political institutions encourage (and often reward) the formation of such costly coalitions?*

In sum, the 2008 financial crisis in Spain was the result of macroeconomic factors, the monetary policies of the ECB, inadequate supervision, the underestimation of risk, and political interference. Yet, as we have seen throughout the book, while the banks/*cajas* managers blamed the real estate bubble for the problems of their institutions, in reality these very deficient managers took advantage of the *bank bargains* that we have described (and of the inadequate and insufficient supervision from the Bank of Spain) to assume risks that not only led to the collapse (or near collapse) of many of their institutions, but also created a systemic risk for the Spanish banking system overall that led to a financial bailout and put the whole Spanish economy at risk. This outcome was not predetermined. As we have seen, other institutions faced a similar environment and made the right choices. The *banks bargains* led to the appointment of bad managers as well as the establishment of a supervisory framework characterized by lack of adequate supervision and controls (for instance, when nonperforming loans were hidden by these banks as refinanced loans, or when these banks approved loans without even appraising the real estate properties) and regulatory lapses (for instance, regarding capital provisions, which were inadequate, or when these banks passed their stress test and shortly afterward they had to be intervened), and both factors were also central to explain the recent crisis.

SPANISH BANKING IN THE AFTERMATH OF THE CRISIS

According to the IMF, the implementation of the financial bailout program was "steadfast," and all of the program's specific measures had been completed by 2014, including identifying undercapitalized banks; requiring banks to address their capital shortfalls; boosting liquidity; adopting plans to restructure or resolve state-aided banks within a few years; reforming the country's frameworks for bank resolution, regulation, and supervision to facilitate a more orderly cleanup and promote financial stability and protect the taxpayer. All these efforts reduced threats emanating from banks to the rest of the economy and strengthened the system's capital, liquidity, and loan-loss provisioning. Notwithstanding this substantial progress, the IMF still stressed the challenges remaining for the financial sector, including the decline in core pre-provision profits, the continuing rising NPL ratio, and the challenges from the private-sector deleveraging and fiscal consolidation (see IMF 2014).

The costs in financial aid to the financial sector since the crisis, according to the latest data from the Bank of Spain, had reached €65,725 million through December of 2018. This amount included €42,561 million provided by the state through the *Fondo de Restructuración Ordenada Bancaria* (FROB), and an additional €23,164 million provided by other Banks through the *Fondo de Garantía de Depósitos* (FGD). Of this amount, only 22.5% (or €14,785 million) have been recovered so far, but there is hope that the eventual privatization of *Bankia* will lower the total costs (see Table 8.1). The first institution that collapsed was CCM in March 2009 with a cost to the estate of €2475 million. It was infamous for its ruinous investment on project like the building

Table 8.1 State aid to banks ten years after the crisis (FROB, FGD and support to Sareb)

	Amount (million euros)	Percentage over assets	Percentage over risk-weighted assets
BFA-Bankia	22,424	7.5	12.3
Catalunya Bank	12,599	16.6	25.9
CAM	12,474	18.5	29.6
NCG	9404	14.1	18.2
B. Valencia	6103	25.9	37
CCM	4215	16.2	24
Sareb	2192		
Unnim	1997	10.2	17.2
BMN	1645	2.4	4.1
Ceiss	1559	3.4	6.2
CajaSur	1192	7.7	11.3
B. Cívica	977	1.4	2.1
Caja 3	407	2	3
B. Gallego	245	5.4	7.8
Liberbank	124	0.2	0.4

Assets' data as of December 31, 2010
Source Bank of Spain

of a new airport in Ciudad Real, which has been barely used. That one institution used all the deposit guarantees funds that the banks, *cajas*, and cooperatives had accumulated during the previous two decades. It was purchased by *Liberbank*.[5]

In December of 2019, the FROB (now renamed the *Autoridad de Resolución Ejecutiva* or the Executive Resolution Authority) celebrated its tenth year anniversary and it issued a full report of its activities since the crisis. According to that report, while *Bankia* is the bank that has received the largest amount of state aid (€22,424 million), other entities received much more support in proportion to their size, particularly *Banco de Valencia* (which was at some point a subsidiary of *Bankia*) and the *Caja de Ahorros del Mediterraneo* (CAM). Four examples will illustrate the magnitude of the problem: *Banco de Valencia* needed aid for the equivalent of 25.9% of their total balance sheet, and for 37% of its assumed risks, in other words, one of three loans could not be recovered and lacked sufficient capital to cover the risk. In the end, the Bank of Valencia received €6103 million and the state only recovered €42 million (it was sold to *CaixaBank* in November 2012 for one euro, and the bank faces 11 judicial proceedings still pending, as of January 2020, on Spanish courts). *Novagalicia Banco* accumulated 18.2% of loans that could not be recovered and received €9404 million of which only €873 have been paid back by *Abanca* after it bought it. Furthermore, at the CAM, the percentage of loans over risk-weighted assets that could not be recovered reached 29.6% of all their loans, and in this case, the FGD (e.g., the other banks) had to assume those losses. The accumulated losses have already reached €7386 million (€693 more than what the FROB expected). The CAM was sold to the *Banco Sabadell* in November 2011 for one euro. Finally, *Catalunya Banc* needed €12,599 million of which only €881 have been recovered, mostly paid by the BBVA, which purchased it. All these banks have achieved dubious recognition for their immense losses that placed the Spanish economy at risk and cost millions of euros to Spanish taxpayers. According to the FROB report, as of June 2019, the FROB was participating in 24 criminal proceedings against managers from all these institutions and was claiming €3705 million.[6]

---

[5] For the reforms in the Spanish financial system after the crisis, see IMF (2014).

[6] From "El Banco de Valencia y la CAM, peores que Bankia," *El País*, Sunday, January 12, 2020, p. 43.

Moreover, while the overall health of the banking sector is not questioned (as of January 2020), significant challenges remain. Spanish banks are being dragged by low interest rates (see Fig. 8.1), low profitability, reputational damage, adverse judicial decisions (mostly around the contractual conditions of some mortgages), and a digitalization process that threaten to reduce even more profit margins. All these obstacles are impacting the performance of their stock: Eight Spanish banks publicly traded (*Santander*, BBVA, *CaixaBank*, *Bankia*, *Bankinter*, *Sabadell*, *Unicaja*, and *Liberbank*) have lost almost a quarter of their stock market capitalization between August 2018 and August 2019, reducing their value in €40,000 million in that period, from €156,000 million to €116,000 million. Consequently, they have embarked in a process to reduce costs: They eliminated 5473 jobs in 2018, and since September of 2008, they have closed 43.1% of their offices, from 45,707 to 26,011 offices. Since 2009, there has been a 30% reduction in the number of banking entities, yet the banking sector remains very competitive according to the Bank of Spain. And the consolidation process continues as banks continue to seek to reduce costs (one of the few levers that they still have available to improve their financial results): There are questions about a possible merge between Bankia and BBVA, *Unicaja* and *Liberbank* have been in discussions to merge (they announced it in January of

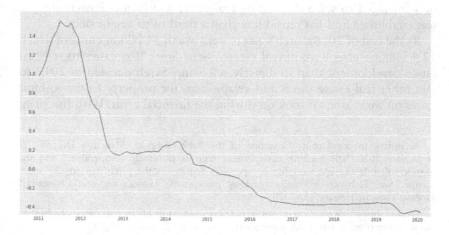

Fig. 8.1 Short-term interest rates. Total, % per annum, Jan 2011–Feb 2020 (*Source* OECD Main Economic Indicators: Finance)

2019 but later retracted it), and further mergers are expected. The Bank of Spain is demanding that any new fusion "generates synergies, and that the potential partners present a viable business plan, that is coherent and credible."[7]

Has Sareb Succeeded?[8]

As part of the rescue deal, Sareb (*Sociedad de Gestión de Activos procedentes de la Restructuración Bancaria,* or *Company for the Management of Assets proceeding from Restructuring of the Banking System*), a 'bad bank' owned by Spain's banks and the state bailout fund FROB, took over more than 50 billion euros ($57 billion) in real estate and other toxic assets from nine Spanish savings banks in 2012. It took on 200,000 assets for more than 50 billion euros, 80% of which were loans buying the value of loans and foreclosed assets at the time at an average discount of 46 and 63%, respectively.

When Sareb was established, it targeted a return on equity of around 14–15%. However, contrary to other 'bad banks' set up in Europe after the financial crisis, such as in Britain and Ireland where property markets have rebounded,[9] Spain's Sareb has been struggling due to a slump in Spanish real estate prices, which has depressed the value of the loans and foreclosed assets it took on. Consequently, by the end of 2018, it had only managed to sell a third of its real estate and financial assets since it was established and has repaid less than a third of its senior debt.

By the end of 2018, Sareb's assets were worth 34 billion euros of which 12.4 billion euros were in real estate assets. Since it is easier to sell real estate development than to directly sell loans, Sareb decided in 2019 to join other real estate funds and swaps loans for property to try to limit losses on toxic assets it took on during the financial crisis. With this plan,

---

[7] According to the Deputy Governor of the Bank of Spain, Margarita Delgado, as of January 2020 "the banking environment remains extremely competitive," and she considers that "there is a worrying competition that makes the profit margins are super narrow." See "El banco de España cree muy 'preocupante' la alta competencia bancaria," *El País,* Tuesday, January 14, 2020.

[8] From: https://www.reuters.com/article/us-spain-badbank-analysis/spains-bad-bank-transforms-into-real-estate-fund-as-it-tries-to-stem-losses-idUSKCN1TQ1XH.

[9] Britain's 'bad bank' set up to manage more than 100 billion pounds ($127 billion) of assets repaid the last of its government loans in June 1019 and Ireland, NAMA's 32 billion euro deleveraging program was 94% complete at the end of 2018.

they were trying to get more liquidity on those assets by transforming up to 18 billion euros (from a total of 22 billion euros) in outstanding loans to real estate developers into assets in order to gain access to the collateral and be able to sell it afterward. As part of this plan, Sareb is planning to speed up filing lawsuits against those who don't fulfill their payments, targeting a total of around 8 billion euros in loans. So far, Sareb says it has transformed 5.8 billion euros into assets. It is also shedding assets since 2017, with discounts ranging from 50% on foreclosed assets to over 70%.

Nevertheless, market conditions in Spain will most likely mean that the lender will not be able to deliver a positive return. In 2018, Sareb reported a loss of 878 million euros, while revenues fell by 5%. Even before the COVID-19 crisis, the ratings agency S&P anticipated a slow-down in the Spanish real estate market, from an estimated 4.5% rise in house prices in 2019 to 3.4% in 2020. Hence, few expect that Sareb will be able to offload its assets by the end of 2027, as planned.

The Crisis of Banco Popular

When everyone thought that the worst was behind, *Banco Popular* (Popular Bank), Spain's fifth-largest bank with 91 years of history, with over $100 billion in loans, collapsed in June 2017, forcing it into the arms of its rival, *Banco Santander*, which purchased it for the nominal sum of one euro after depositors withdrew money massively and the bank's stock price plunged.[10] The root of the problem was, yet again, the nonperforming loans on the bank's books, which were a major cause of its collapse. They can be traced back to the real estate bubble of over a decade ago, as toxic home loans had been festering on its books for years. As a result, *Popular's* stock had lost 95% of its market value over the previous five years, weighed down by its toxic real estate portfolio (see ECB 2017).

Popular's collapse was not merely a problem of lack of liquidity, the capital increases that took place in 2012 and 2016 were insufficient to cover the losses from the real estate mortgages, and they just served to gain more time. The experts (*peritos*) agreed that the bank was solvent when it collapsed but it was very close from falling below the minimum capital requirements. In the end, liquidity problems, combined with

---

[10] See Gretchen Morgeson, "Lessons from the Collapse of Banco Popular," *New York Times*, June 23, 2017.

governance problems and severe management mistakes (like the decision from the bank's chairman, Emilio Saracho, during the August 2017 shareholders' meeting to announce that the bank needed additional capital, which precipitated a run on the bank deposits and a 10% drop of the bank's stock), ultimately led to its collapse.[11]

The collapse of *Popular* illustrated, yet again, the limits of supervision and safeguard instruments, as the stress tests that had been designed by regulators to assess how resilient bank balance sheets will be during downturns failed. Indeed, in 2016, *Banco Popular* conducted a stress test in cooperation with the European Banking Authority. Based on the results of the tests, *Popular's* common equity Tier 1 capital stood at 10.2% of assets, which was below the 12.6% average among 51 big European banks, but not the worst on the list, and even in an adverse scenario, according to the stress tests Popular would still have excess capital of 6.6%. In April of 2017, that capital cushion vanished almost overnight, when top *Banco Popular* officials said they needed to raise capital and the institution began to experience a run. In early June, the bank received $4 billion in emergency assistance from the Bank of Spain but this aid was consumed in two days, and the deal with *Santander* quickly followed.

While there was a sense of comfort based on the fact that the mechanisms that were created after the 2008 crisis worked and no taxpayer money was involved in the takeover, there were significant losers including investors in both the bank's stock and in the $1.4 billion in debt-like instruments Banco Popular issued prior to its collapse to provide a capital cushion (thousands of judicial demands are still pending in Spain, Belgium, and the United States), as well as Santander's shareholders because their holdings were diluted by the bank's decision to raise $7.9 billion in equity to shore up its balance sheet.

The *Popular*'s collapse showed the limits of stress tests and reaffirmed the need for regulators to require banks to maintain higher leverage ratios. According to the 2016 tests, Popular had a leverage ratio of 5.68%. Under the stress scenario, it was supposed to be 3.99%. Yet, the bank's capital proved inadequate.

Two years after the *Popular*'s collapse, the judicial process is still pending with a number of demands from all parties involved. It will be an

---

[11] Saracho had replaced Angel Ron as chairman of Popular in February 2017, a few months before the bank collapsed. Ron, who became chairman of Popular in 2004, was forced to step down in the face of an intense backlash from investors and analysts.

incredibly complex process that will take place in several countries and will involve shareholders, investors, auditors (PwC), the bank managers (for their management of the bank and their compensation[12]), bondholders, regulators (for intervening in the bank), and even *Santander*. So far, there have been 1063 claims and 262 appeals against the FROB.

In the end, the new complex resolution rules that had been established after the crisis in the EU did not protect the system from the *Popular's* collapse. Now, it will be up to the pending trials to determine if it was just an issue of poor management or there was also malfeasance, and whether the regulators (including the supervisor Bank of Spain) and the PwC auditors did their job.

LESSONS FROM THE SPANISH EXPERIENCE

Financial Stability Cannot Be Divorced from Economic Policy

As we have seen throughout the book, the Spanish government (and the ECB) failed to cope with the asset bubble and its imbalances. Hence, the Spanish experience shows that financial stability cannot be divorced from economic policy and macroprudential supervision; while regulation matters, macroeconomic factors do too. The recurrent excuse from politicians and regulators that their hands were tied up because monetary policy was in the hands of the ECB (which was true) was a convenient justification for their inaction. They had other options to curve the bubble: For instance, the government should have eliminated housing tax breaks and/or established higher stamp duty on property sales, or higher capital gains tax on second properties; and the Bank of Spain could have imposed higher provisions and/or capital requirements. Those options, however, would have been politically costly, and it was easier to stand still.

Moreover, the Spanish experience provides an interesting insight into the pitfalls of integration into an incomplete monetary union (one not backed by a political union): Lower interest rates and the loosening of credit will likely lead to a credit boom, driven by potentially overoptimistic expectations of future permanent income, which in turn may

---

[12] Ron was paid €1.47 million in 2016, the same as the year before, and he retired on a pension worth €1.1 million a year, and Saracho was hired with a. $4.5 million hiring bonus. See "Mission Impossible: Saving Banco Popular Too Much for Saracho," *Bloomberg*, June 8, 2018 and "Emilio Saracho Replaces Angel Ron as Banco Popular Chairman," *Financial Times*, February 20, 2017.

increase housing demand and household indebtedness, as well as lead to overestimations of potential output and expansionary fiscal policies. The boom will also lead to higher wage increases, caused by the tightening of the labor market, higher inflation, and losses in external competitiveness, together with a shift from the tradable to the non-tradable sector of the economy, which would have a negative impact on productivity (see Royo 2013). In addition, the crisis also showed that fiscal discipline matters in a monetary union, but it is not enough. Indeed, prior to the crisis as we examined in Chapter 3, Spain was perceived as one of the most fiscally disciplined countries in Europe, and initially, fiscal surpluses allowed the country to use fiscal policy to be used in a countercyclical way to address the global financial crisis. However, although Spain entered the crisis in 2008 in an apparent excellent fiscal position, the country's structurally or cyclically adjusted deficit turned out to be much higher than its actual deficit. As a result of the crisis, as we examined in Chapter 3, the country's fiscal performance collapsed by more than 13% of GDP in just two years. This shows that Spain's structurally or cyclically adjusted deficit was much higher than its actual deficit, and illustrates how difficult it is to know the structural position of a country.

In order to avoid these risks, countries should develop stringent budgetary policies in the case of a boom in demand and/or strong credit expansion. At the same time, they should guard against potential overestimation of GDP and measure carefully the weight of consumption on GDP, because they may inflate revenues in the short term and create an unrealistic perception of the budgetary accounts, as in the case of Spain. It is also important that they use fiscal policies in a countercyclical way to be prepared for recessions. Finally, higher revenues, as in Spain prior to the crisis, should not drive budget surpluses. On the contrary, governments need to address the structural reasons for the deficits and avoid one-off measures that simply delay reforms but do not address the long-term budgetary implications.

Furthermore, to avoid unsustainable external imbalances, countries should also carry out the necessary structural reforms to increase flexibility and productivity, as well as improve innovation in order to allow their productive sectors to respond to the increasing demand and to ensure that their economies can withstand the pressures of increasing competition. They should also set wages based on Eurozone conditions to ensure wage moderation, instead of on unrealistic domestic expectations and/or domestic inflation (Abreu 2006, pp. 5–6). Countries should also take the opportunity presented in boom years to move into higher value-added and faster growth sectors toward a more outward-oriented production structure.

However, the crisis also showed that the economic reform process is a domestic responsibility that countries need to undertake to fully adapt to the challenges (and opportunities) of a single market and a monetary union. Somehow there was an expectation that membership on its own would force structural reforms, and this (naturally) did not happen. On the contrary, the years prior to the crisis showed the limits (and also adverse incentives) of EU/EMU membership in imposing institutional reforms in other areas (e.g., the labor market, the financial sector, or competition policy) and to balance domestic and external economic objectives. Since the crisis, while the EU has gained some leverage to force countries to act in the case of bailouts (as it happened to Spain and other countries that received financial assistance), that leverage is limited once the countries exit the bailout. This remains a domestic process that has to be undertaken at the national level (see Royo 2013).

Moreover, in the context of a monetary union, it is also crucial that countries address current account deficits and losses of competitiveness. During the Eurozone crisis, the focus was largely on the fiscal challenges that countries faced. Yet, it is essential to note that we were also dealing with a crisis of competitiveness. In Spain, EMU membership fostered a false sense of security among private investors, which brought massive flows of capital to the periphery. As a result, costs and prices rose, which in turn led to a loss of competitiveness and large trade deficits. Indeed, below the public debt and financial crisis, there was a balance of payment crisis caused by the misalignment of internal real exchange rates.

Indeed, between 2000 and 2010, there was a significant deterioration of competitiveness in Spain vis-à-vis the Eurozone: 4.3% if we take into account export prices, and 12.4% if we take into account unitary labor costs in the manufacturing sector. In this regard, the experience of Spain within EMU showed that there were lasting performance differences across countries prior to the crisis. These differences can be explained at least in part by a lack of responsiveness of prices and wages, which did not adjust smoothly across sectors, and which, in the case of Spain, led to accumulated competitiveness losses and large external imbalances (see Fig. 8.2). While Germany (and other EMU countries) implemented supply-side reforms in the first years of EMU to bring labor costs down, through wage restraint, payroll tax cuts, and productivity increases,

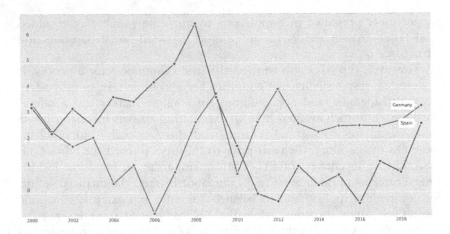

Fig. 8.2 Labor compensation per hour worked in Spain and Germany. Total, Annual growth rate (%), 2000–2019 (*Source* OECD Productivity Statistics: GDP per capita and productivity growth)

making it the most competitive economy with labor costs 13% below the Eurozone average, Spain continued with the tradition of indexing wage increases to domestic inflation rather than the ECB target, and it became one of the most expensive ones with labor costs going up to 16% above average (Portugal led with 23.5%, Greece with 14%, and Italy with 5%).[13] Hence, a lesson for EMU members has been that it is critical to set wages based on Eurozone conditions, and not on unrealistic domestic expectations, to ensure wage moderation (Abreu 2006, 5–6).

Indeed, a crucial problem for Spain was the dramatic erosion of its comparative advantage. The emergence of major new players in world trade, like India and China, as well as the Eastern enlargements of the EU, was damaging some European economies like Spain because those countries have lower labor costs and compete with some of our traditional exports (as exporters of relatively unsophisticated labor-intensive products), leading to losses in export market shares (aggravated by the appreciation of the euro and the increase of unit labor costs relative to those in its trading competitors). Yet, while this was particularly true for

[13] Stefan Collignon, "Germany Keeps Dancing as the Iceberg Looms," *Financial Times*, January 20, 2009, p. 13.

other countries such as Portugal, Italy, and France, in Spain the problem was compounded by the fact that too few companies exported prior to the crises, and that those that exported had differentiated products because they were the large multinationals. At the same time, Spain's attempt to specialize in medium- and higher-technology products was also hindered by the accession of the Eastern European countries into the EU, which were already moving into those sectors specializing in these products.

Furthermore, in order to avoid unsustainable external imbalances, countries should also carry out the necessary structural reforms to increase flexibility (particularly internal flexibility which may be even more important for companies to allow them to deploy effectively their human capital, than the external one, despite the traditional fixation in Spain on dismissal costs) and improve productivity. This would be the most effective way to allow countries' productive sectors to respond to increasing demand and to ensure that their economies can withstand the pressures of membership to a single market. Finally, countries should also take the opportunity presented by the boom to move into higher value-added and faster growth sectors, toward a more outward-oriented production structure.

Last, in regard to economic policies, much has been said about the response to the crisis and the focus on austerity (see Royo 2013; Blyth 2013; Alesina et al. 2019). Suffice to say here that the aftermath of the crisis has shown that discipline and austerity are not enough. The response to the crisis was a real-life experiment to try to prove that an expansionary fiscal contraction can work, and in the end, the obsession with austerity had painful consequences across the European periphery. The problem for countries like Spain was the feeble outlook for growth, as austerity further contributed to the contraction of the Spanish economy and deteriorating fiscal conditions. As a result, Spain's sovereign debt was repeatedly downgraded throughout the crisis; unemployment reached record levels at over 24%; public debt grew from 36% in 2007 to nearly 100% in 2020. The country was trapped in a so-called doom loop: a negative spiral that happens when banks hold sovereign bonds and government bailout banks. Bailing out banks puts pressure on sovereign bonds as investors, concerned about a possible default, sell them. That pressure in turn leads to increase in interest rates, making it more expensive for countries to fund their deficits/debt, which in turn worsens the fiscal outlook for the country and makes a default more likely. Yet, if they do not bail out the banks, countries risk a financial meltdown. This was a trap countries in the European periphery (Greece, Portugal, Ireland, and Spain) found themselves in, which ultimately precipitated their bailouts.

In this regard, the contrast with the United States was striking. Between 2007 and 2010, the US Congress passed the equivalent of three stimulus bills:

- A bipartisan $158 billion package of tax cuts signed by President George W. Bush in early 2008.
- A $787 billion bill pushed by President Obama as he took office in 2009 in the wake of the financial system's collapse.
- A tax cut and unemployment fund extension agreement reached by President Obama and Congressional Republicans in December 2010.

Many studies show that these measures were a key reason why the unemployment rate did not reach double digits in the United States (see Prasad and Sorkin 2009).

Macroprudential Instruments Are Important to Ensure Financial Stability

One of the important lessons from this crisis is that a microprudential approach, even when combined with stable output and inflation, is not enough to ensure financial stability. We still need macroprudential instruments, because they help bolster financial stability and mitigate systemic risk. These instruments have been used more extensively following the crisis, including the use of contemporaneous financial variables as proxies for systemic risk; the use of early warning indicators to gauge financial risks; measures of credit, house prices, and bank balance sheets (e.g., capitalization, profitability, maturity, and currency mismatches); monitoring the financial health of borrowers using detailed firm- or consumer-level data; using housing price and asset-pricing models, as well as a statistical analysis to assess overvaluation in the housing market.

Stress tests also stand out a systemic risk indicator. They are particularly helpful, because they are forward-looking and can consider various extreme scenarios. However, the Spanish financial crisis also shows the shortcomings of these tests, as macrostress tests carried out prior to the financial crisis did not point to any significant risk in the banking sector (nor did they prevent the collapse of *Banco Popular* in 2017).

Furthermore, additional work remains to be done in this area as no consensus exists about the definition of a "macroprudential policy stance," nor regarding the best methods used to measure it. On the one hand, this policy stance can be viewed as the values taken by macroprudential instruments, irrespective of current financial conditions. Alternatively, we could also define the policy stance as conditional on financial developments, e.g., how binding the instruments are at a given time. In addition, it is challenging to determine the best methods to measure it, as it is difficult to aggregate different instruments with potentially very different effects on financial risk.

Finally, some studies have shown the limits of these macroprudential tools. For instance, Montalvo and Raya (2018) have shown that, contrary to other countries, the introduction of regulatory penalties on high loan-to-value (LTV) ratios (one of the most frequently used tools of macroprudential policy) for residential mortgages used by Spanish banking regulators before the onset of the housing crisis of 2008 did not reduce the feedback loop between credit and house prices. The reason was that in Spain appraisal companies were mostly owned by banks, which led to a situation in which the LTV limits were used to generate appraisal values adjusted to the needs of the clients, rather than trying to appropriately represent the value of the property. This caused a tendency toward over-appraisals, which in turn produced important externalities in terms of a higher than otherwise demand for housing, and the intensification of the feedback loop between credit and house prices.

Address EMU's Institutional Deficiencies

While EMU is not the focus of this book, it is important to note that the crisis exposed institutional deficiencies in the European Monetary Union that still need to be addressed. Indeed, the crisis showed that the EMU is a flawed construction. Mario Draghi, president of the ECB, acknowledged as much when he noted that it was like a "bumblebee" and declared "it was mystery of nature because it shouldn't fly but instead it does. So the euro was a bumblebee that flew well for several years." It has not been flying well, and according to him, the solution should be "to graduate to a real bee."[14] The crisis in Spain further illustrated the EMU's institutional shortcomings: Spain had a huge bubble that crashed with the crisis. The "bumblebee" flew for a while and convinced investors that they could invest (and lend) massively within the country; thus, money poured into Spain. However, when the crisis hit, the country could not count on the EU support to guarantee the solvency of its banks or to provide automatic emergency support. And when unemployment soared and revenues plunged, the deficits ballooned. As a result, investors' flight followed and drove up borrowing costs. The government's austerity measures and structural reforms contributed to deepen the country's slump. If anything, the crisis exposed the shortcoming of EMU institutions and showed the fragility of an institutional framework that tried to balance fiscal sovereignty with a monetary union. This model failed to combine flexibility, discipline, and solidarity. Fear has been largely what is keeping it all together. But is fear enough to hold it together in the long term?

Be Humble: This Time It May Not Be so Different

The experience of Spain showcases the need to address deficiencies in the policy-making process and to challenge the dominant paradigm. As we examined in Chapter 4, prior to and during the crisis, there was strong consensus in Spain among economic elites, as well as among Conservatives and Socialists leaders, regarding fiscal consolidation and the balanced budget objective. Indeed, prior to the crisis, Spain presented itself as the model of a country applying the budget surplus policy mantra. This consensus may have worked well in the short term, as it contributed to the credibility of the government policies and allowed the country to become a founding member of EMU, but a more accommodating policy stance would have positively contributed to upgrading the productive base of the country with investments in necessary capital infrastructure and human capital that would have contributed to a faster transition from an economic growth model based largely on low costs, toward one based more on higher value-added and higher productivity, as well as reduced dependency on the construction sector. These investments would have contributed to change the growth model, diversify the economy, and regain

---

[14] Paul Krugman, "Crash of the Bumblebee," *The New York Times*, July 30, 2012.

competitiveness, thus preparing the Spanish economy to better confront the crisis.

Furthermore, more humbleness would have gone a long way in overcoming the overconfidence on the strengths of the Spanish banking sector that was so pervasive during the initial stages of the global financial crisis. Indeed, the Spanish banking crisis also shows that systemic crisis does not only originate from problem of large financial institutions. We also need to pay close attention to the performance of smaller institutions (Garicano 2012). It was precisely the strength of the largest banks (e.g., *Santander*, BBVA, and *La Caixa*) that created a lull and a false sense of overconfidence during the initial stages of the crisis, while the problems of the small institutions, like the *cajas*, were minimized. As we have seen, the Spanish *cajas* sector proved to be the canary in the coalmine. Had more attention been paid to the *cajas*, the regulator would have been able to anticipate earlier the systemic problems and act more timely and decisively.

In addition, the country also overestimated the strengths of its regulatory system based on the countercyclical policies and the dynamic provisions of the Bank of Spain. The crisis exposed the shortcomings of that system, particularly its inability to withstand the political pressures that were at that heart of the "bank bargains" examined throughout the book, and it illustrates the need to build institutional mechanisms that allow the supervisor and regulators to stand up to politicians. As we have seen, regulators in Spain failed to challenge regional politicians, political parties, and unions. This proved to be disastrous. Regulators not only need the supervisory instruments, access, and authority to know the ins and outs of banking institutions, but also they need to courage to be as intrusive as necessary. While individual actors may matter in shaping outcomes because of their ability to act decisively or identify opportunities, we should not just rely on individual actors, but rather develop institutional mechanisms that limit the rent-seeking interest of populist coalitions (like the ones that took over most of the *cajas*).

Moreover, the complacency about the strength of the financial system and its ability to weaver the crisis proved to be disastrous and it prevented the country from taking decisive action earlier, which would have mitigated the impact of the crisis. Indeed, the timing of the decisions is also crucial: Dynamic provisioning worked initially as expected, but it delayed decisive action thus making the crisis larger and deeper than it should have otherwise been. As we have seen, during the initial stages of the global financial crisis, there was consensus in Spain that the stern regulations of the Bank of Spain played a key role in the initial positive performance of Spanish banks, because it forced banks to set aside during the good years "generic" bank provisions in addition to the general provisions for specific risks. In addition, it made it so expensive for them to establish off-balance sheet vehicles that Spanish banks stayed away from such toxic assets. But this consensus led to complacency and hubris. Indeed, the provisions' size—3% of GDP at their highest point (2004)—was simply not of a magnitude commensurate with the credit losses accrued during the crisis. The experience of the Spanish financial sector shows that it is impossible for banks not to be affected from a collapsing bubble in real estate (the Bank of Spain announced in 2012 that bad loans on the books of the nations' commercial banks, mostly in the real estate sector, reached 7.4% of total lending). In the end, Spain suffered a property-linked banking crisis exacerbated by financing obstacles from the international crisis and the delay in taking action proved to be very damaging.

Finally, we should not think that 'this time is different' (Reinhart and Rogoff 2009) and we should be prepared to learn from traditional financial crises. As we have seen, the financial crisis in Spain did not involve subprime mortgages, collateralized debt obligations, structured investment vehicles, or even investment banks. In many ways, some of the main lessons from the Spanish financial crisis are not be so different from those from a traditional banking crisis. First, as noted above, monetary policy has to address asset bubbles and central bankers should be proactive in bursting the bubbles before it is too late. The global financial crisis showed that financial stability will not follow automatically if monetary policy delivers steady growth and low inflation. On the contrary, central banks should try to prevent bubbles from inflating. The crisis has finally shown that it is far more expensive and painful to 'clean up' after asset price bubbles that have burst. At the same time, banks should not lend excessively to property developers and governments. Second, bankers should recognize that retail banking is not a low-risk activity and should avoid overconcentration in property loans, and finally, governments and central bankers should avoid any complacency (as it happened in Spain) and instead need to be vigilant and proactive to avoid the mistakes of the past and to anticipate all possible scenarios, including the most negative ones. In Spain, as noted before, the misplaced and excessive confidence on the strength of the financial sector, and the almost unquestioned belief in the regulatory and oversight prowess of the Bank of Spain, led to hubris.

In sum, in Spain, the collapse of the real estate markets eventually led to a traditional banking crisis fueled by turbocharged lending (largely from international wholesale funding). When the crisis hit, international and wholesale funding dried up and it affected bank lending. In this regard, the main "toxic" assets held by Spanish banks were domestic bad loans and mortgages. In this sense, the Spanish banking crisis has been labeled as a traditional bank crisis. This time it was not so different!

It Is the Politics, Stupid

Throughout the crisis, the focus for most analysts was largely on the economic dimension of the crisis, as well as on its economic causes and consequences. As we have seen, however, it would be a mistake to underplay the political dimensions of the crisis, and not just at the Spanish national level, but also at the European and global ones. This was as much a political crisis as an economic one, and as much a failure of the markets, as a failure of politics. Indeed, political decisions led to the crisis and marked the course of the crisis.

A central argument of this book has been that politics matter, and that political factors are central to understanding why Spain has suffered repeated banking crises. Politics influence bankers' decisions, their operations, and the regulatory framework in which they operate. And political institutions and politics structure the incentives of actors involved in banks, from bankers to shareholders, depositors, debtors, to regulators.

Indeed, political circumstances influence *bank banking bargains*, and they in turn define the types of banks that emerged in any given country (see Calomiris and Haber 2014). As we have seen throughout the book, banking systems are an outcome of politics, and the interplay between politics and banking has been crucial to account for the performance of the Spanish banking system. At the same time, while banking systems shape politics, they also influence the coalitions that bargain and affect the bargaining power of the parties that participate in the bargain. Finally, political circumstances shape the development of new financial services and instruments.

Ensure Conservative Risk Management

The Great Recession of 2008 made it clear that Conservative risk management will be crucial, and hence, banks must be required to manage their risk prudently, hold sufficient levels of capital, and recognize losses in a timely manner. Furthermore, they should be prevented from free riding on safety-net/too-big-to-fail protection. Bankers should also stress the need to be obsessive about credit quality, they should largely avoid non-core activities such as commodities trading, they should absorb past experiences, and they should learn that geographical diversification may help limit the banks' exposure to the weak economic performance of a particular market.

Hence, some of the main lessons from the crisis are old lessons that banks should have already heeded: Banks need to maintain lending standards to ensure banking stability; they need to maintain sufficient provisions and capital to cover expected and unexpected losses; and they have to ensure alignment between the incentives of managers, shareholders, and stakeholders (including taxpayers and deposit guarantee funds). Moreover, supervisors and regulators need to enforce intrusive supervision and ensure that financial institutions do not engage in unsustainable business models/initiatives (see Roldan and Saurina 2012).

In addition, the 2008 crises also showed that safety nets can be destabilizing because they intensify moral hazard: The more the generous, the more unstable the banking system (Calomiris and Haber 2014, pp. 461–62). The most stable banking systems combine a credible system of prudential regulation based on accounting transparency, substantial capital requirements, and limited safety-net protections. And paradoxically safety nets, such as deposit insurance, which are outcomes of political bargains, tend to destabilize banking systems. In Spain, those guarantees and the expectation of bailouts promoted (and even rewarded) the reckless behavior that led to the collapse of so many *cajas*. Hence, it is crucial to overcome political resistance to stricter capital requirements.

Implement Financial Reforms (While Recognizing How Difficult That Process Will Be)

We still need further financial reforms at the national and international level to ensure increasing financial stability. Yet, we need to recognize that reforms are difficult because the coalitions that underpin the "banks bargains" described throughout the book are entrenched, and we cannot simply expect incumbent politicians or regulators to prevent the next banking crisis because they are likely part of the coalition that may cause it (Calomiris and Haber 2014, pp. 278, 281). For this reason, improvements in banking systems would be more lasting if they take place in the context of broader reforms of the political system, fiscal policy, trade policy, industrial policy, and corporate governance that reduce corruption.

As noted by Calomiris and Haber (2014, p. 504), the implementation of reforms hinges on the ability to assemble a winning coalition, and even more importantly, we need *persistent support for good ideas*. Crisis offers windows of opportunity to effect change. However, more often than not, powerful interests succeed in using moments of crisis to strengthen their power. For instance, despite the establishment following the crisis of resolution mechanisms, dynamic provisioning, increasing capital requirements, and other safeguards, moral hazard remains a significant problem across the Western world in the aftermath of the recent Great Recession. Indeed, one of the main outcomes of the crisis in Spain and other countries has been the greater concentration of national banking systems (yet another instance of the political bargains that shape banking), which has only aggravated the risk of 'too-big-to-fail.'

Supervision and an appropriately tough resolution regime must go hand in hand. But it is also important to recognize that smart regulation alone cannot count for a country's banking success. If that was the case, it could be easily replicated. The initial complacency regarding the strength of the dynamic provisioning system that we examined in Chapter 5 proved to be unwarranted in light of the subsequent crisis. On the contrary, banking success is based on the "*bank bargains*" that we have examined throughout the book, and those are much harder to emulate. Countries with stable banking systems that provide abundant credit (i.e., Australia, Canada, New Zealand) have the following characteristics: They were part of the British Empire; they have long-standing democracies; and they have institutions "that limit the opportunities for bankers and populists to form rent-seeking coalitions" (Calomiris and Haber 2014, p. 459). The Spanish experience confirms that premise. Indeed, one of the main lessons from Spain is that in order to avoid banking crises it is crucial to develop strong institutional mechanisms that limit rent-seeking. In Spain, an apparently robust regulatory and institutional framework that was supported by a dynamic provisioning regime was undermined by a coalition of bankers and populists (particularly in the *cajas'* sector) that prioritized their own rent-seeking interests, at the expense of the stability of the system.

Furthermore, we need to accept the premise that banking systems are the result of political compromises. Hence, they can be used as instruments of redistribution and have often provided politically palatable options to governments to avoid harder choices. Governments essentially have three main ways to redistribute income (Calomiris and Haber 2014, p. 445): They can change the tax burden; they can transfer resources to the poor; or they can use subsidized lending through the banks to effect implicit transfers to the poor. Sometimes governments use all three, but this choice has enormous implications for the stability of banking system. Countries cannot expect a 'free lunch.'

Finally, it is also necessary to acknowledge that although political pre-conditions impact banking-system outcomes, these outcomes can also have very significant political consequences, and sometimes these effects are unintended and/or unanticipated. For instance, governments can resort to inflation-tax banking to avoid increasing taxes, but this may eventually undermine the legitimacy or electoral support of that government and lead to a political change.

Be Wary of Capital Inflows

While capital inflows have many benefits for countries because they provide funding for infrastructure and investment, or help offset trade deficits, they can also be a source of instability because they may trigger a boom, as it happened in Spain, which caused an asset price bubble that led to an erosion of Conservative risk management from banks when the drop in interest rates from the ECB led to a search for higher yields. This has led many authors to emphasize the negative impact of global capital and to attribute financial crises to capital inflows (see Chinn and Frieden 2011). In Spain, the real estate market boomed in the year that preceded the 2008 financial crisis, and this led to higher commodity prices and record stock market yields. Most banks tried to capitalize on this bubble and made choices that they came to regret. While capital controls have a mixed record in terms of effectiveness and would not be desirable, capital flows can contain early warning signals of an upcoming crisis; hence, it makes sense for the European Stability Mechanism (ESM), as the crisis resolution mechanism, of the Eurozone to carefully monitor these data.

However, it is important to recognize that not all foreign capital is destabilizing for the banking sector. While countries like Spain, the UK, and the United States experienced large current account deficits in the years prior to the 2008 global financial crisis, other countries, as we mentioned in Chapter 1, like Australia or New Zealand did as well, but did not experience financial crises. This puzzle has to be explained. According to Copelovitch and Singer (2020), while foreign capital inflows can be the fuel for a destabilizing financial boom that can end in a financial crisis, and large international capital inflows are strongly correlated with banking crisis, the key conditioning variable is the domestic financial market structure. It is the relative degree of competition between banks and securities markets, measured by the ratio of stock market size to bank credit that determines the risk proneness of financial systems. This is so because when banks face competition from large, well-developed security markets they take on more risks and these risks are magnified by capital inflows. In other words, they show that capital inflows in countries with more securitized financial markets affect the quality of bank lending and the composition of balance sheets (not necessarily the volume of bank credit) (p. 44). Therefore, it is crucial to examine the size and depth of securities markets in order to understand (and prevent) financial crises.

Consider the Size and Depth of Securities Markets

As we outlined in Chapter 1, Copelovitch and Singer (2020) focus on the political decisions that shape the structure of financial markets and the international capital flows that make some countries more vulnerable to financial crises. They recognize that capital inflows amplify this risk and increase the chance of a banking crisis but argue that banks engage in riskier behavior when they compete alongside well-developed national securities markets. They examine the political origins of financial market structure, looking at the key political decisions that define the structure of these financial markets (from the process of granting and operating license to banks to the rules that determine their operations; to the rules that establish the terms to disclose information, fraud, supervision, and exchange for securities). According to them, banks in countries with large financial markets are likely to feel more pressure to take on more risk to maintain their market share and profits. In other words, their Conservative bias will be eroded as security markets get stronger in their countries. From this perspective, the size and depth of securities markets are key for bank stability and have important implications of risk management. Hence, in order to understand financial crises, we need to delve into the structure and characteristics of national financial markets, which varies significantly from country to country: from universal banks to specialized banks; from developed to underdeveloped securities markets; from national to regional banks; from fragmented to centralized...

However, it is important to recognize that changing the structure of financial markets will not be easy because they are historically contingent and thus difficult to change. Indeed, their development is marked by path-dependent financial trajectories and stickiness (Thelen 2004), which, for instance, makes it harder for countries with underdeveloped securities markets to develop new and riskier financial products whereas it is easier for those with well-developed securities markets, and thus amplifies the risk of financial crises.

This is an important finding that fell outside of the scope of this book's analysis, but it still needs to be explored further and should be in the agenda for future research. As Copelovitch and Singer (2020) show, differences in the size and structure of securities markets and the behavior that they have on banks have also impacted the stability of the Spanish financial system. While this book has focused largely on the structure of the Spanish banking sector and the rules for allocating credit as the key explanation for banking instability in Spain, it has overlooked the role of securities markets in explaining the incidence of financial crises in Spain. Indeed, more work needs to be done to analyze the relationship between banking systems, securities markets, and financial stability in Spain. In particular, it would be important to examine carefully the historical policy decisions and political bargains that helped to determine the relative strength of traditional banking versus securities markets in the country, and analyze in great detail the relationship between banking and securities markets. These political decisions have shaped Spanish financial markets over the long term by allowing room for securities markets to mature, which in turn have made the country more vulnerable to financial crises. This is an omission of broader aspects of Spanish financial markets that still needs to be addressed.

Spanish Banking in Comparative Perspective

Some have compared the Spanish case with the experience of Japan (Kamikawa 2013). Is Japan the future of Spain? While there were some similarities on the assets and liabilities side, there were also key differences; in Spain, the purchase of sovereign bonds depended to a significant degree on international markets (this was not so much the case in Japan, where domestic savings were a major source of funding). In Japan (unlike Spain), during the lost decade citizens still invested in banks despite the bubble, and there was limited capital flight. Finally, Spain has a much larger funding gap, hence a larger dependence on wholesale markets.

Others have compared Spain to Ireland (Dellepiane and Hardiman 2011). Fortunately for Spain, its real estate bubble was smaller than in Ireland: Property prices rose about three times in Spain in the mid-1990s, compared with 4.5 times in Ireland, and real estate loans peaked at 77% of the Irish economy, compared with 29% in Spain. Moreover, the crisis largely affected the second tier of Spanish banks. Santander and BBVA far more diversified and with major international operations, were affected to a far lesser degree. At the same time, Spain unlike Ireland had established a recapitalization instrument and process based on the FROB. The problem, of course, is that it lacked enough funds, which led to a financial bailout in June 2012 (see Chapter 1). However, the Spanish bailout was also 9% of its economy, compared with 63 billion euros, or 43%, in Ireland.

In the end, the Spanish government (as well as the BoS and the ECB) failed to cope with the asset bubble and its imbalances. Hence, the Spanish experience shows that financial stability cannot be divorced from economic policy; while regulation matters, macroeconomic factors do too. As late as 2012, Spanish banks were still dealing with their toxic assets. They should have heeded the lesson from other countries who acted more decisively to clean them up, not just merged into a bigger problem. Bankia is a perfect example of that failure. In the end, Spain did not heed the lessons from previous crises: do not lend excessively to property developers; burst the bubble before it is too late; recognize that retail banking is not a low-risk activity; avoid over-concentration in property loans; and remember what happened before.

Theoretical and Normative Implications

From a theoretical standpoint, the Spanish financial sector's response to the global financial crisis during the crisis shows that cross-national differences persist. While financial capitalist states have converged as a result of the combined processes of globalization and European integration rendered the "Mediterranean" financial model far less distinct from other models than before, in the case of Spain, the crisis led to extensive regulatory intervention that served to reinforce the pre-existing model;

changes in the years immediately preceding the financial crisis have not been reversed (see Hall 2016).

The analysis of the Spanish case shows that institutions do not stand still. On the contrary, they evolve through their ongoing adaptation in response to changes in the political, social, economic, and international (i.e., the EU) environments. These changes have illustrated the ways in which the functions and roles of these institutions have evolved over time. Yet, it calls into question the presumption that increasing economic integration into EU/EMU will force the institutions of member states into convergence on a common model. Indeed, Spanish institutions have proved to be remarkably resilient in the face of significant exogenous shocks and EU/EMU membership did not radically reshape or disrupt previous patterns. Nor it did generate the degree of institutional innovation that could have been expected. The financial bailout forced significant reforms but since 2015 the political paralysis and fragmentation have led to stagnation and complacency. To understand why, future research should explore further, as we have tried in this book for the banking sector, the political coalitions under which these institutions have evolved.

Furthermore, the focus of the VoC literature has been on how national institutional differences condition economic performance, public policy, and social well-being; and whether national institutions will survive the pressures for convergence generated by the crisis. This book contributes to this literature by highlighting how national institutions have conditioned economic performance in Spain (see Royo 2008).

The initial response to the global financial crisis showed that cross-national differences persisted. While financial capitalist states converged as a result of the combined processes of globalization and European integration rendered the "Mediterranean" model far less distinct from other models than before, in the case of Spain, the crisis initially led to extensive regulatory intervention that served to reinforce the pre-existing model, and changes in the years immediately preceding the financial crisis were not reversed. However, it is likely that the recent restructuring of the Spanish financial system and labor regulations will accelerate its convergence towards a more liberal market economy model, more based on the markets.

Moreover, despite the grouping of Spain in the 'Mediterranean' variety of capitalism, there are important differences in how the crisis played out among those countries. A coherent variety of Mediterranean capitalism is missing and certain domestic political economy institutions—namely, banks—are key to explain the outcome of the sovereign debt crisis. In light of recent events, the literature on varieties of capitalism, and the literature in political economy more generally, should pay more attention to banks and national banking systems (Quaglia and Royo 2015).

Yet, it is important to stress that the process of institutional change is not linear and that there is also strong path dependency; therefore, it is still too premature to confirm any definitive outcomes (Hall 2016). In addition, the analysis of the Spanish experience with the crisis confirms the thesis that coordination is a political process and that strategic actors with their own interests design institutions (Thelen 2004). Institutional change is a political matter because institutions are generated by conflict, they are the result of politics of distribution, and, hence, they are politically and ideologically construed and depend on power relations. In other words, institutional change is driven by politics. In this regard, the crisis is having a profound effect on power relations and the interests of actors. The (yet undetermined) outcome(s) of these changes will, in turn, influence the process of institutional change. But it is still too premature to make definite conclusions, and that is a limitation of this book.

The implications of policy-making are also significant. The experience of Spain shows that economic convergence is not sustainable in the absence of institutional convergence. This casts the fundamental challenge of policy-making in a new light, suggesting that policy-makers should not only focus on policy reform, but also on institutional one, as well.

The creation of the Single Monetary Union and the conferral of monetary powers to the ECB were not accompanied by the centralization of prudential institutions in the Eurozone area, which apart from regulatory efforts to harmonize prudential requirements in the internal market largely remained in the hands of national supervisors. As we have seen, the financial crisis brought to the fore the vulnerabilities of the European banking system and had a great impact on prudential supervision of credit institutions (Lo Schiavo 2017). As a result, in 2014, European institutions established a new framework of banking supervision for the Europe area: the Single Supervisory Mechanism (SSM), which places considerable supervisory powers in the hand of the ECB including biding supranational macroprudential powers (including the power to 'apply higher requirements for capital buffers' and also 'apply more stringent measures aimed at addressing systemic or macroprudential risks at the level of credit institutions,' which could be higher than the requirements applied by national governments of the member states in which the banking institutions are established) under certain conditions (art. 5 of the SSM Regulations).

These regulatory changes go a long way in addressing Schoenmaker's financial stability trilemma (2011), according to which financial stability, financial integration, and national financial policies are incompatible. Any two of the three objectives can be combined but not all three; one has to give. As we have seen throughout the book, Spain tried to achieve all three and failed. The establishment of a European-based system of financial supervision that moves powers for financial regulation, supervision, and stability, as well as crisis management operations to maintain financial stability further to the European level, helps address the balance between financial integration and national financial autonomy (e.g., as financial integration increases national policies become less effective) (see Goodhart and Schoenmaker 2009). This development would have had a tremendous impact on the "bank bargains" described throughout the book, as the locus of decision making would have been transferred to European institutions, thus making it far harder for the coalitions that underpinned those bargains to get their way. This new framework of banking regulation will likely set a new structure for the emergence of new supranational coalitions and new "bank bargains." Yet, it is still important to highlight that so far there has been limited progress toward a banking union and much remains to be done.

Finally, the crisis also exposed the weaknesses of the EMU design, as well as the absence of adequate instruments to correct asymmetries among member countries, particularly persistent current account deficits in a monetary union that lacks a fiscal union. If anything, the crisis has shown that we need a more balanced response to any crisis at the EU level. The lack of coordination may have lessened the effects of the measures taken to deal with the crisis.

CHALLENGES FOR SPAIN

The last ten years of Spain's history (2009–2019) have been a period of brutal lows during one of its worst crisis in modern history, but also some impressive highs as the country emerged from it. Those who still argue that the economic crisis in Spain was caused by EMU's institutional deficiencies, the subprime crisis in the United States, neoliberal policies, or deregulation fail to take into account the domestic institutional dimension. As we have seen, the crisis exposed an unsustainable economic model that had no long-term prospects. The emperor had no clothes (see Muñoz Molina 2013). While Spaniards were rightly outraged by the actions of bankers and politicians, and by the amorality of the whole situation and the complete disregard from the majority of the country's elite of the impact that their actions could have on people, most people had a sense of what had been going on and tolerated it. As noted on Chapter 7, there seemed to be a complete disconnect between actions and consequences. The Spanish ruling class largely remains a community interwoven by personal and/or financial connections that have worked to their advantage, often at the expense of regular citizens. That interweaving generated clientelistic networks that have provided protection and a safety net that did not let them fall. Elites became complacent, they had been coasting for a long time, and they could not grasp the unsustainability of their course of action. Their own success skewed them in an optimistic direction and kept them from seeing what was happening around them. For them, actions and consequences did not apply. They did not seem to appreciate that their actions could make any difference, or that they had responsibilities over the consequences of their actions.[15] When the crisis hit the country, they were out of answers. Their complacency took the place of serious thinking about the country's future.

At the time of writing (January 2020), Spanish citizens still demand answers. The economic crisis has been followed by a political crisis that has made it harder to act: The country has experienced four general elections in four years between 2015 and 2019, and regional tensions have intensified as a result of the crisis in Catalonia, where divisions over the illegal independence push in 2017 and the nationalist backlash that it triggered in the rest of the country have intensified the effects of political fragmentation and polarization. Not surprisingly, as we have seen in Chapter 7, Spanish citizens are still upset about the corruption and impunity, the consequences of the severe budget cuts, the taxes, the dramatic increase in unemployment, the pessimism, and desperation that has spread across the country following the crisis. But it would be too easy to just blame the political and economic elites. That account leaves aside the responsibility of people who were not part of the political elite, but also invested in the property pyramid scheme that fueled the bubble. They also seemed to forget that their actions had consequences.

---

[15] See "The Psychology of and Irish Meltdown," *The New York Times*, July 27, 2013.

Indeed, a central problem has been the lack of accountability: In the years prior to the crisis, there was (in Spain and many other countries) a feeling of impunity that came from non-punishment. As of 2020, the list of those who have been sent to jail for their part in the housing bubble and all that followed it is remarkably thin. The result of the *Bankia* trial, in which its main administrators have been accused of embezzlement and falsifying accounts, is still pending at the time of publishing, and just a handful of corruption cases have resulted in convictions or reached any conclusion to date. Certainly, if there is nothing criminal in the conduct of the managers and/or the regulators, it must be because the criminal law is defective in that area.

It is essential that the culture impunity that characterized the years prior to the crisis changes. In this regard, one of the few positive outcomes of the crisis seems to be that Spanish society seems to be less tolerant of corruption. This will be crucial to prevent future crisis: Bankers and politicians must know that if they break the rules they could go to prison. The country needs simple rules, timely judicial decisions and strong enforcement, and accountability must also extend to politicians. Yet, while Rajoy's government was toppled in 2018 over the corruption of the PP, trials still take way too long, few people are convicted, and the prison terms and/or financial penalties are not always severe or come too late.

Spain seems to conform to Mancur Olson's institutional sclerosis thesis (1982), according to which over time all political systems succumb to sclerosis because of rent-seeking activities by organized interest groups, which lead to cronyism and corruption, and thus an erosion in the rule of law. The solution, therefore, must come from civil society. Indeed, civil society needs to be strengthened. Historically, Spaniards (as other continental Europeans) have preferred equality to liberty, which led to the development of strong states, but conversely weak civil society (Ferguson 2013). We need a better balance.

Furthermore, deep-seated structural weaknesses are still holding back growth in Spain and weighting on market assessment: overregulated product and labor markets, poor productivity, and low education achievement in international tests. Spain still has much ground to cover vis-a-vis its wealthiest European counterparts (see Fig. 8.4). In the absence of devaluations, Spain still needs further structural reforms to improve productivity and move toward a higher value-added model. The crisis has shown that 'internal devaluations' through a decrease in prices and salaries are not politically sustainable in the long term and too costly socially.

Indeed, while Spain has regained competitiveness after the crisis, it did so through a very painful internal devaluation with enormous social costs. Carabaña (2016) shows that Spain had roughly the same inequality in 2016 than in 1993: a Gini coefficient of disposable income of around 0.34 (see Fig. 8.3). In comparative perspective, Spain's inequality got closer to the EU-15 average at the peak of its real estate bubble in 2006–2007 but it fell back again in the aftermath of the crisis. In constant 2013 euros, the average income for the very poor went from 1.443 euros in 1993 to 729 euros in 2012, with even negative incomes between 2007 and 2009. This is also reflected in the distribution of overall income in Spain. In 1994, the poorest 20% of the population received 7% of the income; in 2013, this share was just 5.9% (Otero-Iglesias 2019). And the crisis was particularly brutal for young people, and they have not yet fully recovered: A head of family younger than 35 had in 2016 (the last available data) an income 18% lower than someone of that age in 2010 (27,700 euros in 2010 vs. 22,800 euros in 2016), and this despite the fact that by 2016 all overall incomes had recovered the level of 2010. But governments have other options, including the reduction of Social Security contributions

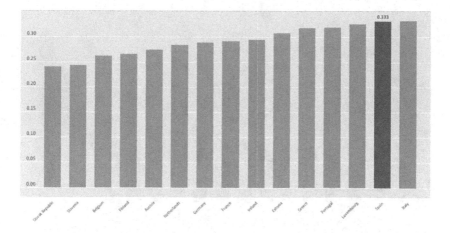

Fig. 8.3 Income Inequality. Euro Area (2017) Gini coefficient, 0 = complete equality; 1 = complete inequality (*Source* OECD Social and Welfare Statistics: Income distribution)

and/or the increase in the value-added tax; the reduction of other non-salary costs such as the energy and infrastructure ones; and the increase in productivity and labor quality.

As of January 2020 (the time of writing), a new leftist government coalition between the Socialist Party and the leftist populist *Unidas Podemos* that emerged from the November 2019 election is coming to power with a progressive agenda that, while committed to fiscal discipline, seeks significant tax increases for the wealthy, higher corporate taxes, the establishment of a basic income and a labor reform to eliminate some of the more neoliberal provisions of Rajoy's 2012 labor reform. This is the first coalition government and the first time in which the Communist Party is part of a coalition government since the 1930s. It comes to power with a very slim and fragile majority in Congress (it won the tightest of investiture votes in history—167 vs. 165) and at a time in which the country has slipped back to the polarization and personal animosity that had characterized other dark periods of Spanish modern history (Fig. 8.4).

Yet, the 2008 crisis has resulted in significant shifts in the business sector and the economy as a whole (see Table 8.2), and prospects for the country seem positive despite challenges. On the one hand, the last election that took place in November 2019 produced an inconclusive result, unemployment remains stubbornly high at 14% (still about half the rate of

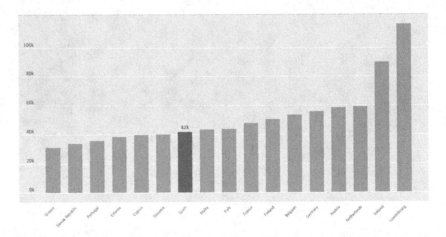

Fig. 8.4 Gross Domestic Product (GDP) Total (US dollars/capita, 2019) (*Source* OECD. Aggregate National Accounts, SNA 2008 [or SNA 1993]: Gross domestic product)

Table 8.2   Economic performance Spain. 2014–2020

Subject Descriptor	2014	2015	2016	2017	2018	2019	2020
GDP, constant prices. % change	1.4	3.7	3.2	3.0	2.6	2.2	1.8
GDP, current prices. US dollars	1379.1	1199.7	1238.0	1317.1	1427.5	1397.9	1440.4
Total investment. % of GDP	19.5	20.4	20.4	21.1	21.9	22.2	22.3
Gross national savings. % of GDP	20.5	21.6	22.7	22.9	22.9	23.1	23.3
Inflation, avg. consumer prices. % change	−0.2	−0.5	−0.2	2.0	1.7	0.7	1.0
Volume of exports of goods and services. % change	4.3	4.2	5.2	5.2	2.3	2.4	3.3
Unemployment rate. % of total labor force	24.4	22.1	19.6	17.2	15.3	13.9	13.2
General government structural balance. % GDP	−1.9	−2.4	−2.8	−2.5	−2.3	−2.3	−2.3
General government gross debt. % of GDP	100.4	99.3	99.0	98.1	97.1	96.4	95.2
Current account balance. % of GDP	1.1	1.2	2.3	1.8	0.9	0.9	1.0

Source International Monetary Fund, World Economic Outlook. October 2019
Estimates after 2018

the peak of the crisis), job security is still an issue with more than a quarter employees on temporary contracts, the quality of many jobs and the level of some wages leave much to be desired, and the country is already experiencing lower rates of growth: The economy is no longer expanding at the 3% rate of 2015–2016, but is expected to reach about 2% in 2019, still above the EU average, which has already slowed down job creation from an annual rate of 3.2% earlier in 2019 (596,900 new additional jobs) to 1.7% by the end of the year (346,300 jobs). Still, the main concern is the dysfunctional state of the country's politics: The dispute between Catalan separatists and the rest of the country remains tense, and four elections in four years have produced inconclusive results and weak governments that have not lasted long. Political fragmentation seems entrenched: The fourth general election in as many years resulted, yet again, in an inconclusive result in November 2019 that made the formation of a stable government even more difficult.

However, there were good reasons for optimism prior to the COVID-19 crisis.[16] Following the 2008 crisis, Spain moved away from an economy that was overly dependent on residential constructions and adjusted magnificently to an export-oriented economy. Indeed, despite political uncertainty, innovation and exports were flourishing and the country was experiencing an expansionary cycle with a positive current account cycle (something quite exceptional in recent history as Spain experienced current account deficits of over 10% prior to the crisis). Indeed, Spain is still benefitting from its extensive connections abroad, particularly in the EU and Latin America, and it is capitalizing from cost advantages compared to other Western European countries, as well as from an impressive infrastructure network that includes the EU's most extensive high-speed train, motorway, and fiber-optic networks. The crisis led a massive deleveraging of the private sector and the reforms that followed the crisis (including a labor reform that further liberalized the labor market and shifted away from sectoral wage bargaining) have resulted in competitiveness gains driven by lower costs and wages (labor costs are between 20 and 40% below France, Germany, and the UK, the outcome of a brutal internal devaluation) (see Fig. 8.5), which have led to a surge of exports (cars, chemicals, industrial equipment, food products) in the last decade, from 22% of GDP to 35%. While Spanish workers bore the brunt of the crisis, salaries are finally growing again: In 2019, salaries negotiated in collective bargaining agreements have grown 2.3%, public salaries have also increased 2.5%, and the government increased the minimum wage of 22.3% to 900 euros (the highest increase since the establishment of democracy in the country). All this has led to an average wage increase of approximately 2% (1876.95 euros).

The new leftist coalition government, which came to power in Spain in January 2020, is in unchartered territory, as the country has never had a coalition government since the transition to democracy more than four decades ago, plus they do not have enough votes in parliament and will need to support from other parties to pass legislation. Moreover, its leftist agenda has raised some alarms among markets and investors. The two parties have agreed to higher taxes, swifter reductions in carbon emissions, and a return to sector-level collective bargaining (from firm-level

[16] See "Companies Ride Out Political Turbulence," *Financial Times: Spain Business and Innovation*, Tuesday, December 10, 2019, pp. 1–2.

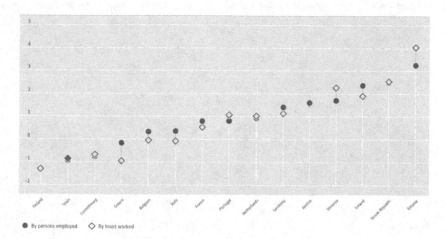

Fig. 8.5   Unit labor costs by persons employed/by hours worked, Percentage change, previous period, 2016 (*Source* OECD. Labor: Unit labor cost—quarterly indicators—early estimates)

wage bargaining). The government is also committed to engage in dialogue with the pro-independence Catalan leaders (in exchange for the most pragmatic of Catalonia's secessionist parties, *Esquerra Republicana de Catalunya*, ERC's abstention in the investiture vote) to try to ease tensions and mend divisions (for which he has been accused of treachery by the Conservative parties). Yet, there are concerns that these policies will deny Spanish companies the ability to adapt swiftly to new market conditions and opportunities, and that raising corporate taxes of bigger groups may deter the investment that the country desperately needs to raise productivity (although the recent data in productivity growth is promising, see Fig. 8.6). Finally, the new government faces significant fiscal constraints as public debt is hovering at around 100% of GDP, and it comes to power at a time when the Spanish economy has cooled and it has to reconcile its spending promises with the EU demands for Spain top rein in its structural deficit (and it lacks a majority in parliament to pass a budget). The implosion of the COVID-19 crisis, which took place as this book goes to the publisher in March 2020, with its devastating costs in lives as well as its social and economic consequences, will make things extraordinarily harder.

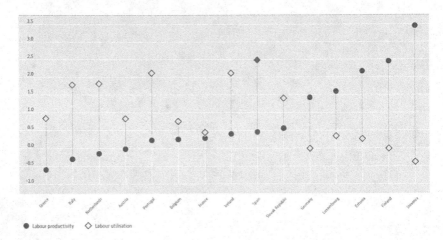

Fig. 8.6 Labor productivity and utilization. Labor productivity/Labor utilization, Annual growth rate (%), 2016 (*Source* OECD Productivity Statistics: GDP per capita and productivity growth)

However, the recent experience of Portugal shows that a government committed to reform and fiscal discipline can turn around investor sentiment. Indeed, while the degree to which the Portuguese Socialist government that came to power in November 2015 overturned austerity was not dramatic, small policy changes that sought to combine fiscal discipline with a fairer distribution of the economic costs and benefits were enough to restore market confidence and increase growth because they lifted confidence, the great driver of economic recovery, thus propelling economic activity. Prime Minister Costa challenged the prevalent austerity dogma based on the imposition of deflationary policies that ended up deepening recessions and increasing unemployment and the probability of defaults. His policies have shown that it is possible to respect common rules on deficit and public debt, while achieving a fairer distribution of economic benefits and promoting economic growth to reduce unemployment and increase people's incomes.

THE IMPLICATION OF BANK BARGAINS FOR DEMOCRATIC POLITICS

At a time in which there are growing calls in many countries for further deregulation and the undoing of some of the safeguard mechanisms that emerged after crisis (dynamic provisions, greater capital requirements, resolution mechanisms, stronger supervision, the Volcker Rule...), it is important to emphasize again the crucial role of *bank bargains* in banking systems. Not only governments have been historically the largest demanders of credit, but they also define the property-right systems that structure banking; establish and regulate banks; enforce credit contracts; and allocate losses among creditors in the case of bank failures. This book has emphasized the political deals that determine which banking rules are passed and which groups are in charge, and it has shown that banking systems are the result of a partnership between governments and bankers that is shaped by the institutions that determine the distribution of power in the political system. The supervision and regulation of banks may be based on technical criteria, but those criteria are the outcome of a political process of deal making, Calomiris and Haber's *Game of Bank Bargains*. While the rules that determine who is part of the government-banker partnership are set by political institutions, coalitions among the actors determine the rules governing bank entry, access to credit, and the allocation of profits and losses. And these decisions are not merely technical decisions based on some efficiency criterion, but rather the outcome of political deals that are guided by the logic of politics (Calomiris and Haber 2014, p. 13) and had substantive economic and political consequences.

Democratic governments in Western Europe and the United States responded to the 2008 *Great Recession* by bailing out failed financial institutions, and most of them did it while they were implementing severe austerity policies that were exceptionally painful for millions of their citizens. And this happened while there was limited accountability (if any) against the leaders of those financial institutions who were also responsible for the crisis. These governments' inadequate responses to that crisis have had disastrous social, economic, and political consequences, and it has made the allocation of losses among creditors in the event of bank failures one of the most contentious issues of the day and a challenge to our democratic politics. To this day, we are still suffering the consequences of those decisions.

The crisis provided a window of opportunity to address the self-interested policies that had led to such catastrophic consequences for many banking systems (and their citizens!). But we seem to have quickly forgotten the terrible consequences of previous policy choices. It is important to emphasize the political nature of this process because at this precise time, powerful self-interested coalitions in countries across the world are looking for opportunities to reversing the safeguards established after the crisis and lower, again, underwiring standards.

Indeed, while in the nineteenth and early twentieth century governments responded to systemic banking crisis with either minimalistic policies or simply stayed aside, the increasing financialization of the last few decades has led to governments' public-funded bailouts in the most recent financial crises of 2007–2008. As Chwieroth and Walter (2019) show, democratic institutions have not been able to constraint governments' propensity to implement taxpayer-funded rescues of the financial system, which now seem to have become the norm. According to them, this may reflect the evolving interest of middle-class voters, who are exercising pressure from below and 'forcing' governments to implement those bailouts to ensure electoral success.[17] This era of growing financialization has increased the middle class's expectations that their governments will act to protect their wealth, which has enlarged the constituency supportive of bailouts.

However, the most recent crisis has shown the limits of this approach as governments have been finding it increasingly difficult to meet the demands of that middle class (whose investment interests largely align with those of the elites) for costly taxpayer-funded bailouts while mitigating the impact of the crisis on the majority of the population, because these bailouts have had an asymmetric impact on the distribution of wealth. They contribute to larger fiscal deficits (and to less resources to support those adversely impacted by them) and to rising economic volatility and lower growth (Reinhart and Rogoff 2009, pp. 145–47). Moreover, they inherently increase moral hazard and in turn foster greater financialization and leveraging, thus making financial system more fragile.

---

[17]Yet, less than 14% of American households own corporate stock directly and the middle class only owns 8% of all stock (by comparison, the top 1% owns almost 40%) (Wolff 2017; Holmberg 2018); and the share the middle-class shares are decreasing even as incomes overall are rising in most Western European countries at the U.S. (Kochhar 2017).

The result of the crisis (amplified by the continued impact of globalization and technological change) has been rising inequality, job losses, and lower quality of jobs. The entrenched perception among millions of citizens is that these bailouts have left middle class and poorer households relatively worse off. All this, in turn, has been turning citizens against traditional parties (particularly leftist parties who have intervened in favor of the financial sector, as it happened to Zapatero's Socialists in Spain, see Rodríguez Zapatero 2013) and democratic institutions, and has contributed to the growth of populism and anti-system parties across the world (Pappas 2019). This presents a growing dilemma for our democratic politics because financialization and financial instability are likely here to stay.

Growing populism has not been a unique Spanish development. On the contrary, it has become a global phenomenon emerging all over the world. As we have seen, the *Great Recession* was a return to zero-sum politics, and one of its main consequences was the erosion of collective interest, solidarity, and cooperation, as well as the re-emergence of a new age of nationalism. The new politics of the day have been mostly marked everywhere by transactional arrangements and not enlightened self-interest. While societal upheaval was driven by social change, as well as technological and economic disruptions, the crisis led to the growth of populism because the traditional elites did not have effective solutions. Indeed, the liberal regimes failed to address the economic and social consequences that arose from liberal policies like tax cuts, deregulation, or fiscal consolidation. Those policies resulted in financial instability, inequality, deteriorating living conditions for low income people and stalled social mobility and wages across the world. And the response from the traditional elites to the crisis was largely austerity *rather* than accountability. Indeed, their obsession with austerity only intensified the resentment and the anger against the elites and their policies. That sense of insecurity was intensified by immigration, as well as low trust and low expectations for the future. All this provided a fertile ground for the emergence of populism in Spain and elsewhere.

What can we learn from this experience? History has shown that the main political drivers for people are fear and hope. Indeed, history moves by stories of identity, sovereignty, and self-respect, not just economic factors or rationality. While technological changes and automation have been crucial in the dislocations experienced by our societies, the growing inequalities are fueling a fury that is seeking someone to blame: elites and immigrants; and the *Great Recession* has made social, economic, and geographical disparities and living conditions intolerable. Foa and Mounk (2016) have shown that as inequality rises, citizens are less likely to believe that their government is democratic, which undermines the legitimacy of the system. There is an enormous fear of losing control over the future and losing status vis-à-vis one's neighbor, which generates a backlash. At the same time, the traditional monopoly of elites over access to elected office has been eroded by new technologies that open up communication and fundraising (Levitsky and Zibblatt 2018). Finally, we need to heed the lessons from the past: As Paxton (2004) has eloquently shown, the rise of extremism in Europe was aided by businesses worried more about the possibility of wealth redistribution than about the threat of political extremism. Indeed, fascist parties would not have been able to approach power without the complicity of Conservatives willing to sacrifice the rule of law for security. That trade-off was disastrous and cannot be repeated. Passivity against current challenges is not an option.

We are living in a context in which the politics of destruction are increasingly defined by opposition: to the status quo, to the establishment, and to the other side. There is growing polarization, partisan rancor, intransigence, tribalism, distrust of institutions, and destructiveness across the world; and more instability driven by fractures within political parties, fragmentation, and elections that reject incumbents. This is all leading toward a growing tear-it all-down ethos characterized by *negative partisanship* (Abramowitz and Webster 2016) in which people vote based on fear and distrust of the other side, rather than support for one's side. This empowers those who want to destroy the other side and make crisis of governability more common, thus making the sense of political powerlessness more pervasive. In this context, outrage and distrust dominate our politics.

In this challenging environment, *how do we respond to the populists' fears?* Two important lessons from the past decade are that moderation and gradualism will not be enough, and also that institutions alone will not contain this threat. We must address urgently and effectively the forthcoming challenges, such as the threats (and opportunities) that will arise from artificial intelligence; the increasing inequality and poverty; racism; or the economic dislocations of climate change. And all this will be much harder after the extraordinary dislocations and pain caused by the COVID-19 crisis throughout the world (in Spain in less than a month, the crisis had led to the destruction of 304,000 jobs!).

This will require credible plans to address the security and economic concerns of our citizens. And in order to deliver and implement them, we need to overcome the growing fragmentation and polarization, and take effective action. For this, we need a politics built on hope to fill the vacuum, and we also need to develop a positive narrative that focuses on opportunities to address the fears of the day. In sum, we need to offer real solutions to citizens' problems that will allow our citizens to regain their confidence, while embracing diversity and the larger identities that emerge from engaging with one another.

In order to escape this dilemma, we need to engage the public at large to regain faith in our democratic institutions and ensure that they fulfill their solemn obligation to the public interest. This is something that we seemed to forget in Spain in the previous decades when our institutions rather than constraining the vices and treacheries of people within them seemed to encourage it and mask it. This institutional dereliction converted many Spanish institutions into platforms for prominence and enrichment. Instead, institutions must play a formative role shaping the people who populate them to serve the public interest and be trustworthy. And we need to leave behind the tendency to look to the other side and/or blame others, because we all have a role to play in our institutions to ensure they are trustworthy. Rather than using them for our own personal gain, we need to serve them to rebuild the bonds of trust that are so crucial to our societies.[18] We also need to build new institutions that can integrate all citizens into decision making.

Finally, in the financial realm, which is the main focus of this book, we must learn from previous crises and work to bolster support for democratic institutions that shape partnerships between governments, bankers, and citizens that will limit the emergence of populist coalitions that have proved to be so detrimental to the establishment of stable banking systems, and that will rather work to establish robust banking regulations that avoid banking crises and implement policies that will make countries less vulnerable to crises. This is not an impossible task or a lost cause. Countries such as Canada have been able to do it. We should hope that participants in the game identify ways to build better-suited coalitions and institutions to make beneficial changes happen. Too much is at stake.

---

[18] See Yuval Levin, "How We Lost Faith in Everything," *The New York Times*, January 19, 2020, p. 4.

References

Abramowitz, Alan, and Steven Webster. "The Rise of Negative Partisanship and the Nationalization of U.S. Elections in the 21st Century." *Electoral Studies* 41 (March 2016): 12–22.

Abreu, Orlando. "Portugal's Boom and Bust: Lessons for Euro Newcomers." *ECOFIN Country Focus* 3, no. 16 (2006), 1–6.

Alesina, Alberto, Carlo Favero, and Francesco Giavazzi. *Austerity: When It Works and When It Doesn't.* Princeton, NJ: Princeton University Press, 2019.

Blyth, Mark. *Austerity: The History of a Dangerous Idea.* New York: Oxford University Press, 2013.

Boix, Carles. *Partidos Políticos, Crecimiento e Igualdad.* Madrid: Alianza Editorial, 1995.

Calomiris, Charles W., and Stephen H. Haber. *Fragile by Design: The Political Origins of Banking Crises & Scarce Credit.* Princeton: Princeton University Press, 2014.

Carabaña, Julio. *Ricos y Pobres.* Madrid: Los Libros de la Catarata, 2016.

Chinn, Menzie D., and Jeffry A. Frieden. *Lost Decades: The Making of America's Debt Crisis and the Long Recovery.* New York: W. W. Norton, 2011.

Chwieroth, Jeffrey, and Andrew Walter. *The Wealth Effect: How the Great Expectations of the Middle Class Have Changed the Politics of Banking Crises.* New York: Cambridge University Press, 2019.

Copelovitch, Mark, and David A. Singer. *Banks on the Brink: Global Capital, Securities Markets, and the Political Roots of Financial Crises.* New York: Cambridge University Press, 2020.

Coterill, Joseph. "What Is Up with Spanish Mortgage Lending?" *FT Alphaville,* July 20, 2010.

Cuñat, Vicente, and Luís Garicano. "Did Good Cajas Extend Bad Loans? Governance, Human Capital and Loan Portfolio." Working Papers 2010-08, FEDEA, 2010.

Dellepiane, Sebastián, and Niamh Hardiman, "Governing the Irish Economy." UCD Geary Institute Discussion Series Chapters, Geary WP2011/03. Dublin: University College, February 2011.

Ekaizer, Ernesto. *El Libro Negro. Como Fallo el Banco de España a los Ciudadanos.* Madrid: Espasa, 2018.

Ellul, Andrew, and Vijay Yerramilli. "Stronger Risk Controls, Lower Risks: Evidence from U.S. Bank Holding Companies." National Bureau of Economic Research Working Paper 16178, 2010.

European Central Bank (ECB). "Failing or Likely to Fail." Assessment of Banco Popular Español. Frankfurt, 2017.

Ferguson, Neil. *The Great Degeneration.* New York: Penguin Press, 2013.

Fernández Ordóñez, Miguel. *Economistas, Políticos y Otros Animales.* Madrid: Ediciones Península, 2016.

Fishman, Robert. "Rethinking the Iberian Transformations: How Democratization Scenarios Shaped Labor Market Outcomes." *Studies in Comparative International Development* 45, no. 3 (2010): 281–310.

Foa, Roberto, and Yascha. Mounk. "The Democratic Disconnect." *Journal of Democracy* 27, no. 3 (July 2016): 5–17.

Garicano, Luís. "Five Lessons from the Spanish Cajas Debacle for a New Euro-Wide Supervisor." October 16, 2012. http://www.voxeu.org/article/five-lessons-spanish-cajas-debacle-new-euro-wide-supervisor.

Goodhart, C., and D. Schoenmaker. "Fiscal Burden Sharing in Cross-Border Banking Crises." *International Journal of Central Banking* 5 (2009): 141–65.

Hall, Peter. "Varieties of Capitalism in Light of the Euro Crisis." Paper presented at the Annual Meeting of the American Political Science Association, Philadelphia, August 2016.

Hoffman, Phillip, Giles Postel-Vinay, and Jean-Laurent Rosenthal. *Surviving Large Losses: Financial Crises, the Middle Class and the Development of Capital Markets*. Cambridge, MA: Belknap Press, 2007.

Holmberg, Susan. "Who are the Shareholders?" *Roosevelt Institute Report*. June 2018.

IMF. *Spain: Financial Sector Reform—Final Progress Report*. Washington: IMF, 2014.

Jiménez, Gabriel, Steven Ongena, José-Luis Peydró, and Jesús Saurina. "Hazardous Times for Monetary Policy: What Do Twenty-Three Million Bank Loans Say about the Effects of Monetary Policy on Credit Risk-Taking?" *Econometrica* 82, no. 2 (2014): 463–505.

Kamikawa, Ryunoshin. "Market-Based Banking in Japan: From the Avant Garde to Europe's Future?" In *Market-Based Banking, Varieties of Financial Capitalism and the Financial Crisis*, edited by Iain Hardie and David Howarth. New York: Oxford University Press, 2013.

Kochhar, Rakesh. "7 Key Findings on the State of the Middle Cass in Western Europe." Pew Research Center. April 24, 2017. https://www.pewresearch.org/fact-tank/2017/04/24/7-key-findings-on-the-state-of-the-middle-class-in-western-europe/.

Levitsky, Steven, and Daniel Zibblatt. *How Democracies Die*. New York: Crown, 2018.

Linz, Juan, and Alfred Stepan. *Problems of Democratic Transition and Consolidation: Southern Europe, South America, and Post-Communist Europe*. Washington, DC: John Hopkins University Press, 1996.

Linz, Juan, et al. *Informe Sociologico sobre el Cambio Politico en España, 1975–1981*. Madrid: Fundación FOESSA, 1981.

Lo Schiavo, Gianni. *The Role of Financial Stability in EU Law and Policy*. The Netherlands: Kluwer Law International, 2017.

Lukauskas, Arvid. *Regulating Finance*. Ann Arbor: Michigan University Press, 1997.

Montalvo, José G., and Josep M. Raya. "Constraints on LTV as a Macroprudential Tool: A Precautionary Tale." *Oxford Economic Papers* 70, no. 3 (July 2018): 821–45.

Muñoz Molina, Antonio. *Todo lo que era Sólido*. Madrid: Seix Barral, 2013.

Olson. Mancur. *The Rise and Decline of Nations*. New Haven: Yale University Press, 1982.

Otero-Iglesias, Miguel. "Inequality in Spain: Let's Focus on the Poor." *Elcano Blog* (2019). https://blog.realinstitutoelcano.org/en/inequality-in-spain-lets-focus-on-the-poor/.

Pappas, Takis. *Populism and Liberal Democracy: A Comparative and Theoretical Analysis*. New York: Oxford University Press, 2019.

Paxton, Robert. *The Anatomy of Fascism*. New York: Vintage, 2004.

Pérez, Sofia A. *Banking on Privilege*. New York: Cornell University Press, 1997.

Prasad, Eswar, and Isaac Sorkin. *Assessing the G-20 Economic Stimulus Plans: A Deeper Look*. Washington, DC: Brookings, 2009.

Quaglia, Lucia, and Sebastián Royo. "Banks and the Political Economy of the Sovereign Debt Crisis in Italy and Spain." *Review of International Political Economy* 22, no. 3 (2015): 485–507.

Reinhart, Carmen M., and Kenneth Rogoff. *This Time Is Different: Eight Centuries of Financial Folly*. New York: Princeton University Press, 2009.

Rodríguez Zapatero, José Luís. *El Dilema: 600 Días de Vértigo*. Barcelona: Planeta, 2013.

Roldan, José María and Jesús Saurina. "Old and New Lessons of the Financial Crisis for Risk Management." In Banks at Risk: Global Best Practices in an Age of Turbulence, ed. Peter Hoflich. New York: Wiley, 2012.

Royo, Sebastián. *From Social Democracy to Neoliberalism*. New York: St. Martin's Press, 2000.

Royo, Sebastián. *Varieties of Capitalism in Spain*. New York: Palgrave, 2008.

Royo, Sebastián. *Lessons from the Economic Crisis in Spain*. New York: Palgrave, 2013.

Royo, Sebastián. "A 'Ship in Trouble': The Spanish Banking System in the Midst of The Global Financial System Crisis: The Limits of Regulation." In *Market-Based Banking, Varieties of Financial Capitalism and the Financial Crisis*, edited by Iain Hardie and David Howarth. New York: Oxford University Press, 2013a.

Schoenmaker, Dirk. "The Financial Trilemma." *Economics Letters* 111 (2011): 57–59.

Stallings, Barbara. *Finance for Development: Latin America in Comparative Perspective*. Washington, DC: Brookings Institution Press, 2006.

Thelen, Kathleen. *How Institutions Evolve: The Political Economy Skills in Germany, Britain, the United States and Japan.* New York: Cambridge University Press. 2004.

Tortella, Gabriel, and José Luís García Ruiz. *Spanish Money and Banking. A History.* New York: Palgrave, 2013.

Villena Oliver, Andrés. *Las Redes de Poder en España.* Madrid: Rocaeditorial, 2019.

Wolff, Edward. 2017. "Household Wealth in the United States, 1962 to 2016: Has Middle Class Wealth Recovered?" National Bureau of Economic Research. Working Paper Series, No. 24085. www.nber.org/papers/w24085.

BIBLIOGRAPHY

Abramowitz, Alan, and Steven Webster. "The Rise of Negative Partisanship and the Nationalization of U.S. Elections in the 21st Century." *Electoral Studies* 41 (March 2016): 12–22.

Abreu, Orlando. "Portugal's Boom and Bust: Lessons for Euro Newcomers." *ECOFIN Country Focus* 3, no. 16 (2006): 1–6.

Acemoglu, Daron, and James Robinson. *Why Nations Fail: The Origins of Power, Prosperity and Poverty*. New York: Random House, 2012.

Acemoglu, Daron, and James Robinson. *The Narrow Corridor: States, Society and the Fate of Liberty*. New York: Penguin Press, 2019.

Ahamed, Liaquat. *The Lords of Finance*. New York: Penguin Books, 2009.

Albert, Michael. *Capitalism against Capitalism*. London: Whurr, 1992.

Alesina, Alberto, Carlo Favero, and Francesco Giavazzi. *Austerity: When It Works and When It Doesn't*. Princeton, NJ: Princeton University Press, 2019.

Alesina, Alberto, and Francesco Giavazzi. *The Future of Europe: Reform or Decline*. Cambridge, MA and London: MIT Press, 2006.

Allen, Franklin, and Kenneth Rogoff. "Asset Prices, Financial Stability and Monetary Policy." In *The Riksbank's Inquiry into the Risks in the Swedish Housing Market*, edited by Per Jansson and Mattias Persson, 189–218. Stockholm: Sveriges Riksbank, 2011.

Almarcha Barbado, A., ed. *Spain and EC Membership Evaluated*. New York: St. Martin's Press, 1993.

Alston, Lee J., Thrain Eggertsson, and Douglass C. North. *Empirical Studies in Institutional Change*. New York: Cambridge University Press, 1996.

Alvarez-Miranda, Berta. *El Sur de Europa y la Adhesion a la Comunidad: Los Debates Politicos*. Madrid: CIS, 1996.

© The Editor(s) (if applicable) and The Author(s) 2020
S. Royo, *Why Banks Fail*,
https://doi.org/10.1057/978-1-137-53228-2

Amable, Bruno. *The Diversity of Modern Capitalism.* Oxford: Oxford University Press, 2003.

Amri, Puspa D., Greg M. Richey, and Thomas D. Willet. "Capital Surges and Credit Booms: How Tight Is the Relationship." *Open Economics Review* 27, no. 4 (2016): 637–70.

Analistas Financieros Internacionales. "Guia del sistema financiero español" [Guide to the Spanish Financial System]. Madrid: AFI, 2005.

Aoki, M. *Toward a Comparative Institutional Analysis.* Cambridge: MIT Press, 2001.

Armigeon, Klaus, and Lucio Baccaro. "The Sorrows of Young Euro." In *Coping with the Crises,* edited by Nancy Bermeo, and Jonas Pontusson, pp. 162–98. New York: Russell Sage Foundation, 2012.

Asensio Menchero, Maria. *El Proceso de la Reforma del Sector Publico en el Sur de Europa: Estudio Comparativo de España y Portugal.* Madrid: Instituto Juan March, 2001.

Avgouleas, Emilios, and Charles Goodhart. "Critical Reflections on Bank Bail-ins." *Journal of Financial Regulation* 1, no. 1 (March 2015): 3–29.

Banco, de España. *Report on the financial and banking crisis in Spain, 2008–2014.* Madrid, 2017.

Banerjee, Abhijit V., and Esther Duflo. *Good Economics for Hard Times.* New York: Public Affairs, 2019.

Bank of Spain. *Financial Stability Review.* Madrid, March 2010.

Barrón, Iñigo. *El Hundimiento de la Banca.* Madrid: Catarata, 2012.

Barry, Frank. "Economic Integration and Convergence Process in the EU Cohesion Countries." *Journal of Common Market Studies* 41, no. 5 (2003): 1–25.

Bentolila, Samuel, and Juan J. Dolado. "Labour Flexibility and Wages: Lessons from Spain." *Economic Policy* 9, no. 18 (1994): 55–99.

Bentolila, Samuel, J. Segura, and L. Toharia. "La Contratación Temporal en España." *Moneda y Crédito* 193 (1991): 225–65.

Bermeo, Nancy. *Unemployment in the New Europe.* New York: Cambridge University Press, 2002.

Bermeo, Nancy, and Jonas Pontusson. *Coping with the Crises.* New York: Russell Sage Foundation, 2012.

Blanchard, Olivier J., J. Andrés, C. Bean, E. Malinvaud, A. Revenga, D. Snower, G. Saint-Paul, R. Solow, D. Taguas, and L. Toharia. *Spanish Unemployment: Is There a Solution?* Madrid and London: Consejo Superior de Cámaras de Comercio, Industria, y Navegación/Center for Economic Policy Research, 1995.

Blanchard, Olivier J., and Juan Jimeno. "Structural Unemployment: Spain Versus Portugal." *American Economic Review* 85, no. 2 (1995): 212–18.

Blanco, Roberto. "The Securization Market in Spain: Past, Present and Future." In *Working Arty on Financial Statistics: Proceedings of the Workshop on Securitisation*, edited by M. Chavoix-Mannato. OECD Statistics Working Papers (2011/03), OECD Publishing, 2011.

Blyth, Mark. "An Approach to Comparative Analysis or a Sub-discipline within a Sub-field? Political Economy." In *Comparative Politics: Rationality, Culture and Structure*, 2nd ed., edited by M. I. Lichbach and A. S. Zuckerman, 193–219. Cambridge: Cambridge University Press, 2002.

Blyth, Mark. *Austerity: The History of a Dangerous Idea*. New York: Oxford University Press, 2013.

Boix, Carles. *Partidos Políticos, Crecimiento e Igualdad*. Madrid: Alianza Editorial, 1995.

Boix, Carles. *Political Parties, Growth and Equality: Conservative and Social Democratic Strategies in the World Economy*. New York: Cambridge University Press, 1998.

Botella, Joan, Richard Gunther, and Josè Ramón Montero. *Democracy in Modern Spain*. Yale University Press: New Haven and London, 2004.

Botín, Emilio. "Sage Advice." *Financial Times*, October 16, 2009a.

Botín, Emilio. "Why Banks Must Adopt a 'Back-to-Basics Approach.'" *Financial Times: Future of Finance*, 9, October 19, 2009b.

Bover, Olympia, Pilar Garcia-Perea, and Pedro Portugal. "Labour Market Outliers: Lessons from Portugal and Spain." *Economic Policy* 31 (2000): 381–428.

Cabana, Francesc. *Historia del Banc de Barcelona 1844–1920*. Barcelona: Edicions 62, 1978.

Caja Madrid. *Annual Report*. Madrid, 2007.

Calavita, Kitty. *Immigrants at the Margins: Law, Race, and Exclusion in Southern Europe*. Cambridge: Cambridge University Press, 2005.

Calomiris, Charles W., and Stephen H. Haber. *Fragile by Design: The Political Origins of Banking Crises & Scarce Credit*. Princeton: Princeton University Press, 2014.

Cameron, David. "Unemployment, Job Creation, and Economic and Monetary Union." In *Unemployment in the New Europe*, edited by Nancy Bermeo, 7–51. Cambridge: Cambridge University Press, 2001.

Cameron, David. "European Fiscal Responses to the Great Recession." In *Coping with the Crises*, edited by Nancy Bermeo, and Jonas Pontusson, 91–129. New York: Russell Sage Foundation, 2012.

Carabaña, Julio. *Ricos y Pobres*. Madrid: Los Libros de la Catarata, 2016.

Carballo Cruz, Francisco. "Causes and Consequences of the Spanish Economic Crisis: Why the Recovery Is Taken so Long?" *Panoeconomicus* 3 (2011): 309–28.

Carreras, Albert, and Xavier Tafulell. *Historia Económica de la España Contemporánea (1789–2009)*. Barcelona: Crítica, 2004.

Chinn, Menzie D., and Jeffry A. Frieden. *Lost Decades: The Making of America's Debt Crisis and the Long Recovery*. New York: W. W. Norton, 2011.

Chwieroth, Jeffrey, and Andrew Walter. *The Wealth Effect: How the Great Expectations of the Middle Class Have Changed the Politics of Banking Crises*. New York: Cambridge University Press, 2019.

Closa, Carlos, and Paul Heywood. *Spain and the European Union*. New York: Palgrave, 2004.

Copelovitch, Mark, and David A. Singer. *Banks on the Brink: Global Capital, Securities Markets, and the Political Roots of Financial Crises*. New York: Cambridge University Press, 2020.

Coterill, Joseph. "What Is Up with Spanish Mortgage Lending?" *FT Alphaville*, July 20, 2010.

Crouch, Colin. *Capitalist Diversity and Change*. New York: Oxford University Press, 2005.

Cuñat, Vicente, and Luís Garicano. "Did Good Cajas Extend Bad Loans? Governance, Human Capital and Loan Portfolio." Working Papers 2010-08, FEDEA, 2010.

De Juan, Aristobulo, Francisco Uría, and Iñigo De Barrón. *Anatomía de una Crisis*. Barcelona: Deusto, 2013.

De la Dehesa, Guillermo. "Spain." In *The Political Economy of Policy Reform*, edited by J. Williamson, 123–140. Washington, DC: Institute for International Economics, 1994.

De la Dehesa, Guillermo. *La Primera Gran Crisis del Siglo XXI*. Madrid: Alianza Editorial, 2009.

De la Fuente, Angel, and Rafael Demenech. "Ageing and Real Convergence: Challenges and Proposals." In *Spain and the Euro: The First Ten Years*, edited by J. F. Jimeno, 191–273. Madrid: Banco de España, 2010.

Deeg, Richard, and Sofía Pérez. "International Capital Mobility and Domestic Institutions: Corporate Finance and Governance in Four European Cases." *Governance* 13, no. 2 (2000): 119–53.

Deeg, Richard, and Susanne Luetz. "Internationalisation and Financial Federalism: The United States and Germany at the Cross Roads?" *Comparative Political Studies* 33, no. 3 (2001): 374–405.

Della Sala, Vincent. "The Italian Model of Capitalism: On the Road between Globalization and Europeanization?" *Journal of European Public Policy* 11, no. 6 (December 2004): 1041–57.

Dellepiane, Sebastián, and Niamh Hardiman. "Governing the Irish Economy." In *UCD Geary Institute Discussion Series Chapters*, Geary WP2011/03. Dublin: University College, February 2011.

Diamond, Douglas W., and Raghuram G. Rajan. "Illiquid Banks, Financial Stability, and Interest Rate Policy." *Journal of Political Economy* 120 (2012): 552–91.

Duran Munoz, Rafael. Contencion y Transgresion: *Las Movilizaciones Sociales y el Estado en las Transiciones Espanola y Portuguesa.* Madrid: Centro de Estudios Politicos y Constitucionales, 2000.

Dut, A. K., and J. Ross. "Aggregate Demand Shocks and Economic Growth." *Structural Change and Economic Dynamics* 18, no. 1 (2007): 75–99.

The Economist. "The Party's over." November 8 (a special report), 2008.

Editorial Board. "The Economic Policies of the Zapatero Government." *Revista de Fomento Social* 247 (2007).

Eichengreen, Barry. *The European Economy since* 1945. Princeton, NJ: Princeton University Press, 2007.

Ehremberg, Richard. *Capital and Finance in the Age of the Renaissance: A Study of the Fuggers.* New York: Harcout Brace, 1928.

Ekaizer, Ernesto. *El Libro Negro. Como Fallo el Banco de España a los Ciudadanos.* Madrid: Espasa, 2018.

Ellul, Andrew, and Vijay Yerramilli. "Stronger Risk Controls, Lower Risks: Evidence from U.S. Bank Holding Companies." National Bureau of Economic Research Working Paper No. 16178, 2010.

Esping-Andersen, Gosta. *Social Foundations of Postindustrial Economies.* Oxford: Oxford University Press, 1999.

Esping-Andersen, Gosta. "Who Is Harmed by Labor Market Regulations?" In *Why Deregulate Labor Markets?* edited by G. Esping-Andersen and M. Regini, 66–98. Oxford: Oxford University Press, 2000.

Esping-Andersen, Gosta, and Marino Regini, eds. *Why Deregulate Labor Markets?* Oxford: Oxford University Press, 2000.

Estefanía, Joaquín. *La Larga Marcha: Medio Siglo De Política (Económica) Entre La Historia Y La Memoria.* Barcelona: Ediciones Península, 2007.

Estrada, A., J. F. Jimeno, and J. L. Malo de Molina. "The Performance of the Spanish Economy in EMU: The First Ten Years." In *Spain and the Euro: The First Ten Years*, edited by J. F. Jimeno, 83–138. Madrid: Banco de España, 2010.

Etchemendy, Sebastián. "Revamping the Weak, Protecting the Strong, and Managing Privatization: Governing Globalization in the Spanish Takeoff." *Comparative Political Studies* 37, no. 6 (August 2004): 623–51.

Etchemendy, Sebastián. *Models of Economic Liberalization: Business, Workers, and Compensation in Latin America, Spain, and Portugal.* New York: Cambridge University Press, 2012.

European Central Bank (ECB). "Failing or Likely to Fail." Assessment of Banco Popular Español. Frankfurt, 2017.

European Commission. *Joint Employment Report 2018*. Adopted by the EPSCO Council on 15 March 2018.

Ferguson, Neil. *The Great Degeneration*. New York: Penguin Press, 2013.

Fernández Méndez De Andes, Fernando, ed. *La Internacionalización de la Empresa Española: Aprendizaje y Experiencia*. Madrid: Universidad Nebrija, 2006.

Fernández Ordóñez, Miguel. "The Challenges to the Spanish Banking System in the Face of the Global Crisis." Lecture on the Occasion of the 50th Anniversary of ESADE. Barcelona, October 30, 2008.

Fernández Ordóñez, Miguel. *Economistas, Políticos y Otros Animales*. Madrid: Ediciones Península, 2016.

Field, Bonnie. *Spain's 'Second Transition'?: The Socialist Government of Jose Luis Rodriguez Zapatero*. New York: Routledge, 2011.

Fishman, Robert M. "Rethinking State and Regime: Southern Europe's Transition to Democracy." *World Politics* 42 (1990a): 422–40.

Fishman, Robert M. *Working Class Organization and the Return of Democracy in Spain*. London: Cornell University Press, 1990b.

Fishman, Robert M. *Democracy's Voices*. Ithaca: Cornell University Press, 2004.

Fishman, Robert M. "Rethinking the Iberian Transformations: How Democratization Scenarios Shaped Labor Market Outcomes." *Studies in Comparative International Development* 45, no. 3 (2010): 281–310.

Fishman, Robert M. "Democratic Practice after the Revolution: The Case of Portugal and beyond." *Politics & Society* 39, no. 2 (2011): 233–67.

Fishman, Robert M. "Anomalies of Spain's Economy and Economic Policy Making." *Contributions to Political Economy* 31, no. 1 (2012): 67–76.

Fishman, Robert. *Democratic Practice: Origins of the Iberian Divide in Political Inclusion*. New York: Oxford University Press, 2019.

Fishman, Robert M., and Anthony Messina, eds. *The Year of the Euro*. Indiana: University of Notre Dame Press, 2006.

Foa, Roberto, and Yascha. Mounk. "The Democratic Disconnect." *Journal of Democracy* 27, no. 3 (July 2016): 5–17.

Fondo de Restructuración Ordenada Bancaria (FROB). http://www.frob.es/index_en.html.

Frieden, Jeffry, and Ronald Rogowski. "The Impact of the International Economy on National Policies: An Analytical Overview." In *Internationalization and Domestic Politics*, edited by Robert O. Keohane and Helen V. Milner, 108–136. New York: Cambridge University Press, 1996.

Führer, Ilse Marie. *Los Sindicatos en España*. Madrid: CES, 1996.

Galí, J. Comments on "The Performance of the Spanish Economy in EMU: The First Ten Years." In *Spain and the Euro: The First Ten Years*, edited by J. F. Jimeno, 139–146. Madrid: Banco de España, 2010.

García Agustín, O., and Marco Briziarelli. *Podemos and the New Political Cycle.* New York: Palgrave, 2018.

Garicano, Luís. "Five Lessons from the Spanish Cajas Debacle for a New Euro-Wide Supervisor." October 16, 2012. http://www.voxeu.org/article/five-lessons-spanish-cajas-debacle-new-euro-wide-supervisor.

Garicano, Luís. *El Dilema de España.* Barcelona: Ediciones Península, 2014.

Garrett, Geoffrey. *Partisan Politics in the Global Economy.* New York: Cambridge University Press, 1998.

Germain, Randall. "Governing Global Finance and Banking." *Review of International Political Economy* 19, no. 4 (2012): 530–35.

Gerschenkron, Alexander. *Economic Backwardness in Historical Perspective.* Cambridge, MA: Harvard University Press, 1962.

Goldthorpe, John A., ed. *Order and Conflict in Contemporary Capitalism.* New York: Oxford University Press, 1984.

González, A., and E. Gutiérrez. "Spain: Collective Bargaining and Wage Determination." In *Wage Policy in the Eurozone*, edited by P. Pochet, 217–238. Brussels: Peter Lang, 2002.

Goodhart, Charles A. E., Anil K. Kashyap, Dimitrios P. Tsomocos and Alexandros P. Vardoulakis. "Financial Regulation in General Equilibrium." NBER Working Papers Series. Working Paper No. 17909, March 2012. http://www.nber.org/papers/w17909.

Goodhart, C., and D. Schoenmaker. "Fiscal Burden Sharing in Cross-Border Banking Crises." *International Journal of Central Banking* 5 (2009): 141–65.

Goodhart, Charles A. E., Pojanart Sunirand, and Dimitrios P. Tsomocos. "A Model to Analyse Financial Fragility." *Economic Theory* 27, no. 1 (2006): 107–42.

Goodhart, Charles A. E., Dimitrios P. Tsomocos, and Alexandros P. Vardoulakis. "Modeling a Housing and Mortgage Crisis." In *Financial Stability, Monetary Policy, and Central Banking*, edited by Rodrigo A. Alfaro. Santiago, Chile: Central Bank of Chile, 2010.

Gourinchas, Pierre-Olivier, and Maurice Obstfeld. "Stories of the Twentieth Century for the Twenty-First." *American Economic Journal: Macroeconomics* 4 (2012): 226–65.

Güell, Joan. "Las Cajas de Ahorro en el Sistema Financiero Espaōl. Trayectoria Histórica y Realidad Actual." Paper presented at the Universidad de Zaragoza, during the *Jornadas sobre La singularidad de las cajas de ahorros españolas*, May 28, 2001.

Guillén, Mauro. *The Rise of Spanish Multinationals.* Cambridge: Cambridge University Press, 2005.

Guitart, Joan. "Zapatero: Left in Form, Right in Essence." *IV Online Magazine*, IV402, July 2008.

Gunther, Richard, Jose Ramon Montero, and Joan Botella. *Democracy in Modern Spain.* New Haven: Yale University Press, 2004.

Gunther, Richard, Giacamo Sani, and Goldie Shabad. *Spain after Franco: The Making of a Competitive Party System.* Berkeley: University of California Press, 1986.

Haggard, Stephan, and Robert Kaufman, eds. *The Politics of Economic Adjustment.* Princeton: Princeton University Press, 1992.

Haggard, Stephan, and Robert Kaufman, eds. *The Political Economy of Democratic Transitions.* Princeton: Princeton University Press, 1995.

Haggard, Stephan, and Jongryn Mo. "The Political Economy of the Korean Financial Crisis." *Review of International Political Economy* 7, no. 2 (2000): 197–218.

Hall, Peter. "The Eurocrisis and Beyond: The Challenges for Germany and Europe." Presented at the Annual Conference of the International Association for the Study of German Politics, London, May 16, 2011.

Hall, Peter, and Danbiel Gingerich. "Varieties of Capitalism and Institutional Complementarities in the Macroeconomy: An Empirical Analysis." MPIfG Discussion Paper No. 04/5. Cologne: Max Plank Institute for the Study of Societies, September 2004.

Hall, Peter. "Varieties of Capitalism in Light of the Euro Crisis." Paper presented at the Annual Meeting of the American Political Science Association, Philadelphia, August 2016.

Hall, Peter, and David Soskice. *Varieties of Capitalism.* New York: Oxford University Press, 2001.

Halleberg, M., R. Strauch, and J. von Hagen. "The Design of Fiscal Rules and Forms of Governance in European Union Countries." ECB Working Paper No. 419, December 2004.

Hamann, Kerstin. *The Politics of Industrial Relations: Labor Unions in Spain.* New York: Routledge, 2012.

Hancké, Bob. "The Political Economy of Wage-Setting in the Eurozone." In *Wage Policy in the Eurozone,* edited by P. Pochet, 131–148. Brussels: Peter Lang, 2002.

Hancké, Bob, Martin Rhodes, and Mark Thatcher, eds. *Beyond Varieties of Capitalism: Contradictions, Complementarities, and Change.* Oxford: Oxford University Press, 2007.

Hardie, Ian. "How Much Can Governments Borrow? Financialization and Emerging Markets Government Borrowing Capacity." *Review of International Political Economy* 18, no. 2 (2011): 141–67.

Hardie, Iain, and David Howarth. "Market-Based Banking and the Financial Crisis." Mimeo: Paper present at the University of Victoria, 2011.

Hardie, Iain, and David Howarth, eds. *Market-Based Banking, Varieties of Financial Capitalism and the Financial Crisis*. New York: Oxford University Press, 2013.

Harrison, Joseph. *An Economic History of Modern Spain*. New York: Holmes & Meir, 1978.

Harrison, Joseph, and David Corkill. *Spain: A Modern European Economy*. Burlington: Ashgate, 2004.

Hassel, Anke, and Bernhard Ebbinghaus. "From Means to Ends: Linking Wage Moderation and Social Policy Reform." In *Social Pacts in Europe—New Dynamics*, edited by Giuseppe Fajertag and Phillipe Pochet, 61–84. Brussels: European Trade Union Institute, 2000.

Heclo, Hugh. "Ideas, Interests and Institutions." In *The Dynamics of American Politics: Approaches and Interpretations*, edited by Lawrence C. Dodd and Calvin Jillson, 366–92. Boulder, CO: Westview, 1993.

Hoffman, Phillip, Giles Postel-Vinay, and Jean-Laurent Rosenthal. *Surviving Large Losses: Financial Crises, the Middle Class and the Development of Capital Markets*. Cambridge, MA: Belknap Press, 2007.

Holmberg, Susan. "Who Are the Shareholders?" *Roosevelt Institute Report*. June 2018.

Huber, Evelyne, and John D. Stephens. Development and Crisis of the Welfare States: Parties and Politics in Global Markets. Chicago: University of Chicago Press, 2001.

Huber, Evelyne, and John D. Stephens. *Democracy and the Left*. Chicago: University of Chicago Press, 2012.

Iglesias, Pablo. *Politics in a Time of Crisis: Podemos and the Future of Democracy in Europe*. New York: Verso, 2015.

International Monetary Fund (IMF). *World Economic Outlook*. Washington, DC: IMF, Various Years.

International Monetary Fund (IMF). IMF Spain—Staff Report for the 2011 Article IV Consultation. Country Report No. 11/215, July 2011.

International Monetary Fund (IMF). Spain: The Reform of Spanish Savings Banks Technical Notes. IMF Country Report No. 12/141, 2012a.

International Monetary Fund (IMF). *The Good, the Bad and the Ugly: 100 Years of Dealing with Public Debt Overhangs*. Washington: IMF, October 2012b.

International Monetary Fund (IMF). *Spain: Financial Sector Reform—Final Progress Report*. Washington: IMF, 2014.

Iversen, Torben. *Capitalism, Democracy and Welfare*. New York: Cambridge University Press, 2005.

Iversen, Torben, and David Soskice. *Democracy and Prosperity: Reinventing Capitalism through a Turbulent Century*. Princeton: Princeton University Press, 2019.

Jabko, Nicolas, and Elsa Massoc. "French Capitalism under Stress: How Nicolas Sarkozy Rescued the Banks." *Review of International Political Economy* 19, no. 4 (2012): 562–95.

Jackson, Gregory, and Richard Deeg. "How Many Varieties of Capitalism? From Institutional Diversity to the Politics of Change." *Review of International Political Economy* 15, no. 4 (2008): 679–708.

Jenson, Jane. "Ideas and Policy: The European Union Considers Social Policy Futures." American Consortium on European Union Studies, ECAS Cases, No: 2010/2, 2010. Available at transatlantic.sais-jhu.edu/bin/y/d/2010.2_ACES_Cases_Jenson.pdf.

Jenson, Jane, and Frédéric Mérand. "Sociology, Institutionalism and the European Union." *Comparative European Politics* 8, no. 1 (2010): 74–92.

Jiménez, Gabriel, Steven Ongena, José Luís Peydró, and Jesús Saurina. "Hazardous Times for Monetary Policy: What do Twenty-Three Million Bank Loans Say about the Effects of Monetary Policy on Credit Risk-Taking?" *Econometrica* 82(2), (2014): 463–505.

Jordana, Jacint. "Reconsidering Union Membership in Spain, 1977–1994: Halting Decline in a Context of Democratic Consolidation." *Industrial Relations* 27 (1996): 211–24.

Kahler, Miles, and David Lake. *The Great Recession in Comparative Perspective.* Ithaca: Cornell University Press, 2013.

Kamikawa, Ryunoshin. "Market-Based Banking in Japan: From the Avant Garde to Europe's Future?" In Iain Hardie and David Howarth, eds. *Market-Based Banking, Varieties of Financial Capitalism and the Financial Crisis.* New York: Oxford University Press, 2013.

Kaminsky, Graciela, and Carmen Reinhart. "The Twin Crises: The Causes of Banking and Balance-of-Payments Problems." *The American Economic Review* 89, no. 1 (1999): 473–500.

Katzenstein, Peter. *Small States in World Markets.* Ithaca: Cornell University Press, 1985.

Kesselman, Mark. *The Politics of Globalization.* Boston: Houghton Mifflin, 2011.

Kindleberger, Charles. *The World in Depression 1929–1939.* Berkeley: University of California Press, 1973.

Kindleberger, Charles. *A Financial History of Western Europe.* New York: Routledge, 1984.

Kindleberger, Charles. *Manias, Panics, and Crashes: A History of Financial Crises.* New York: Wiley, 1996.

King, Robert G., and Ross Levine. "Finance and Growth: Schumpeter Might Be Right." *Quarterly Journal of Economics* 108 (1993): 717–37.

Kochhar, Rakesh. "7 Key Findings on the State of the Middle Class in Western Europe." Pew Research Center. April 24, 2017. https://www.pewresearch.org/fact-tank/2017/04/24/7-key-findings-on-the-state-of-the-middle-class-in-western-europe/.

Levitsky, Steven, and Daniel Zibblatt. *How Democracies Die*. New York: Crown, 2018.

Linz, Juan, and Alfred Stepan. *Problems of Democratic Transition and Consolidation: Southern Europe, South America, and Post-Communist Europe*. Washington, DC: John Hopkins University Press, 1996.

Linz, Juan, et al. *Informe Sociologico sobre el Cambio Politico en Espana, 1975–1981*. Madrid: Fundacion FOESSA, 1981.

Lipset, Seymur Martin. *Political Man*. New York: Anchor Books, 1960.

Lipset, Seymur Martin, and Stein Rokkan. *Party Systems and Voter Alignment*. New York: Free Press, 1967.

Locke, Richard M. *Remaking the Italian Economy*. Ithaca: Cornell University Press, 1995.

López, Julia. *Un Lado Oculto de la Flexibilidad Salarial: El Incremento de la Judicializacion*. Albacete: Bomarzo, 2008.

Lo Schiavo, Gianni. *The Role of Financial Stability in EU Law and Policy*. The Netherlands: Kluwer Law International. 2017.

Lukauskas, Arvid. *Regulating Finance*. Ann Arbor: Michigan University Press, 1997.

Mabbet, Deborah, and Waltraud Schelke. "What Difference Does Euro Membership Make to Stabilization? The Political Economy of International Monetary Systems Revisited." *Review of International Political Economy* 22, no. 3 (June 2015): 508–34.

Macedo, Jorge Braga de. "Portugal's European Integration: The Limits of External Pressure." In *Portugal: Strategic Options in a European Context*, edited by J. A. Tavares, F. Monteiro, M. Glatzer, and A. Cardoso, 61–97. Lanham, MD: Lexington Books, 2003a.

Macedo, Jorge Braga de. "Portugal's European Integration: The Good Student with a Bad Fiscal Institution." In *Spain and Portugal in the European Union*, edited by Royo Sebastián and Paul Manuel, 169–94. Portland: Frank Cass, 2003b.

Macedo, Jorge Braga de. "A Mudança do Regime Cambial Português: Um Balanço 15 anos Depois de Maastricht." UNL WP 502, 2007.

Maddaloni, Angela, and José Luís Peydró. "Bank Risk-Taking, Securitisation, Supervision and Low Interest Rates: Evidence from the Euro-area and the U.S. Lending Standards." *The Review of Financial Studies* 24, no. 6 (2011): 2121–65.

Mahoney, James, and Kathleen Thelen. *Explaining Institutional Change: Ambiguity, Agency, and Power*. New York: Cambridge University Press, 2009.

Major, Aaron. "Neoliberalism and the New International Financial Architecture." *Review of International Political Economy* 19, no. 4 (2012): 536–61.

Mallet, Victor. "Prudence Pays Off for Big Banks." In "Investing in Spain," special report, *Financial Times*, October 2, 2009, p. 3.

Malo de Molina, José Luís, and Pablo Martin-Aceña, eds. *The Spanish Financial System: Growth and Development Since 1900*. New York: Palgrave, 2012.

Maravall, Jose Maria. "Politics and Policy: Economic Reforms in Southern Europe." In *Economic Reforms in New Democracies: A Social-Democratic Approach*, edited by Luiz Carlos Bresser Pereira, Jose Maria Maravall, and Adam Przeworski, 77–131. Cambridge: Cambridge University Press, 1993.

Maravall, Jose Maria. *Regimes, Politics and Markets*. New York: Oxford University Press, 1997.

Marks, M. *The Formation of European Policy in Post-Franco Spain*. Avebury: Brookfield, 1997.

Martin, Cathy, and D. Swank. *The Political Construction of Business Interests: Coordination, Growth, and Equality*. New York: Cambridge University Press, 2012.

Martín Aceña, Pablo. "Desarrollo y Modernización del Sistema Financiero, 1844–1935." In *La Modernización Económica de España*, edited by Nicolás Sánchez Albornoz. Madrid: Alianza Editorial, 1985.

Martín Aceña, Pablo. "Universia Business Review-Actualidad Económica: 150 Aniversario." *Banco de Santander*, November 5, 2007.

Martín Aceña, Pablo. "The Spanish Banking System From 1900 to 1975." In *The Spanish Financial System: Growth and Development Since 1900*, edited by José Luís Malo de Molina and Pablo Martin-Aceña. New York: Palgrave, 2012.

Martinez-Mongay, Carlos, and Luís Angel Maza Lasierra. "Competitiveness and Growth in EMU: The Role of the External Sector in the Adjustment of the Spanish Economy." Economic Papers No. 355, October 2009.

Mauro, Filippo, and Katrin Forster. "Globalisation and the Competitiveness of the Euro Area." Occasional Paper Series, No. 97, European Central Bank, 2008.

McDonough, Peter, Samuel Barnes, and Antonio López Pina. *The Cultural Dynamics of Democratization in Spain*. Cambridge: Harvard University Press, 1998.

McGuire, Patrick, and Goetz von Peter. "The US Dollar Shortage in Global Banking." *BIS Quarterly Review*, March, 2009.

Menz, Georg. *Varieties of Capitalism and Europeanization: National Response Strategies to the Single European Market*. New York: Oxford University Press, 2005.

Minsky, Hyman. *Can 'It' Happen Again: Essays on Instability and Finance*. New York: M.E. Sharpe, 1982a.

Minsky, Hyman. "The Financial-Instability Hypothesis: Capitalist Processes and the Behavior of the Economy." In *Financial Crises: Theory, History, and Policy*, edited by Charles Kindleberger and jean Pierre Laffargue. New York: Cambridge University Press, 1982b.

Muñoz Molina, Antonio. *Todo lo que era Sólido*. Madrid: Seix Barral, 2013.

Molina, Oscar, and Martin Rhodes. "Conflict, Complementarities and Institutional Change in Mixed Market Economies." In *Beyond Varieties of Capitalism*, edited by B. Hancké, M. Rhodes, and M. Thatcher, 223–53. Oxford: Oxford University Press, 2007.

Molinas, César. *Qué Hacer con España*. Madrid: Imago Mundi, 2013.

Montalvo, José G., and Josep M. Raya. "Constraints on LTV as a Macroprudential Tool: A Precautionary Tale." In *Oxford Economic Papers* 70, no. 3 (July 2018): 821–45.

Moore, Barrington. *Social Origins of Dictatorship and Democracy: Lord and Peasant in the Making of the Modern World*. New York: Beacon Press, 1967.

Morlino, Leonardo. "The Europeanisation of Southern Europe." In *Southern Europe and the Making of the European Union*, edited by António Costa Pinto and Nuno Severiano Teixeira, 237–60. New York: Columbia University Press, 2002.

Morlino, Leonardo, and José Ramón Montero. "Legitimacy and Democracy in Southern Europe." In *The Politics of Democratic Consolidation*, edited by Richard Gunther, P. Nikiforos Diamandouros, and Hans-Jürgen Puhle, 231–260. Baltimore: Johns Hopkins University Press, 1995.

Mosley Layna. "Attempting Global Standards: National Governments, International Finance, and the IMF's Data Regime." *Review of International Political Economy* 10, no. 2 (2003): 331–62.

Norris, Floyd. "Spain's Banking Mess." *New York Times*, September 23, 2011, p. B1.

OECD. *Economic Surveys: Spain*. Paris: OECD, 2010.

Olson, Mancur. The Logic of Collective Action Public Goods and the Theory of Groups. Cambridge, MA: Harvard University Press, 1965.

Olson, Mancur. *The Rise and Decline of Nations*. New Haven: Yale University Press, 1982.

Ortega, Andrés, and Angel Pascual-Ramsay. *¿Qué nos ha pasado?* Madrid: Galaxia Gutemberg, 2013.

Otero-Iglesias, Miguel. "Inequality in Spain: Let's Focus on the Poor." *Elcano Blog* (2019). https://blog.realinstitutoelcano.org/en/inequality-in-spain-lets-focus-on-the-poor/.

Otero-Iglesias, Miguel, Sebastián Royo, and Federico Steinberg. "War of Attrition and Power of Inaction: The Spanish Financial Crisis: Lessons for the European Banking Union." *Revista de Economía Mundial*. Number 46, September 2017.

Packer, George. *The Unwinding*. New York: Farrar, Straus and Giroux, 2013.

Pappas, Takis. *Populism and Liberal Democracy: A Comparative and Theoretical Analysis*. New York: Oxford University Press, 2019.

Paxton, Robert. *The Anatomy of Fascism*. New York: Vintage, 2004.

Pereira-Zazo, Oscar, and Steven Torres. *Spain after the Indignados/15M Movement*. New York: Palgrave, 2019.

Pérez, Sofía A. *Banking on Privilege*. New York: Cornell University Press, 1997.

Pérez, Sofía. A. "Systemic Explanations, Divergent Outcomes: The Politics of Financial Liberalization in France and Spain." *International Studies Quarterly* 42 (1998): 755–84.

Pérez, Sofía A., and Philippe Pochet. "Monetary Union and Collective Bargaining in Spain." In *Monetary Policy and Collective Bargaining in the New Europe*, ed. Philippe Pochet. Brussels: Peter Lang, 1999.

Pérez, Sofía A., and Jonathan Westrup. "Finance and the Macroeconomy: The Politics of Regulatory Reform in Europe." *Journal of European Public Policy* 17, no. 8 (2010): 1171–92.

Pérez Díaz, Victor. *The Return of Civil Society*. Cambridge: Harvard University Press, 1993.

Piketty, Thomas. *Capital e Ideologia*. Bilbao: Deusto, 2019.

Plender, John. "Respinning the Web." *Financial Times*, June 22, 2009, 5.

Pontusson, Jonas. *Inequality and Prosperity*. Ithaca: Cornell University Press, 2005.

Porter, Michael. *The Competitive Advantage of Nations*. New York: Free Press, 1990.

Portes, Richard. "Global Imbalances." In *Macroeconomic Stability and Financial Regulations: Key Issues for the G20*, edited by Mathias Dewatripont, Xavier Freitas and Riichard Portes, vol. 19, 19. London: CEPR, 2009.

Poveda, Raimundo. "Banking Supervision and Regulation over the Past 40 Years." In *The Spanish Financial System: Growth and Development since 1900*, edited by José Luís Malo de Molina and Pablo Martín Aceña. New York: Palgrave, 2012.

Prasad, Eswar, and Isaac Sorkin. "Assessing the G-20 Economic Stimulus Plans: A Deeper Look." Washington, DC. Brookings, 2009.

Pridham, Geoffrey. "European Integration and Democratic Consolidation in Southern Europe." In *Southern Europe and the Making of the European Union*, edited by António Costa Pinto and Nuno Severiano Teixeira, 183–207. New York: Columbia University Press, 2002.

Quaglia, Lucia, and Sebastián Royo. "Banks and the Political Economy of the Sovereign Debt Crisis in Italy and Spain." *Review of International Political Economy* 22, no. 3 (2015): 485–507.

Rajan, Raghuram G. "Has Finance Made the World Riskier?" *European Financial Management* 12 (2006): 499–533.

Regini, Marino. "Still Engaging in Corporatism? Recent Italian Experience in Comparative Perspective." *European Journal of Industrial Relations* 3, no. 3 (1997): 259–78.

Rhodes, M. "Globalization, Labour Markets and Welfare States: A Future of Competitive Corporatism?" In *The Future of European Welfare*, edited by M. Rhodes and Y. Meny, 178–203. New York: St. Martin's Press, 1998.

Reinhart, Carmen M., and Kenneth Rogoff. *This Time Is Different: Eight Centuries of Financial Folly.* New York: Princeton University Press, 2009.

Reinhart, Carmen M., and Kenneth Rogoff. "From Financial Crash to Debt Crisis." *American Economic Review* 101 (August 2011): 1676–1706.

Rodríguez Zapatero, José Luís. *El Dilema: 600 Días de Vértigo.* Barcelona: Planeta, 2013.

Rogowski, Ronald. *Commerce and Coalitions: How Trade Affects Domestic Political Alignments.* Princeton, NJ: Princeton University Press, 1990.

Rojo, Luís Angel. "Spain's Membership of EMU: Lessons for 2009." In *Spain and the Euro: The First Ten Years,* edited by Juan Francisco Jimeno, 27–30. Madrid: Banco de España, 2010.

Roldan, José María and Jesús Saurina. "Old and New Lessons of the Financial Crisis for Risk Management." In Banks at Risk: Global Best Practices in an Age of Turbulence, ed. Peter Hoflich. New York: Wiley, 2012.

Rothstein, Bo. *Social Traps and the Problem of Trust.* Cambridge: Cambridge University Press, 2005.

Royo, Sebastián. *From Social Democracy to Neoliberalism.* New York: St. Martin's Press, 2000.

Royo, Sebastián. *A New Century of Corporatism?* Westport: Praeger, 2002.

Royo, Sebastián. "The 2004 Enlargement: Iberian Lessons for Post-Communist Europe." In *Spain and Portugal in the European Union: The First 15 Years,* edited by Sebastián Royo and Paul Manuel, 287–313. Portland, OR: Frank Cass, 2003.

Royo, Sebastián. *Varieties of Capitalism in Spain.* New York: Palgrave, 2008.

Royo, Sebastián. "After the Fiesta: The Spanish Economy Meets the Global Financial Crisis." In "Southern Europe and the Financial Earthquake: Coping with the First Phase of the International Crisis," South European Atlas special issue, *South European Society & Politics* 14, no. 1 (March 2009a): 19–34.

Royo, Sebastián. "Reforms Betrayed? Zapatero and Continuities in Economic Policies." In "Spain's 'Second Transition'? The Socialist Government of Jose Luis Rodriguez Zapatero," Special Issue, *South European Society & Politics* 14, no. 4 (December 2009b).

Royo, Sebastián. "Portugal and Spain in the EU: Paths of Economic Divergence (2000–2007)." *Análise Social* 45, no. 195 (2010): 209–54.

Royo, Sebastián. "Lessons from Portugal and Spain in the EU after 25 Years: The Challenges of Economic Reforms." In *Spain and the European Union: The first Twenty-Five Years (1986–2011),* edited by Joaquín Roy and Maía Lorca-Susino (155–292). Miami: Jean Monnet EU Chair, University of Miami, 2011.

Royo, Sebastián. "Why the Spanish Financial System Survived the First Stage of the Global Crisis?" *Governance* 26, no. 4 (October 2013a): 631–56.

Royo, Sebastián. *Lessons from the Economic Crisis in Spain*. New York: Palgrave, 2013b.

Royo, Sebastián. "A 'Ship in Trouble' The Spanish Banking System in the Midst of The Global Financial System Crisis: The Limits of Regulation." In *Market-Based Banking, Varieties of Financial Capitalism and the Financial Crisis*, edited by Iain Hardie and David Howarth. New York: Oxford University Press, 2013c.

Royo, Sebastián. "Institutional Degeneration and the Economic Crisis in Spain." Special issue of *American Behavioral Scientist. The Economic Crisis from Within: Evidence from Southern Europe*. Anna Zamora-Kapoor, and Xavier Coller (eds.). 58, no. 12 (2014): 1568–91.

Royo, Sebastián. "After Austerity: Lessons from the Spanish Experience." In *Towards a Resilient Eurozone: Economic, Monetary and Fiscal Policies*, ed. John Ryan. New York: Peter Lang, 2015.

Royo, Sebastián. "The Causes and Legacy of the Great Recession in Spain." In *Oxford Handbook of Spanish Politics*, edited by Diego Munro and Ignacio Lago. New York, NY: Oxford University Press, 2020.

Royo, Sebastián, and Federico Steinberg. "Using a Sectoral Bailout to Make Wide Reforms: The Case of Spain." In *The Political Economy of Adjustment Throughout and Beyond the Eurozone Crisis: What Have We Learnt?* edited by Michelle Chang and Federico Steinberg. New York: Routledge, 2019.

Royo, Sebastián, and Paul C. Manuel, eds. *Spain and Portugal in the European Union*. Portland: Frank Cass, 2005.

Ryan, John, ed. *Towards a Resilient Eurozone: Economic, Monetary and Fiscal Policies*. New York: Peter Lang, 2015.

Santillán, Ramón. *Memorias, 1808–1856*. Madrid: Tecnos and Banco de España, 1966.

Scharpf, Fritz, and Vivien Schmidt. *Welfare and Work in the Open Economy: From Vulnerability to Competitiveness*. Oxford: Oxford University Press. 2000.

Schmidt, Vivien. *The Futures of European Capitalism*. New York: Oxford University Press, 2002.

Schmidt Vivien. "Putting the Political Back into Political Economy by Bringing the State Back Yet Again." *World Politics* 61, no. 3 (2009): 516–48.

Schneider, Ben Ross. *Business Politics and the State in Twentieth-Century Latin America*. New York: Cambridge University Press, 2004.

Schoenmaker, Dirk. "The Trilemma of Financial Stability." Paper prepared for CFS-IMF Conference: *A Financial Stability Framework for Europe: Managing Financial Soundness in an Integrating Market*. Frankfurt, September 26, 2008.

Schoenmaker, Dirk. "The Financial Trilemma." *Economics Letters* 111 (2011): 57–59.

Schularick, Moritz, and Alan M. Taylor (2012): "Credit Booms Gone Bust: Monetary Policy, Leverage Cycles, and Financial Crises, 1870–2008." *American Economic Review* 102 (2012): 1029–61.

Sebastián, Miguel. "Spain in the EU: Fifteen Years May Not Be Enough." Paper presented at the conference "From Isolation to Europe: 15 Years of Spanish and Portuguese Membership in the European Union," Minda de Gunzburg Center for European Studies, Harvard University, November 2–3, 2001.

Serra Ramoneda, A. *Los Errores de las Cajas: Adiós al modelo de las cajas de ahorro*. Barcelona: Ediciones Invisibles, 2011.

Skocpol, Theda. *States and Social Revolutions*. New York: Cambridge University Press, 1979.

Solbes, Pedro. *Recuerdos*. Bilbao; Deusto, 2013.

Stallings, Barbara. *Finance for Development: Latin America in Comparative Perspective*. Washington, DC: Brookings Institution Press, 2006.

Steinmo, Sven, Kathleen Thelen, and Frank Longstreth, eds. *Structuring Politics: Historical Institutionalism in Comparative Analysis*. New York: Cambridge University Press, 1992.

Stiglitz, Joseph. "Capital Market Liberalization, Economic Growth, and Instability." *World Development* 28, no. 6 (2000): 1075–86.

Streeck, Wolfgang, and Kathleen Thelen, eds. *Beyond Continuity: Institutional Change in Advanced Political Economies*. New York: Oxford University Press, 2005.

Swank, Duane. *Global Capital, Political Institutions, and Policy Change in Developed Welfare States*. New York: Cambridge University Press, 2002.

Taylor, Alan. "On the Costs of Inward-Looking Development: Price Distortions, Growth, and Divergence in Latin America." *Journal of Economic History* 58 (2009): 1–28.

Tedde de Lorca, Pedro. "La Evolución del Sistema Bancario Español en el Siglo XX." In *Guía de Archivos Históricos de la Banca en España*, edited by María de Inclán Sánchez, Elena Serrano García, and Ana Calleja Fernández. Madrid: División de Archivos y gestión Documental del Banco de España, 2019.

Thatcher, Mark. *Internationalisation and Economic Institutions: Comparing the European Experience*. Oxford: Oxford University Press, 2007.

Thelen, Kathleen. *How Institutions Evolve: The Political Economy Skills in Germany, Britain, the United States and Japan*. New York: Cambridge University Press. 2004.

Tilford, Simon, and Philip Whyte. *The Lisbon Scorecard X: The Road to 2020*. Brussels: Centre for European Reform, 2010.

Toral, Pablo. *The Reconquest of the New World: Multinational Enterprises and Spain's Direct Investment in Latin America*. Great Britain: Ashgate, 2001.

Torcal, Mariano. "Political Dissatisfaction in New Democracies: Spain in Comparative Perspective." Ph.D. Dissertation, Ohio State University, 2002.

Torrero, Antonio. "La formación de los tipos de interés y los problemas actuales de la economía española." *Economistas*. Madrid, no. 39, 1989.

Torres, Francisco. "A Convergéncia para a Uniáo Económica e Monetária." In *Em Nome da Europa. Portugal em Mudança, 1986–2006*, edited by Marina Costa Lobo and Pedro Lains, 97–120. Cascais: Principia (ICS), 2007.

Tortella, Gabriel. *Banking, Railroads and Industry in Spain*. New York: Arno Press, 1977.

Tortella, Gabriel, and José Luís García Ruiz. *Spanish Money and Banking. A History*. New York: Palgrave, 2013.

Tovias, Alfred. "The Southern European Economies and European Integration." In *Southern Europe and the Making of the European Union*, edited by António Costa Pinto and Nuno Severiano Teixeira, 159–81. New York: Columbia University Press, 2002.

Traxler, F. "The Logic of Social Pacts." In *Social Pacts in Europe*, edited by G. Fajertag and P. Pochet, 27–36. Brussels: European Trade Union Institute, 1997.

Urquizu, Ignacio. *La crisis de representación en España*. Madrid: Los Libros de la Catarata, 2016.

Vicens Vives, Jaume. *An Economic History of Spain*. Princeton, NJ: Princeton University Press, 1969.

Villena Oliver, Andrés. *Las Redes de Poder en España*. Madrid: Rocaeditorial, 2019.

Villoria, Manuel, and Fernando Jiménez. "La Corrupción en España (2004–2010): Datos, percepción y efectos." *REIS* 138 (abril-junio 2012): 109–34.

Vives, Xavier. "The Spanish Financial Industry at the Start of the 21st Century: Current Situation and Future Challenges." In *The Spanish Financial System: Growth and Development since 1900*, edited by José Luís Malo de Molina and Pablo Martín Aceña. New York: Palgrave, 2012.

Weiner, Richard, and Iván López. *Los Indignados: Tides of Social Insertion in Spain*. Winchester, NY: Zero Books, 2018.

Weiss, Linda. *States in the Global Economy: Bringing Domestic Institutions Back in*. Cambridge: Cambridge University Press, 2003.

Williamson, J., ed. *The Political Economy of Policy Reform*. Washington, DC: Institute for International Economics, 1994.

Wolf, Martin. *Why Globalization Works*. New Haven: Yale University Press, 2004.

Wolff, Edward. 2017. "Household Wealth in the United States, 1962 to 2016: Has Middle Class Wealth Recovered?" National Bureau of Economic Research. Working Paper Series, No. 24085. www.nber.org/papers/w24085.

The World Economic Forum. *Global Competitiveness Report*. https://www.weforum.org/.

Zysman, John. *Governments, Markets, and Growth: Financial Systems and Politics Industrial Change*. Ithaca, NY: Cornell University Press, 1983.

Index

A

access to capital, 6, 167
accountability, 15, 127, 223, 234, 296, 303, 305
accountable, 17, 214, 219, 220, 223, 225, 235, 241
accounting, 81, 84, 102, 106, 109, 110, 114, 115, 145, 146, 154, 157, 183, 286
acquisitions, 89, 91, 92, 102, 108, 109, 113, 123, 148, 155, 206, 229
agriculture, 74, 80, 83, 124
alliance, 54, 55, 97, 113, 184, 250, 251
apolitical, 219
asientos, 57, 58
asset based securities (ABS), 153, 154, 211
assets, 12, 13, 19, 27–29, 31, 37, 39, 44, 60–62, 64, 65, 71, 79, 85, 86, 89, 94, 95, 98, 103, 105, 106, 108, 110, 114–116, 134, 136, 144–146, 151–154, 157, 159, 160, 163, 164, 167, 169,
171, 174, 177, 179, 183–186, 188, 191, 194–198, 200–202, 209–212, 249, 261, 263, 269, 270, 272–275, 284, 285, 288, 290, 291
assistance, 6, 26, 42, 89, 101, 136, 142, 143, 177, 193, 202, 274, 277
austerity, 137, 138, 195, 217, 242, 279, 282, 302, 303, 305
autarchy, 91
autarkic, 88, 90
autocratic regime, 12, 54, 96, 248, 251

B

bad loan(s), 159, 163–165, 176, 180, 196–198, 201, 212, 284, 285
bailout(s), 3, 4, 8, 17, 19, 26, 28, 40–43, 46, 53, 54, 67, 137–139, 152, 171, 178, 193, 196, 211, 216, 217, 224, 226, 264, 268, 272, 277, 279, 280, 286, 291, 292, 304, 305

balance of payment, 22, 24, 82, 91, 138, 277

balance sheet, 11, 39, 110, 115, 136, 144, 151, 153, 160, 174, 195, 198, 210, 211, 227, 270, 274, 280, 289

bancarization, 116

Banco Bilbao Vizcaya Argentaria (BBVA), 112, 113, 136, 137, 142, 143, 148, 153, 156, 167, 169, 185, 190, 194, 196, 198, 204, 228, 237, 264, 270, 271, 283, 291

bankarization, 157

bank bargains, 2, 3, 24, 31, 44–46, 69, 72, 83, 134, 137, 138, 247, 250, 257, 259, 264, 266, 268, 283, 285–287, 294, 303

bank charters, 6, 55, 59, 72, 97

Bankia, 40, 41, 139, 177, 178, 187–193, 202, 203, 206, 209, 225–227, 257, 260, 262, 263, 269–271, 291, 296

banking culture, 149

banking laws, 55, 65, 66, 68, 72, 84, 88, 92, 104, 251

banking structure, 3, 117, 249, 253

banking system, 1, 2, 4, 6, 8, 10–12, 18, 19, 21, 22, 24, 32, 37, 40, 42, 43, 45, 46, 53–57, 67, 72, 74, 76, 79–81, 85–87, 90, 92, 97, 98, 106–108, 114, 116, 117, 119, 136, 137, 139, 142–144, 150, 151, 156, 157, 162, 168, 175, 176, 193, 199, 213, 237, 247–250, 252, 254, 256, 258, 264, 267, 268, 285–288, 290, 293, 303, 304, 307

bank insiders, 12, 13, 55, 60, 67, 72, 97, 98, 117, 250, 251

Bank of Barcelona, 62, 63, 65, 73, 74, 83, 84, 101

Bank of Saint Charles, 59, 60, 62

Bank of Saint Ferdinand, 63–66, 68

Bank of Spain (BoS), 11, 18–20, 31, 35–37, 40, 46, 59, 68–75, 81–96, 99–103, 106, 108–110, 112, 115, 132, 136, 138, 144–147, 150, 151, 156–160, 162–164, 167, 168, 171, 173, 175–180, 184, 190–193, 196, 197, 199, 200, 202–208, 210, 223, 225, 227–230, 252, 254, 261–263, 265, 267–269, 271, 272, 274, 275, 283, 284, 291

bankruptcies/bankruptcy, 54, 57, 58, 64, 65, 74, 101–103, 147, 163, 257

banks crisis, 1–8, 11, 15, 18, 21, 22, 24, 26, 27, 29–31, 33, 35, 40, 43, 45, 53, 64, 66, 72, 77, 81, 84, 86, 100, 101, 103, 107, 109, 112, 113, 115, 117, 147, 170, 174–176, 196, 210–213, 218, 227, 247–249, 256, 259, 263, 283–287, 289, 304, 307

bargains, 1–4, 8, 10, 20, 34, 35, 53, 55, 67, 72, 77, 97, 117, 119, 138–140, 142, 171, 175, 189, 203, 213, 231, 248–252, 265, 285–287, 290, 294, 300

Big Seven Club, 94

blame, 8, 138, 191, 205, 211, 215–217, 219, 227, 253, 263, 267, 268, 295, 305, 307

board of directors, 70, 83

bond yields, 40, 41, 168

borrow/borrowing, 1, 11, 13, 14, 19–21, 27, 28, 40, 44, 57, 64, 88, 96, 105, 106, 121, 133, 135, 137, 139, 152, 155, 162, 167, 175, 189, 195, 199–201, 203, 205, 216, 259, 260, 265, 282

branches, 63, 68, 69, 74, 80, 82, 87,
89, 91–96, 102, 106, 113, 114,
144, 159–161, 177, 183–186,
192, 198, 251, 264, 266
Brazil, 1, 113, 148, 155, 164, 251
bubble, 9, 11, 16, 17, 27, 30, 31, 37,
46, 128, 129, 133, 135, 136,
138, 139, 153, 170, 171, 175,
196, 203, 205–207, 209–212,
214–217, 227, 228, 237, 263,
268, 273, 275, 282, 284, 288,
291, 295–297

C
cajas, 7, 8, 19, 29, 31, 36, 39,
40, 43, 44, 64, 96, 114, 115,
119, 137–139, 141, 144, 145,
150, 154–168, 170, 176–179,
185, 187, 189–191, 193–195,
197, 198, 200, 202–211, 214,
216–218, 226–229, 238, 248,
257–268, 270, 283, 286, 287
cajas de ahorro, 20, 36, 85, 157,
165, 171, 204
capital, 5, 7–9, 12–15, 20–27, 29–33,
35, 37, 40, 42–44, 55, 59, 60,
63–65, 68, 69, 72, 74–76, 81–
83, 85, 86, 88, 91, 93–95, 97,
102–107, 110, 112–114, 133,
135, 143, 145–148, 150, 156,
158, 159, 161, 162, 165, 167,
170, 177, 179, 181, 183, 184,
187, 190, 192–194, 198, 200,
201, 207, 211, 218, 219, 251,
254, 256–258, 260, 261, 263,
264, 266, 268, 270, 273–275,
277, 279, 282, 286–289, 291,
303
capital controls, 91, 288
capital cushions, 44, 258, 260, 263,
274
capitalization, 3, 116, 200, 271, 280

capital ratios, 29, 66, 160, 164, 177,
259–261
Carlist War, 61, 63, 74, 75
cartelization, 90
central bank, 8, 23, 30, 59, 84–87,
126, 146, 165, 193, 247, 253,
263, 284
circulation, 63, 65, 68, 72, 75, 86
civil society, 9, 219, 296
Civil War, 61, 71, 80, 81, 87, 89,
120, 221, 242, 248, 254
clientelistic, 218, 295
coalition(s), 1–4, 8, 10, 18, 34, 35,
45, 46, 53–56, 60, 67, 68, 76,
81, 97, 106, 117, 132, 211, 213,
224, 238, 240, 241, 247–254,
257–259, 261, 265–268, 283,
285–287, 292, 294, 298, 300,
303, 304, 307
coefficients, 82, 94–96, 167, 206, 297
collapse, 4, 5, 10, 11, 29, 32, 37,
43, 44, 60, 71, 72, 86, 98, 102,
104, 121, 129, 131, 133, 135,
136, 139, 142, 143, 147, 153,
155–157, 162, 171, 174–176,
179–188, 192, 193, 195, 196,
198, 204, 207, 210, 211, 217,
225–227, 236, 238, 256–258,
267–269, 273–276, 280, 281,
285, 286
collateral, 29, 39, 56, 63, 65, 83, 85,
86, 96, 163, 184, 273, 284
collective responsibility, 219
collusion, 218
colonization, 108, 223, 226
commercial, 13, 36, 56–58, 73, 74,
77, 80, 82, 93, 104, 106, 107,
149, 163
commercial banks, 36, 37, 62, 63,
69, 72, 76, 77, 88, 91, 93, 94,
96, 104, 106, 107, 145, 146,
148, 157, 197, 199, 250, 284

competition, 4, 14, 32, 55, 68, 80,
90, 95–99, 103, 105–107, 111,
116, 117, 128, 147, 204, 228,
250, 251, 254, 255, 265, 276,
277, 289
competitiveness, 9, 16–19, 21, 26, 46,
57, 96, 113, 116, 121, 127–129,
131, 134, 135, 137–139, 194,
195, 214, 218, 231–233, 254,
276, 277, 283, 297, 300
complacency, 138, 176, 207, 229,
267, 283, 284, 287, 292, 295
computerization, 96
concentration, 80, 94, 105, 109, 110,
116, 180, 187, 207, 209, 229,
230, 251, 287
confidence, 33, 66, 130, 132, 137,
152, 158, 167, 193, 202, 203,
226, 227, 262, 284, 302, 307
consolidation, 37, 44, 77, 80, 82, 85,
89, 92, 94, 103, 108, 111, 112,
115, 150, 158, 159, 162, 164,
170, 177, 257, 260, 271
Constitution, 9, 75, 84, 120
construction, 18, 54, 68, 71, 84, 110,
123, 124, 128, 130, 131, 135,
139, 141, 153, 159, 160, 163,
174, 175, 180, 190–192, 194,
195, 198, 205, 207, 209, 210,
216, 228, 229, 231, 259, 281,
282, 300
consumption, 4, 18, 124, 129, 131,
132, 135, 204, 215, 228, 276
convergence, 124, 134, 135, 232,
292, 293
convertibility, 91
coordination, 24, 34, 92, 293, 294
corruption, 15, 58, 67, 127, 128,
139, 188, 219, 232, 233, 235,
236, 238–241, 261, 287, 295,
296
cost of capital, 113, 126, 263

cotton, 63, 65, 71, 74
counter-cyclical
counter-cyclical capital regime,
144–147, 168, 170
counter-cyclical provisions, 112
countercyclical, 20
countercyclical capital regime, 206
countercyclical provisions, 19
coverage, 159, 160
coverage ratio, 193
creative destruction, 17, 215, 223
credit, 2, 4, 6, 8, 10–15, 19, 21,
22, 27–30, 33, 36, 37, 39, 40,
45, 46, 53–55, 58, 62, 64–76,
79–83, 85–87, 89–95, 97, 98,
100, 102, 108, 110, 114–117,
127, 132, 138, 139, 143, 146,
150, 153, 157, 159, 160, 163,
170, 171, 174–176, 180, 196,
197, 204, 207, 208, 210, 211,
237, 238, 248, 251–253, 255,
256, 258, 259, 261, 265, 266,
275, 276, 280, 281, 284, 287,
289, 290, 293, 303
credit quality, 154, 286
credit risk, 30, 149, 150, 154
crisis/crises, 1, 2, 4, 5, 7, 9–11,
13–15, 18–26, 28–30, 32,
36, 39, 40, 43–46, 53, 57,
64–67, 71–73, 75–77, 81, 83,
84, 86–88, 97, 98, 100, 101,
103–106, 109–113, 115–117,
119, 121, 127, 128, 130–134,
136–148, 151–156, 159–163,
165, 168, 170, 171, 174–176,
178, 187, 190, 191, 195–198,
200–223, 227–232, 234,
236, 238–244, 247–249, 252,
257–259, 262–265, 267–270,
273–277, 279–298, 300, 301,
303–307
crisis prone, 4, 11, 32

crisis prone, 4, 11, 32
crony, 12, 16, 67, 68, 72, 97, 139,
    179, 214, 217, 223, 296
cross-directorship, 90
cross-shareholding, 90
Crown, 54–58, 62, 67
cultural arguments, 149–150
current account, 9, 27, 30, 68,
    104, 122, 128–130, 134, 135,
    137–139, 175, 195, 277, 288,
    294, 299, 300

D
damaged assets, 143
debasement, 57, 58
debtors, 3, 13, 54, 55, 66, 83, 98,
    155, 249, 285
decentralization, 217, 218
deficit, 4, 9, 13, 14, 21, 26, 27,
    30, 57–60, 62, 72, 85, 91,
    127–130, 133–135, 138, 139,
    165, 175, 194, 195, 216, 247,
    257, 276, 277, 279, 282, 288,
    294, 300–302, 304
degeneration, 15–17, 46, 121, 139,
    213–215, 222, 231, 232, 238,
    244
delegitimization, 221
delinquent rate, 155–157, 159, 160,
    162, 164
democracy, 24, 74, 80, 98, 103, 117,
    120, 121, 217, 221, 230, 235,
    240, 242, 243, 248–250, 252,
    253, 255, 300
demoralization, 220
depositors, 3, 12, 44, 54, 56, 66, 68,
    97, 98, 100, 176, 249, 251, 257,
    265, 273, 285
deposits, 37, 39, 40, 44, 56, 58, 60,
    63, 66, 68, 77, 80, 82, 85, 86,
    88–91, 93, 94, 96–100, 102,
    103, 105, 111, 114, 144, 150,

154, 174, 179, 198, 211, 218,
    250–252, 274
deregulation, 14, 24, 46, 80, 92, 95,
    96, 113, 244, 256, 294, 303,
    305
desamortización, 61, 62, 74
devaluation(s), 21, 113, 296, 297,
    300
development, 3, 5, 11, 14, 18, 19,
    24, 26, 31, 32, 36, 37, 45,
    57, 68, 72, 76, 77, 79, 81, 82,
    90, 92, 95, 96, 98, 99, 102,
    104, 106–109, 112, 116, 120,
    124–126, 129, 150, 155, 162,
    164, 175, 210, 214, 218, 226,
    235, 241, 250, 253, 262, 263,
    272, 281, 285, 290, 294, 296,
    305
development plans, 92, 93
dictatorship, 77, 86–88, 93, 96, 106,
    117, 120, 242, 248, 249, 251
differentiation, 83
disentailment, 61, 68, 71, 74
disintermediation, 107, 114
disruptions, 57, 305
distribution, 4, 35, 54, 72, 88, 251,
    253, 266, 293, 297, 302–304
divergence, 16, 17, 135, 214, 215,
    232, 234
diversification, 77, 92, 148, 155, 164,
    194, 286
diversify, 113, 148, 249, 265, 266,
    282
dividend, 59, 65, 83, 88, 101, 145,
    146, 251
doom loop, 40, 43, 137, 196, 279
dynamic provisioning/dynamic
    provisions, 30, 112, 115, 145,
    146, 283, 287, 303

E

economic crisis, 8, 9, 15, 17, 19, 25, 26, 42, 45, 46, 53, 64, 81–83, 98, 99, 103, 106, 113, 117, 119–121, 126, 129–132, 136, 138, 140, 143, 147, 176, 177, 192, 210, 213–215, 221, 231, 240, 252, 294, 295

economic policies/economic policy, 26, 74, 91, 126, 216, 231, 243, 252–254, 267, 275, 279, 291

economies of scale, 80, 96, 108, 109, 115, 210, 259, 265, 266

economies of scope, 108

education, 25, 116, 129, 217, 218, 224, 233–235, 262, 263, 296

elites, 9, 17, 23, 75, 106, 214, 215, 217–219, 221–223, 232, 237, 244, 250, 252, 253, 266, 282, 295, 304–306

enforcement, 97, 145, 233, 296

equity, 12, 13, 27, 29, 43, 83, 93, 95, 110, 145, 156, 166, 171, 177, 193, 249, 250, 258, 260, 272, 274

Euro, 5, 21, 23, 41, 112, 121, 133–135, 137, 139, 140, 165, 178, 185, 195, 196, 199, 216, 270, 272, 273, 278, 281, 297

Europe, 8, 11, 12, 18–20, 24–26, 28, 40–43, 46, 56–58, 64, 77, 79, 86, 108, 115, 116, 120, 121, 123, 124, 126, 129, 132, 133, 135, 136, 140, 143, 144, 148, 154, 165, 167, 174, 175, 209, 219, 224, 232, 242, 244, 254, 272, 276, 279, 280, 285, 292–294, 296, 303, 304, 306

European Central Bank (ECB), 30, 31, 35, 39, 40, 42, 114, 137, 138, 155, 156, 162, 171, 193, 196, 198, 199, 201, 206, 207, 218, 228, 229, 263, 268, 273, 275, 278, 281, 288, 291, 293

European Economic Community (EEC), 99, 100, 107, 112, 115, 116

European Monetary Union (EMU), 9, 13, 17–20, 37, 46, 126, 127, 135, 137–140, 150, 214–216, 232, 234, 237, 277, 278, 281, 282, 292, 294

European Union (EU), 5, 12, 16–18, 20, 23, 36, 41, 42, 121, 123–125, 127, 128, 136–141, 143, 144, 158, 167, 168, 178, 193, 194, 201, 208, 214, 217, 218, 222, 227, 234, 235, 275, 277, 278, 282, 292, 294, 299–301

exchange rate, 9, 86, 134, 277

exports, 57, 64, 82, 85, 128, 129, 216, 233, 277, 278, 299, 300

expropriated, 55, 56, 67, 100, 102

extractive, 17, 77, 214, 218, 219, 222, 223, 232

F

family, 9, 63, 67, 123, 128, 210, 230, 231, 235, 266, 297

fiduciary, 68, 86

financial crisis, 1, 3–5, 7–11, 16–20, 23, 25, 27, 28, 32, 40, 43, 44, 46, 86, 87, 115, 119–121, 127, 129, 133, 135, 136, 138–143, 154, 173, 174, 176, 195, 196, 198, 199, 209–211, 214, 215, 227, 235, 257, 258, 267, 268, 272, 276, 277, 280, 281, 283, 284, 288, 289, 291–293

financial engineering, 110, 112

financialization, 23, 39, 151, 153, 174, 209, 211, 304, 305

financial liberalization, 14, 22, 23, 28, 37, 80, 99, 150

financial repression, 91, 94
financial system, 4, 6, 10, 11, 14, 17, 18, 20, 21, 23–25, 27, 32, 33, 37–40, 45, 46, 76, 80, 84, 85, 87, 89, 91, 96, 128, 136, 141, 144, 145, 147, 150, 153, 160, 163, 168, 170, 173, 174, 176, 189, 209, 252, 256, 259, 270, 280, 283, 289, 290, 292, 304
First World War, 82
fiscal consolidation, 126, 127, 218, 254, 269, 282, 305
fiscal crisis, 24, 133, 135, 138
Fondo de Restructuración Ordenación Bancaria (FROB), 42, 158, 159, 161, 177, 179–181, 183–185, 187, 191, 192, 200–202, 269, 270, 272, 275, 291
foreign banks, 14, 69, 77, 86, 88, 94, 96, 108, 175
foreign direct investment (FDI), 91, 112, 123
fragile, 10, 45, 54, 72, 81, 127, 153, 198, 251, 258, 304
fragmentation, 5, 132, 238–240, 243, 292, 295, 299, 306, 307
Franco, 77, 87, 88, 91, 96–98, 102, 115, 117, 120, 237, 242, 250–252, 254
Francoism, 219
funding crises, 54, 167, 199

G
Game of Bank Bargains, 1, 2, 98, 138, 189, 248, 253, 303
Germany, 12, 32, 39–42, 58, 77, 123, 133, 135, 136, 142, 143, 154, 176, 196, 218, 277, 278, 300
global crisis, 45, 120, 121, 129, 131, 195, 217, 218
  global financial crisis, 1, 8, 9, 11, 19, 25, 27, 32, 44, 120, 127, 133, 136, 139, 141, 142, 173, 174, 176, 195, 196, 209, 215, 267, 276, 283, 284, 289, 291, 292
Glorious Revolution, 72
government, 1–5, 10, 12, 15, 17–19, 25, 26, 28, 29, 32, 34–37, 39, 41–45, 53–55, 59–62, 64–76, 81, 84–91, 93–99, 101, 103, 104, 106–108, 110, 111, 117, 122, 124, 126, 127, 129–139, 142, 143, 147, 148, 150, 151, 155, 156, 158, 161, 164–168, 170, 171, 174, 176–178, 183, 184, 187–189, 191–193, 195, 196, 200–203, 205–209, 214–218, 220, 222–224, 226–233, 236–243, 248–255, 257–262, 264, 265, 267, 272, 275, 276, 279, 282, 284, 288, 291, 294, 296–304, 306, 307
Great Depression, 58, 86, 87
great recession, 8, 119, 239, 285, 287, 303, 305, 306
Greece, 37, 41–43, 127, 132, 133, 136, 137, 143, 150–152, 167, 195, 196, 209, 210, 239, 242, 278, 280
growth, 5, 11, 12, 14, 17, 20, 24, 27, 28, 30, 61, 68, 69, 74–77, 80, 82–84, 89, 90, 92, 95, 100, 106, 108, 111, 113, 116, 120, 121, 123–125, 127–129, 131, 133, 134, 139, 145, 146, 148, 153, 165, 194–196, 198, 200, 204, 205, 208, 215–217, 223, 228, 229, 231, 252, 254–256, 263, 264, 277–279, 282, 284, 296, 299, 301, 302, 304, 305
guarantee coefficients, 93
guarantee(s), 20, 29, 37, 62, 68, 88, 100, 136, 143, 148, 163, 177,

179, 193, 202, 258, 260, 266, 270, 282, 286
guilds, 58

H
historical learning, 147
housing bubble, 30, 128, 170, 171, 237, 296

I
imbalances, 8, 17, 21, 27, 58, 60, 129, 134, 137, 138, 195, 231, 232, 275–277, 279, 291
immigration, 8, 121, 124, 125, 127, 242, 305
import substitution, 82
impunity, 250, 295, 296
inaction, 138, 158, 208, 227, 229, 261, 262, 275
incentives, 3, 8, 13, 23, 31, 34, 45, 54, 69, 76, 81, 94, 97, 105, 119, 135, 139, 142, 149, 171, 175, 258, 260, 261, 267, 277, 285, 286
inclusive, 222, 223
income, 21, 61, 62, 76, 83–85, 121, 123, 128, 135, 139, 231, 233, 253, 275, 288, 297, 302, 304, 305
indebtedness, 14, 21, 27, 57, 59, 64, 100, 128, 130, 133, 190, 207, 209, 267, 276
industrial banks, 80, 93, 94, 101, 104–106
industrial firms, 63, 77, 82, 90, 105, 251
industry, 57, 61, 63, 64, 76, 80, 82, 83, 85, 92, 111, 127, 149, 163, 170, 251
inflation, 13, 19, 21, 44, 87, 90, 91, 93, 98, 103, 104, 122, 124,

128–130, 133, 196, 231, 252, 254, 276, 278, 280, 284, 299
inflationary impact, 89, 103
initial public offering (IPO), 148
innovate, 17, 215, 223
innovation, 3, 12, 13, 17, 23, 34, 57, 99, 215, 217, 218, 223, 233, 235, 276, 292, 300
insolvency, 2, 33, 44, 54, 77, 138, 256, 257
inspection, 36, 73, 93, 100, 103, 180, 205
institutional change, 24, 35, 216, 221, 293
institutional crisis, 9, 16, 214, 221, 223, 232, 281, 282, 294
institutional degeneration, 15–17, 46, 121, 139, 213–215, 221–231, 232, 238, 244
institutional degradation, 15, 17, 214
institutional divergence, 16, 17, 214, 215, 231–236
institutional framework, 25, 31, 45, 77, 80, 81, 258, 282, 287
institutional setting, 144–145
institutional structure, 38, 138, 253
institutions, 3–5, 8–10, 13, 17–19, 24–27, 31, 33–37, 39, 43, 44, 46, 54, 55, 68, 70–72, 79, 80, 82, 85–89, 92, 95–98, 100, 107, 108, 111, 112, 114–116, 119, 135, 136, 140, 142–148, 150, 152–162, 164, 165, 167, 168, 170, 171, 173, 175, 177, 181, 183, 184, 186, 189–192, 195, 197–199, 201, 202, 208, 209, 211, 213–216, 218, 221–223, 226, 227, 229, 230, 232–235, 237, 242, 243, 248–250, 257, 260, 266–270, 274, 282, 283, 286, 287, 292–294, 303–307
interbank borrowing, 152

interest rates, 8, 14, 18, 19, 21,
    29–31, 44, 55, 58, 64, 66,
    72, 80, 83, 86, 88–91, 93, 94,
    96–100, 103, 114, 116, 121,
    126, 139, 143, 157, 181, 201,
    207, 251, 253, 256, 257, 263,
    271, 275, 279, 288
intermediation, 39, 79, 80, 92, 99,
    116
  intermediation coefficient, 82
internationalization, 26, 96, 113, 116,
    117, 148
intervention/interventionism, 24, 25,
    37, 46, 80, 84, 87, 89, 93, 101,
    105, 110–112, 115, 148, 150,
    165, 177, 180, 187, 206, 228,
    251–253, 266, 291, 292
investment, 4, 5, 7, 12–15, 31, 32,
    39, 43, 55, 64, 68–71, 73, 76,
    80, 82, 83, 85, 87, 89, 91–93,
    97, 104, 107, 108, 110, 113,
    123, 126, 128–130, 132, 133,
    137, 138, 151, 155, 156, 160,
    179–181, 183–185, 190, 195,
    204, 217, 222, 251–254, 256,
    261, 263, 264, 269, 282, 288,
    299, 301, 304
  investment coefficients, 88, 95, 96
investment banks, 68, 70, 80, 89, 93,
    107, 142, 146, 148, 284
Ireland, 11, 40–42, 121, 132, 133,
    136, 143, 152, 174, 195, 209,
    215, 239, 272, 280, 291
Italy, 12, 25, 26, 37, 127, 133, 150,
    151, 195, 209, 210, 242, 278,
    279

J
joint-stock bank, 59, 77, 250
judiciary, 15, 222, 223, 235, 236,
    267

K
king, 54, 57, 58, 60, 230

L
labor market, 9, 21, 120, 123–125,
    129, 131, 244, 255, 276, 277,
    296, 300
Latin America, 112, 113, 148, 155,
    300
legitimacy, 223, 288, 306
lender of last resort, 84, 86, 134, 138
lending, 7, 10, 11, 13, 20, 30, 39,
    44, 45, 54, 55, 57, 60, 62, 63,
    65, 67, 74, 76, 80–82, 85, 88,
    92, 102, 133, 135, 138, 141,
    142, 151, 153, 155, 156, 164,
    174, 175, 197, 198, 208, 210,
    211, 251, 258, 260, 265, 284,
    285, 288, 289
lending standards, 31, 194, 263, 286
liability/liabilities, 11, 21, 39, 40, 64,
    94, 141, 151, 153, 155, 167,
    174, 198–200, 208–211, 290
liberalization, 8, 14, 22, 23, 26, 28,
    37, 77, 80, 91, 92, 95, 99, 100,
    105, 107, 108, 112, 125, 150,
    253, 255
liquidity, 21, 28, 30, 42, 71, 82, 83,
    85, 86, 93, 100, 115, 129, 137,
    138, 143, 152, 154–156, 167,
    170, 181, 191, 196, 199, 202,
    204, 210, 268, 269, 273
loans, 7, 12–14, 21, 29, 37, 39,
    43, 55, 56, 58, 63–65, 67,
    69, 70, 77, 85, 86, 89–91,
    93, 95, 97, 98, 100–102, 106,
    114, 115, 133, 135–138, 143,
    151, 154, 155, 157, 160, 163,
    171, 176–181, 185, 188, 190,
    193–198, 201, 209, 211, 225,
    226, 233, 249–251, 256, 257,

260, 262, 268, 270, 272, 273, 284, 291
loss of competitiveness, 9, 16, 46, 113, 134, 135, 194, 214, 231, 232, 277

M
macro-prudential supervision, 207
management, 22, 44, 68, 75, 80, 83, 85, 90, 95, 96, 100, 104, 105, 109, 110, 114–116, 145, 149, 150, 156, 158, 176, 194, 201, 217, 249, 258, 262, 264, 265, 274, 275, 285, 288, 289, 294
market-based assets, 153
market-based banking (MBB), 10, 39, 141, 143, 151, 152, 155, 174, 175, 204, 208–212
market liquidity, 154
Mediterranean, 37, 101, 150, 183, 184, 291, 292
mercantile societies, 69
merger(s), 74, 87, 89, 92, 108, 109, 111, 112, 137, 147, 156, 158–160, 162, 165, 177, 178, 181, 183, 186, 187, 191–193, 200, 201, 203, 206–208, 259, 260, 263, 272
merging, 6, 111, 112, 156, 160, 164, 170, 191, 192, 200, 203
Miguel Angel Fernández Ordoñez (MAFO), 160, 173, 176, 205, 207, 225, 227–229, 261
mining, 63, 70, 76, 80, 83, 110
mixed banks, 77, 80, 82, 93, 105, 106, 248
modernization, 8, 17, 37, 77, 79, 107, 120, 121, 125, 126, 150, 214, 221, 242
monarchy, 54, 75, 86, 230
monetary, 19, 21, 31, 59, 73, 86, 112, 253

monetary policy, 14, 30, 31, 88, 92, 93, 206, 228, 252, 254, 263, 268, 275, 284
monetary union, 17, 73, 87, 121, 134, 135, 140, 231, 275–277, 282, 294
monopoly, 66, 68, 72, 74, 75, 84, 85, 306
moral hazard, 260, 286, 287, 304
mortgages, 8, 29, 30, 44, 74, 96, 136–138, 141, 142, 145, 152, 153, 155, 157, 165, 170, 171, 174, 175, 178, 184, 190, 194, 195, 197, 198, 204, 211, 212, 228, 257–261, 263, 271, 273, 281, 284, 285

N
nationalist/nationalistic/nationalization, 41, 55, 84, 87, 88, 93, 117, 139, 177, 178, 185, 186, 189–191, 193, 242, 243, 254, 255, 295
network(s), 54, 67, 96, 97, 109, 114, 116, 144, 149, 161, 187, 190, 194, 248, 250, 253, 254, 266, 267, 295, 300
nonperforming loan(s) (NPL), 136, 163, 178, 257, 268, 269
numerus clausus, 87, 91, 92

O
off-balance sheet, 19, 39, 112, 115, 145, 146, 170, 210
official banks, 85, 94, 95
oil crisis, 53, 98, 103
oligopolistic, 14, 23, 37, 88, 90, 91, 97, 99, 117, 150, 251, 253, 256
oligopoly, 97, 117, 252–254, 256, 265
operation(s), 3, 15, 25, 58, 60, 63, 65, 75, 84, 88, 92–96, 99–103,

112, 113, 116, 137, 146, 148,
157, 158, 160, 164, 165, 170,
190, 196, 209, 259, 260, 266,
285, 289, 291, 294
oversight, 178, 207, 227, 258, 260,
265, 284

P

partnership(s), 97, 98, 117, 158, 188,
237, 248, 254, 303, 307
passive rates, 94, 97, 99
peseta, 21, 70, 73, 75, 81, 82, 85–87,
89, 91, 93, 100–103, 105, 109,
110, 113–115
polarization, 86, 132, 240, 295, 298,
306, 307
policy-making, 253, 282
political choices, 7, 11, 15, 22, 45,
54, 76, 81, 249, 258
political institutions, 1–3, 7, 15, 26,
34, 36, 45, 53, 54, 76, 81, 98,
119, 139, 142, 175, 213, 223,
249, 258, 285, 303
politization, 227, 241, 262
portfolio(s), 27, 33, 40, 44, 83, 89,
95, 101, 111, 145, 146, 151,
154, 157, 166, 197, 201, 266,
273
Portugal, 12, 41, 42, 58, 132, 133,
136, 137, 143, 163, 184, 196,
216, 219, 278–280, 302
private debt, 133, 214, 215
private sector debt, 134, 137, 138,
195
productivity, 12, 19, 21, 124, 125,
127–129, 134, 135, 139, 194,
231–233, 276, 277, 279, 282,
296, 298, 301, 302
professionalization, 155, 226, 264
profitability, 8, 13, 90, 104, 117, 146,
149, 155, 254, 271, 280

profit(s), 14, 28, 32, 36, 68, 77, 82,
88, 101, 109, 110, 123, 136,
142, 143, 148, 150, 155, 156,
164, 190, 193, 194, 207, 229,
251, 256, 269, 271, 272, 289,
303
property market bubble, 153
proprietary lending, 106
protectionism, 88
provisioning, 30, 59, 136, 173, 197,
201, 269
provision(s), 10, 36, 40, 59, 69,
88, 94, 105, 109, 110, 115,
144–147, 157, 159, 160, 163,
168, 169, 174, 177, 178, 190,
191, 193, 197, 200, 201, 207,
229, 230, 260, 262, 268, 275,
284, 286, 298
prudential regulation, 22, 30, 33, 43,
80, 81, 107, 117, 211, 249, 250,
257–260, 263, 286
public banks, 84, 91, 93, 95, 97, 112
public debt, 13, 40, 58, 59, 62, 66,
71, 73, 83, 85–89, 107, 127,
133, 195, 277, 279, 301, 302
public deficit, 13, 14, 26, 130, 195
public funds, 29, 88, 89, 91, 95, 304

Q

queen, 61, 64, 67, 68, 73

R

railroad, 53, 63, 64, 67–71, 73, 74,
77
Railroad Law, 68
Rato, Rodrigo, 126, 161, 188–193,
203, 225, 227, 262
real estate, 5, 8, 9, 11, 16–18, 31,
37, 44, 46, 53, 62, 102, 131,
133, 135, 136, 138, 139, 153,
155, 157, 159, 160, 163, 170,

171, 174–181, 183–185, 187,
190, 191, 193–198, 200, 201,
203–205, 207–211, 214–217,
226–229, 238, 257, 259, 262,
263, 268, 272, 273, 284, 285,
288, 291, 297
recapitalization, 26, 183, 196, 202,
203, 228, 256, 291
recession, 8, 14, 22, 72, 103, 129,
132, 136, 178, 194, 195, 198,
207, 211, 252, 253, 256, 258,
276, 302
reformers, 23, 117, 252–254
reform(s), 18, 19, 23, 72, 73, 84, 87,
106, 115, 135, 160, 162, 166,
167, 192, 201, 202, 221, 225,
252, 277, 293, 298, 300, 302
regional government(s), 41, 101, 132,
139, 161, 166, 171, 187, 189,
191, 202, 205, 206, 208, 218,
228, 259, 260, 265
regulation, 3, 10, 18, 19, 22, 23, 28,
30, 37, 45, 53–55, 57, 66, 67,
77, 79–81, 88, 93–95, 105, 106,
114, 139, 145, 147, 150, 159,
166, 167, 210, 232, 233, 251,
258, 260, 261, 265, 268, 275,
283, 287, 291, 292, 294, 303,
307
regulatory framework, 3, 11, 18, 20,
36, 44–46, 53, 77, 80, 81, 98,
105, 119, 136, 137, 139, 142,
144, 145–147, 151, 153, 171,
173, 175, 176, 200, 207, 210,
229, 258, 264, 285
rent-distribution system, 12, 55, 67,
72, 98
rent-seeking, 4, 10, 17, 214, 217,
218, 232, 248, 249, 283, 287,
296
rescue(d), 4, 13, 28, 40–42, 53, 67,
84, 98, 101, 136, 143, 144, 147,

168, 192, 193, 201, 224, 262,
272, 304
reserves, 33, 86–88, 93, 95, 198
resolution, 111, 114, 208, 261, 268,
287
  resolution rules, 275
resolution mechanism(s), 114, 287,
288, 303
responsibilities/responsibility, 30, 37,
89, 90, 149, 150, 193, 215, 217,
218, 220, 223, 225, 227, 277,
295
restoration, 75, 87
retail, 39, 80, 89, 146, 148, 151, 153,
174, 190, 209
  retail banking, 93, 116, 284, 291
return on capital, 55, 97
revenue(s), 54, 55, 60, 64, 71, 98,
133, 134, 148, 156, 157, 159,
164, 195, 205, 250, 252, 273,
276, 282
risk, 7–9, 21, 22, 29, 30, 32, 33,
37, 39, 41, 43, 44, 55, 57, 60,
64, 67, 73, 83, 93, 97, 101,
104, 105, 110, 112, 113, 115,
117, 138, 144–146, 148–150,
152, 153, 157, 158, 160,
161, 167–170, 174, 178–180,
184, 187, 197, 203, 204, 207,
208, 216, 229, 230, 249–251,
256–261, 263, 264, 266–270,
276, 280, 281, 284, 286, 287,
289–291, 293
risk exposure, 3, 153, 263
risk management, 44, 149, 150, 258,
264, 265, 285, 288, 289
rule of law, 15–17, 55, 67, 127, 214,
232–234, 243, 296, 306
rules, 1, 2, 4, 8, 18, 22, 30, 34, 36,
37, 75, 80, 84, 90, 108, 115,
119, 142, 163, 166, 170, 175,
183, 201, 210, 216, 217, 223,

237, 238, 248, 265–267, 289, 290, 296, 302, 303

S

safety net, 4, 22, 249, 260, 286, 295
sanctions, 13, 115
Santander, 40, 63, 69, 74, 76, 80, 89, 93, 94, 110–113, 116, 123, 136, 137, 142, 143, 146, 148–150, 153, 155, 156, 164, 169, 190, 194, 196, 198, 204, 228, 264, 271, 273–275, 283, 291
savings, 36, 60–63, 76, 82, 85, 96, 97, 100, 129, 144, 149, 152, 154, 156, 158, 162, 164, 169, 199, 251, 254, 263, 272, 291, 299
savings and loans, 7, 14, 43, 85, 114, 115, 176, 214
scandal(s), 95, 150, 171, 180, 183, 188, 219, 230, 237, 241
scarce credit, 2, 13, 55, 238, 251
securities, 32, 39, 62, 65, 71–73, 76, 83, 85, 86, 89, 91, 95, 99, 151, 153, 154, 165, 166, 200, 203, 289, 290
securitization, 30, 39, 141, 151–153, 174, 200, 210
securitization market, 152–154, 199
securitized assets, 153
securitized lending, 153
segmentation, 96
shareholder(s), 3, 54–56, 59, 60, 64–67, 97, 117, 149, 166, 249–251, 256, 259, 274, 275, 285, 286
Single Market, 13, 37, 99, 107, 150, 277, 279
social pressures, 255
solidarity, 216, 282, 305

solvency, 57, 64, 70, 83, 88, 114, 115, 117, 143–146, 161, 164, 168, 179, 187, 202, 206, 282
Spanish Bank of Saint Ferdinand, 60, 62, 63
specialization, 76, 83, 89, 92, 108, 114, 116
spending, 14, 19, 21, 133, 135, 152, 165, 216, 235, 301
stability, 5, 15, 27, 28, 46, 75, 97, 112, 115, 117, 126, 240, 248, 249, 268, 275, 280, 284, 286–291, 294
state aid, 193, 269, 270
status quo, 80, 87, 91, 92, 95, 97, 104–107, 111, 219, 241, 306
stress tests, 144, 168, 206, 208, 268, 274, 280
structured investment vehicles (SIVs), 39, 174, 284
supervision, 28, 30, 36, 37, 42, 68, 80, 145, 173, 176, 203, 265, 268, 274, 275, 286, 287, 289, 293, 294, 303
systemic crisis, 5, 6, 31, 45, 53, 80, 87, 283, 304

T

taxpayers, 3, 4, 54, 55, 57, 98, 218, 249, 251, 256–259, 262, 268, 270, 274, 286, 304
tax reform, 84
technological innovation, 13, 99
test(s), 1, 4, 144, 168, 234, 274, 280, 296
toxic, 19, 39, 136, 145, 146, 151, 174, 177, 184, 191, 194, 197, 201, 209, 211, 212, 272, 273, 284, 285, 291
transition to democracy, 98, 103, 117, 121, 217, 221, 240, 249, 250, 252, 253, 255, 300

troubled assets, 136, 197

U
underwriting standards, 43, 211, 257, 258, 260
unemployment, 4, 5, 103, 104, 113, 120, 122, 123, 125, 129–133, 136, 157, 176, 178, 194, 195, 231, 232, 238, 241, 242, 256, 258, 279, 280, 282, 295, 298, 299, 302
United States, 27, 32, 35, 113, 127, 129, 136, 143, 145, 146, 148, 149, 153, 155, 163, 198, 205, 209, 214, 215, 220, 239, 261, 274, 280, 288, 294, 298, 299, 303, 304

V
vales reales, 58, 59
vulnerability, 44, 134, 152, 155, 170, 196, 257, 293

W
wages, 9, 21, 124, 254, 276–278, 299–301, 305
war, 22, 58, 59, 61, 62, 65, 74, 75, 82, 87, 220
wholesale, 10, 39, 40, 151, 152, 171, 174, 176, 183, 184, 198–200, 210, 259, 265, 291
wholesale banking, 93, 175
wholesale financing, 20, 137, 152, 155, 156, 199
wholesale funding, 18, 44, 46, 137, 152, 198–200, 174–176, 179, 210, 211, 257, 285

Printed in the United States
by Baker & Taylor Publisher Services